COMMUNICATIONS, RADAR AND ELECTRONIC WARFARE

COMMUNICATIONS, RADAR AND ELECTRONIC WARFARE

Adrian Graham
Independent Consultant, UK

A John Wiley and Sons, Ltd, Publication

This edition first published 2011

Registered office
John Wiley & Sons Ltd, The Atrium, Southern Gate, Chichester, West Sussex, PO19 8SQ, United Kingdom

For details of our global editorial offices, for customer services and for information about how to apply for permission to reuse the copyright material in this book please see our website at www.wiley.com.

Library of Congress Cataloging-in-Publication Data

Graham, Adrian W.
Communications, radar, and electronic warfare / Adrian William Graham.
p. cm.
Includes bibliographical references and index.
ISBN 978-0-470-68871-7 (hardback)
1. Radio wave propagation. 2. Radio frequency. 3. Wireless communication systems. I. Title.
TK6553.G697 2011
621.384–dc22 2010035687

A catalogue record for this book is available from the British Library.

Print ISBN: 9780470688717 (H/B)
ePDF ISBN: 9780470977163
oBook ISBN: 9780470977170
ePub ISBN: 9780470977149

Typeset in 11/13pt, Times Roman by Thomson Digital, Noida, India

Contents

Preface

I have been fortunate during my career to work with a wide range of designers, managers and operators in the communications, radars and electronic warfare fields. During that time, I have had the opportunity to provide a technical input into their activities via providing consultancy, design of new systems and techniques and training courses at a range of levels. One of my main tasks has been to develop methods suitable for operators based on highly technical materials. This is not always an easy task; it can be difficult to present complex material in a form that does not require the operators to be experts themselves. This is not to denigrate in any way the abilities of operators. They have to perform excellently in their roles, often in very difficult circumstances, and they simply do not have the time to commit to working out how to convert theory into practice in the field. Thus, much of my time has been spent working out ways to simplify practical methods of applying theory for the widest possible range of circumstances. I have tried to adopt the same approach in the preparation of this book. It is also my own personal opinion that there are two types of information available about these subjects. One is highly theoretical and beyond the needs of most workers in the field. The other is slightly too simplified, omitting vital information without which it is impossible to really understand the subject. In this book, I have tried to bridge the gap between these two opposing approaches. I hope I have succeeded.

During my time working with operators, engineers and managers, I have identified a number of specific areas where I believe understanding is limited throughout the industry. These are the areas I have tried to emphasise in this book, and the ones I spend most time on when I design and run training courses. I have also tried to bridge the terminology gap between workers in the three areas covered by this book. Often, there is misunderstanding between these people when they meet, simply based on terminological and learned approaches when they are in fact talking about the same things. I have deliberately used a mix of terminology to blur these lines. I have in some cases also opted not to use terminology used by one group that may not be understood by others. This is often service or allegiance based, and again they can act to cause confusion where the terminology is not shared by others; this book is, after all, aimed at a worldwide market.

I have provided some references and further reading after most of the chapters. I have tried to choose reading material that is not too theoretical such as academic papers where possible. Where no references are provided, the material is based on my own experience.

As the reader can imagine, I could not have created this book without the input of a vast array of input from other people over the years. There are a few in particular I would like to thank. Since they are still working in the field, particularly in EW, most would rather not be identified. However, both they and I know who they are.

I would also like to thank my long-suffering project editor at John Wiley & Sons, who has had to wait far too long for this manuscript. The support given to me by my equally long-suffering special friend Leanne, my brother Jim and my mother Brenda has also been invaluable. Finally, I would like to thank Alan Smith, the best friend anyone could hope for, who has supported me in many ways during the writing of this book and who, when times have been tough, has comforted me with cider.

Adrian Graham

Glossary

AAM	Air to Air Missile
AGA	Air-Ground-Air, usually relating to communications
AGC	Automatic Gain Control
ARM	Anti-Radiation Missile
AM	Amplitude Modulation
ASM	Air to Surface Missile
AOR	Area of Responsibility; area within which a military force element works
APOD	Air Point of Departure
AWGN	Average White Gaussian Noise (a flat response over the band of interest)
Antenna	Device to convert electrical energy to RF energy and the converse
Backhaul	Network used to trunk traffic from a mobile system
Battlespace	Term used to define the battle area, which extends beyond physical bounds (the battlefield)
BER	Bit Error Rate
BSM	Battlespace Spectrum Management (plan); military spectrum plan
Burn-through	Overcoming jamming by the robustness of the target link
CDMA	Code Division Multiple Access
CEW	Communications Electronic Warfare
C/I	Carrier-to-Interference ratio (dB)
CIWS	Close In Weapons System
CME	Coronal Mass Ejection; an eruption on the surface of the sun
COMINT	COMmunications INTelligence
CONUS	Continental United States (of America)
CNR	Combat Net Radio
Combiner	Device to combine more than one radio signal into a single antenna
Connector	Physical electrical connector for RF cables and systems
COTS	Commercial Off The Shelf; standard systems available to buy
CW	Continuous wave; as opposed to periodic pulsed transmissions

Diplexer	Passive device to combine radio signals into a single antenna without loss
dBd	Loss or gain reference an ideal dipole antenna
dBi	Loss or gain reference an ideal isotropic antenna
DEM	Digital Elevation Model
DF	Direction Finder/Finding
DME	Distance Measuring Equipment (aeronautical)
Downlink	(1) Link from a terrestrial fixed radio station to an associated mobile station (2) In satellite systems, from satellite to Earth station
DRDF	Digitally Resolved Direction Finding
DTM	Digital Terrain Model
DVOR	Digital VHF Omni-directional Radio ranging (aeronautical)
EA	Electronic Attack (EW)
ECCM	Electronic Counter-Counter Measures
EIRP	Effective Isotropic Radiated Power, versus a perfect isotropic antenna
EHF	Extra High Frequency (30–300 GHz)
ELF	Extra Low Frequency (0.3–30 kHz)
EMCON	EMission CONtrol; controlling RF emissions to avoid exploitation by the enemy
EM	Electro-Magnetic
EMC	Electro-Magnetic Compatibility
EMI	Electro-Magnetic Interference
EMP	Electro-Magnetic Pulse; damaging RF energy from a nuclear weapon or EMP weapon
EOD	Explosive Ordnance Demolition
EORBAT	Electronic Order of BATtle
EP	Electronic Protection (EW)
ERP	Effective Radiated Power, normally versus a dipole antenna
ES	Electronic support (EW)
EW	Electronic Warfare (EW)
EW	Early Warning (alternative meaning, depending on context)
FAA	Federal Aviation Authority
FDD	Frequency Division Duplex
FDMA	Frequency Division Multiple Access
FEBA	Forward Edge of Battle Area
Feeder	RF cable used to connect RF components together
FFZ	First Fresnel Zone
FH	Frequency Hopping
Filter	Device to condition an electrical signal in the spectral domain
FM	Frequency Modulation

Force Element	Military assets assigned to a specific task
FSL	Free Space Loss; spreading loss only (dB)
GCHQ	Government Communications HeadQuarters (UK)
GCI	Ground Controlled Intercept
GIS	Geographic Information System
GPS	Global Positioning System
GSM	Global System for Mobile Communications
Hardkill	Physical destruction of assets
HF	High Frequency (3–30 MHz)
HME	Home Made Explosive
HND	Host Nation Declaration; response to an SSR
HUMINT	HUMan INTelligence; informants
ICAO	International Civil Aviation Organisation
ICD	Improvised Chemical Device
IED	Improvised explosive device
IF	Intermediate Frequency
IFF	Identification Friend or Foe
IID	Improvised Incendiary Device
ILS	Instrumented Landing Systems
IMINT	Image INTelligence
IND	Improvised Nuclear Device
IMP	Inter-Modulation Product
IRD	Improvised Radiological Device
IRF	Interference Rejection Factor
ITU	International Telecommunications Union
JRFL	Joint Restricted Frequency List
J/S	Jamming to Signal ratio
JSIR	Joint Spectrum Interference Resolution (process) – US interference resolution method
JSR	Alternative form of Jammer to Signal Ratio
LIDAR	LIght Detection And Ranging; high resolution terrain data capture method
LF	Low Frequency (30–300 kHz)
MASINT	Measurement And Signature INTelligence
MBITR	Multi-Band Inter/Intra Team Radio
MCFA	Most Constrained First Assigned; frequency assignment approach
MF	Medium Frequency (300 kHz–3 MHz)
MGRS	Military Grid Reference System
MLS	Microwave Landing System (aeronautical)
MOTS	Mostly Off The Shelf; standard systems that are partially modified
MSR	Main Supply Routes
NDB	Non-Directional Beacon (aeronautical)

NFD	Net Filter Discrimination
NSA	National Security Agency (USA)
OP	Observation Post
OPTEMPO	Level of operational intensity; OPerational TEMPO
ORBAT	ORder of BATtle
OTHT	Over The Horizon Targeting
PIRA	Provisional Irish Republican Army
PM	Pulse Modulation
POD(1)	Probability of Detection
POD(2)	Point of Departure; air (also known as APOD) or port used in military operation
POI	Probability of Intercept
POJ	Probability of Jamming
PRF	Pulse Repetition Frequency
PRI	Pulse Repetition Interval
PSK	Phase Shift Keying
PSO	Probability of Successful Operation; the likelihood that a given link will work
QAM	Quadrature Amplitude Modulation
QPSK	Quadrature Phase Shift Keying
Radio System	Any system that uses RF channels in order to function, including communications, navigation, radars, jammers etc
RCIED	Radio Controlled Improvised Explosive Device
RF	Radio Frequency, as in radio frequency device
SAG	Surface Action Group; naval force element
SAM	Surface to Air Missile
SAR	Synthetic Aperture Radar
SHF	Super High Frequency (3000–30 000 MHz)
SHORAD	Short Range Air Defence system
Short sector	A region where the nominal signal level will not change, but within which the instantaneous level changes due to fast fading
SINAD	Signal In Noise and Distortion
SMM	Simplified Multiplication Method; method of assessing interference from multiple interferers
SNR	Signal to Noise Ratio
Softkill	Disruption or destruction by non-lethal means
SOP	Standard Operating Procedure
Spoofing	A radiating system pretending to be a different system to fool enemies
SSM	Surface to Surface Missile
SSN (1)	Sun-Spot Number (HF)
SSN (2)	Nuclear Submarine (force element)

SSR	Spectrum Supportability Request; request to a host nation for spectrum
TAPS	TETRA Advanced Packet Service
TEL	Transporter, Erector, Launcher – a missile launch platform, usually a large vehicle holding a tactical land or coastal surface-attack missile
TETRA	TErrestrial Trunked RAdio
UGS	Unattended Ground Sensor
UHF	Ultra High Frequency (300–3000 MHz)
UN	United Nations
Uplink	(1) Link from a mobile station to a fixed terrestrial radio station (2) In satellite, from Earth station to satellite
UTM	Universal Transverse Mercator; a data projection
VHF	Very High Frequency (30–300 MHz)
VLF	Very Low Frequency (3–30 kHz)
VOIED	Victim Operated Improved Explosive Device
WGS84	World Geodetic System 1984; geographic datum used by GPS

Part One

Basic Theory

1

Introduction

1.1 The Aim of this Book

This book looks at the subjects of radio communications, radar and electronic warfare. The aim is to provide the reader with a mixture of theory and practical illustrations to explain the way in which these systems are used in practice. The book is aimed at operators, designers and managers operating in these areas. It is designed to provide a detailed overview at a level suitable for this audience. This means that the intention has been to provide explanation of complex theory in as simple manner as possible, and to link the theory to real life as far as possible. One of the main reasons for writing the book is that there is a large body of very in-depth, complex works that are beyond the grasp of the average reader. There are also works that provide simple overviews but without introducing the necessary background theory. Hopefully, this book provides a middle way between these two extremes.

The book has been split into two main sections; theory and practice. The idea is to lay the necessary theoretical groundwork, and then to spend more time in the main, practical part of the book identifying the operational effects of the theory when applied. In this way, the book is designed to bridge the gap of theory to application in a way that makes sense to communications and electronics operators, system designers and managers.

One aim in writing the book has been to provide as compact knowledge as possible in each section. Thus rather than having to find an earlier reference, in some cases the theory has been re-introduced, and some diagrams replicated, in some of the practical sections where they are explicitly required. The reader can therefore easily dip into to individual sections and get most of the information without having to go back to the theoretical sections. Thus, those whose interest is primarily for radar for example, the book has been split up in such a ways as to collate the relevant information into readily located sections. ***To make the book more readable, I have used the term 'radio' to mean any system that uses the RF spectrum, including radars and navigation system.***

Communications, Radar and Electronic Warfare Adrian Graham

The main focus of this book is on the Radio Frequency (RF) part of the system. This is the part between two antennas in a link. However, in order to make practical use of this, it is necessary to examine the parts of the system that are essential to allow the construction of an accurate radio link budget. This means every step from the radio output from the transmitting radio to the output of the receiving radio. The focus will be on those aspects over which operators and developers have some control, such as selected frequency, antenna, location and system configuration.

To those new to the field, I would recommend reading all of the theory section and then the sections relevant to the reader's area of work. More experienced readers may choose to go to the sections that are relevant to their work, with the theoretical sections being available as an easy reference when required.

My hope is that this book finds resonance with those involved in this topical and important area and that it helps such people to improve the state of the art of mission planning and simulation of real-life scenarios.

1.2 Current Radio Technology

1.2.1 Introduction

No one can be unaware of the pervasive nature of radio systems in the modern world. The rise of mobile phone systems has been phenomenal, and this has been matched by other recent developments that have improved the links between mobile phone masts (normally called 'backhaul'), provision of internet access via WiMax and other systems, improvements to broadcast systems brought about by new digital services, and worldwide navigation via GPS.

In the military sphere, similar new technologies have been used to extend system ranges, improve security and to provide information throughout the Battlespace. However, this description could equally be used to describe the developments of civilian systems as well. Increasingly, civilian equipment is becoming more frequently used by armed forces and particularly by insurgents. In some cases, the increasing capabilities of commercial systems are also being exploited by even the most well equipped armies because they are better than their own systems and they can be fielded very much more quickly than new military systems.

Because of these factors, this book includes analysis of modern civilian services as well as military ones. Such systems may be used to provide emergency or short-term communications for military operators, and are also increasingly of interest to electronic warfare operators as legitimate targets since they are used by the opposition. As we will see in the next few pages, the historically distinct fields of military and civilian use of the radio spectrum are in many ways merging into a single set of requirements, at least at outline level.

This section will look at military, civilian and joint technologies. It will look at the commonalities and contrasts and draw conclusions as to how they can be

Figure 1.1 The basic radio link between two terminals. These can be fixed, static locations or can be dynamic links where the terminals move and the link is only present during short communications.

managed by military and other organisations for communications and electronic warfare purposes.

First, we will take an overview of the different types of radio system as they appear to their users.

The simplest configuration is that of a radio link between two defined locations as illustrated in Figure 1.1. In this case, there are two locations with radios, which act as terminals to the link. The arrows at both ends indicate that the link is bi-directional, sending voice or data from either terminal to the other. This type of link is known as a 'point-to-point' or abbreviated to 'point-point'.

Links can be permanently established between two terminals, such as in fixed microwave links, or they may be temporary, such as between a mobile phone base station and a mobile subscriber or between two tactical groups. Single links can be combined into networks as shown in Figure 1.2. The structure shown is typical of the traditional military command and control model (this is a generic model, not built around any particular country's organisation). In this case, the view is in plan form (from above). Each terminal is a black dot and each link is a solid black line. Note that in this model, not all terminals are linked to each other. None of the individual echelon levels (battalion, regiment, brigade, division) talk directly to each other; instead they have to go to a higher level of command until direct links are provided. The network structure is therefore hierarchical.

Networks of point-point links can also take many other forms, ranging from the instantaneous configuration of Personal Role Radio (PRR) networks covering a few hundred metres to national microwave networks.

Apart from point-point radio systems, the other main type of communication system is the mobile network as shown in Figure 1.3. In this case, there is a single fixed base station and a number of mobiles moving through the coverage area, shown as black squares with a track showing where they have been.

It is worth noting that at an instant in time, the mobile network can be described as a point-multi-point network (one base station serving a number of users). Mobile networks normally consist of many base stations to provide coverage over a wide area, for a very large number of subscribers.

Mobile radio systems are often trunked, so calls between parts of the network can be passed to over parts of the network. Figure 1.4 shows an example of a simple trunked network. Fixed base stations are shown as black circles. The coverage of the base stations is shown in grey. The main trunks are shown as thick dotted lines, with feeder

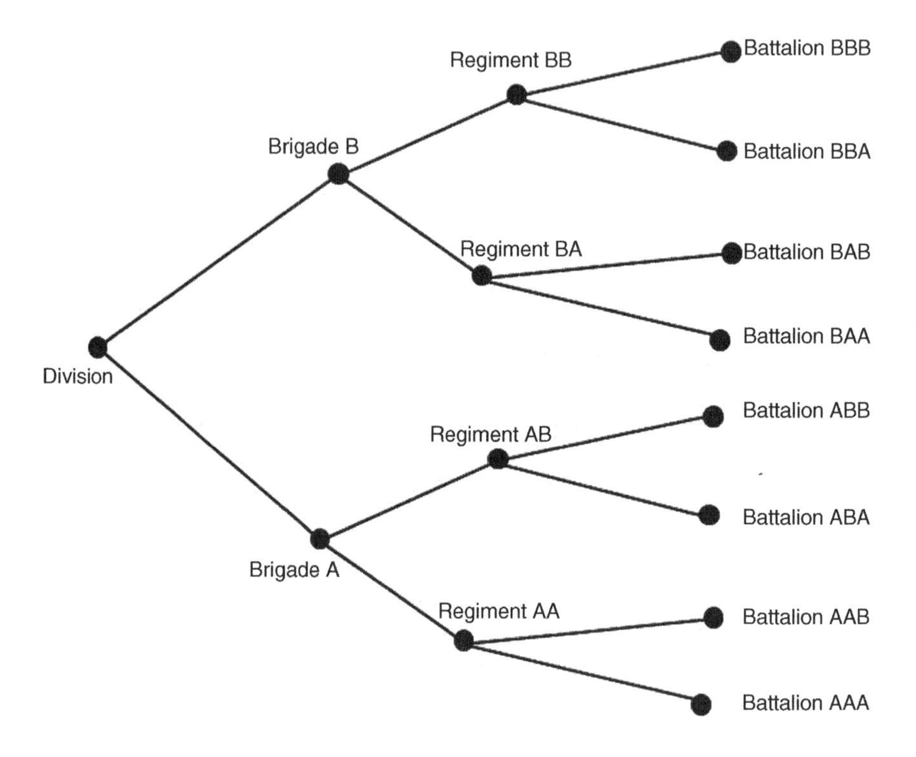

Figure 1.2 A typical hierarchical military command and control radio network. The network design is deliberate so that it supports the way the command structure works.

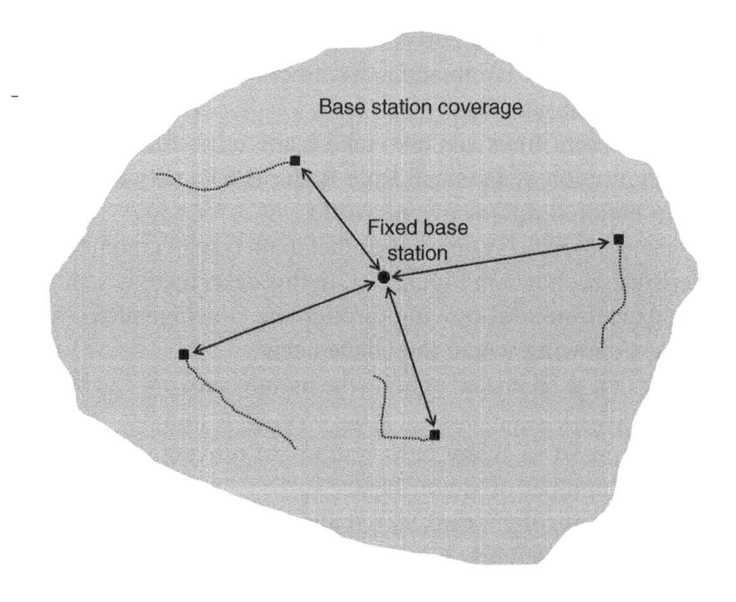

Figure 1.3 A simple mobile network.

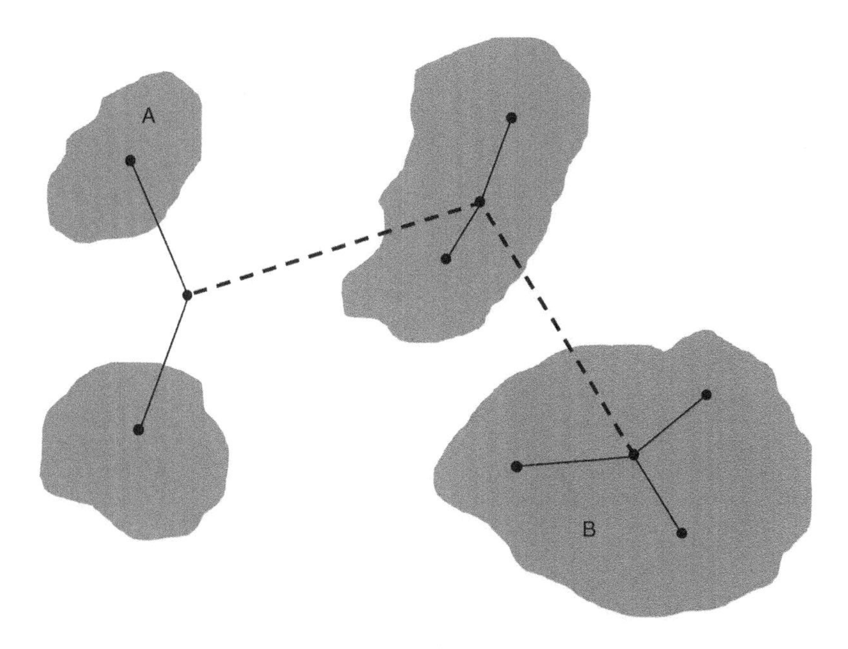

Figure 1.4 A trunked radio network providing mobile coverage to subscribers in the coverage area shown in grey. The thick dashed lines show main trunks and the thin solid lines show trunks between out-stations and the main trunk terminals. This type of network is used in military deployments and also for PMR and mobile phone networks.

links shown in thin solid lines. The trunks provide a link between the different parts of the network, so a caller originating at point A can talk directly to a mobile at point B.

Radar systems provide the means to detect and localise aircraft, ships and battlefield systems. In many cases, such as blue-water maritime scenarios and high altitude aircraft, coverage from radar systems will be circular in form, out to a maximum range for a given type of target at a given altitude. However, in many other cases, radar coverage will be limited by the environment as illustrated in Figure 1.5. This shows the coverage of ship air search radar looking for low altitude targets. Over the sea (left hand side), coverage is circular but over the ground (right hand side) the coverage is limited by hills and ground clutter.

Radars can also be used in networks to provide wide area coverage by a number of linked radar stations. This is shown for a naval group in Figure 1.6, where the composite network coverage from the combined ships is shown in grey.

We will see many more examples of radio systems during the rest of the book.

1.2.2 Military Communications

Military communications have traditionally evolved to meet perceived needs for the battlefield. Since the end of the Second World War, the Western and Eastern blocs

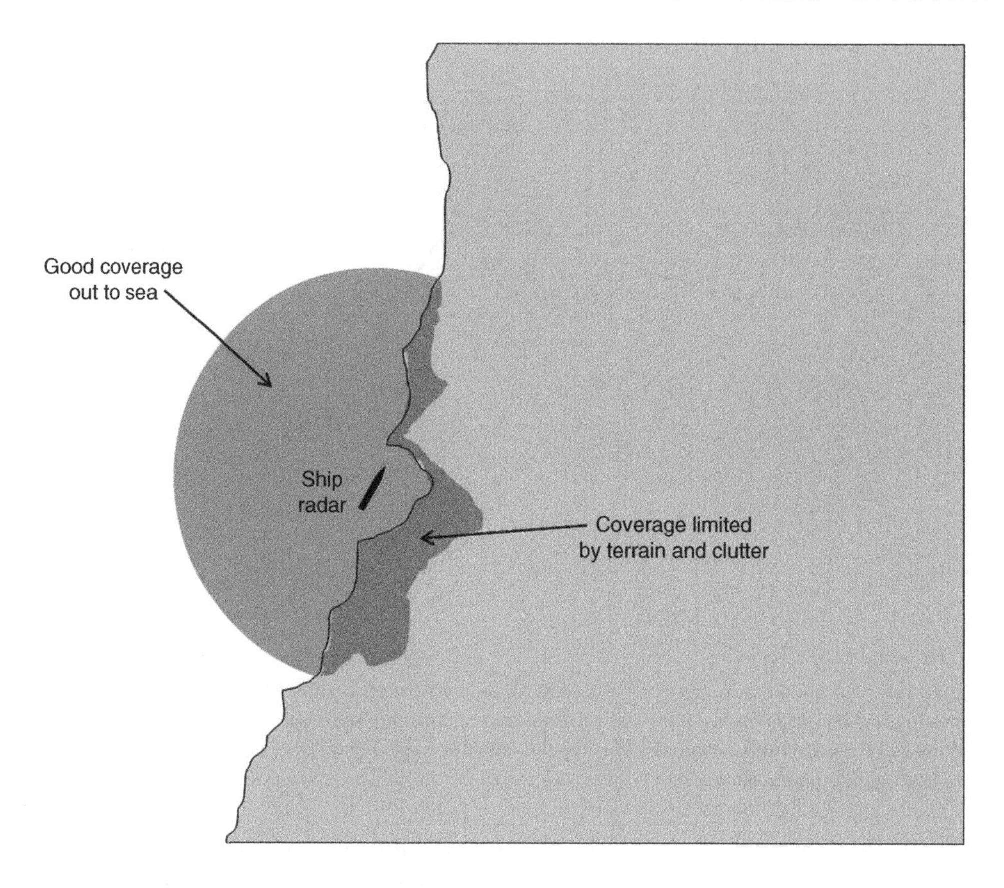

Figure 1.5 Illustration of a ship using radar close to shore. The coverage of the radar over the land is limited by the effect of terrain such as hills and also by radio clutter, which adds noise and obscures genuine targets.

focused entirely on the possibility of a major European or World war that would be essentially nasty, brutish and short.

The major characteristics of this scenario are worth analysis to see how they influenced radio network architecture and radio design. These characteristics included:

- It would take place in a known environment. This was true for the land, air and naval conflict. Although there was the possibility of some variation from the central script, actions and responses were well practiced and known.
- It would be of particularly high intensity. This would have been particularly true in the Forward Edge of Battle Area (FEBA). This would have meant congested airwaves, little time to detect and assess radio targets of interest and a very difficult spectrum management regime. In practice, this meant that dynamic spectrum

Figure 1.6 A naval force with ships shown in black and the coverage of their surface search radars shown in grey. As long as the ships using the radars can communicate with each other, they work as a radar network with all ships benefitting from the composite coverage.

management would have been impossible and thus in general, frequency plans were worked out well in advance of the beginning of a conflict. This meant the system was inflexible and depended on correct usage by all operators. Communications electronic warfare would have been very difficult to use effectively during the conflict due to the expected speed of development on the battlefield.

- The scenario seldom changed over most of the history of the Cold War. The development communications and electronic warfare equipment evolved to meet the perceived needs and from then on, only developed slowly because the original requirements remained the same. Much work was done on improving portability and battery lives and so on, but not to the basic radio requirements.
- If the feared conflict had broken out, the military would have had control over civilian systems, many of which would have been switched off. Thus although military spectrum demand would have been very high they would have had all of the usable radio spectrum to themselves. They would not have to co-exist with civilian users.
- Much of the radio architecture could be made to be semi-fixed or at least deployable to known and tested wartime locations. These would have been well tested before their selection (at least in theory). Thus, the initial action would have been a sprint to deploy communications nodes to their designated locations, using pre-agreed frequency plans.
- Since it was known that communications were vulnerable to detection, interception and jamming as well as hardkill, forces practiced radio communications disruption. Since all NATO planning was for defence purposes, this meant that physical landline

networks could be laid and training for lost comms (communications) procedures carried out. These both helped to negate the operational risk posed by communications disruptions.

Clearly, these factors were crucial in designing the requirements for military communications and electronic warfare equipment and thus they had a major impact on the design of the systems. Some design factors that came out included:

- Radios and other equipment had to be resistant to physical damage. In such a high-intensity conflict, damage would be commonplace.
- Radio range requirements could be based on fairly well-known deployment strategies. This was particularly the case for terrestrial tactical VHF.
- Most land radio would be vehicle-mounted and thus weight and power consumption was not of primary concern.
- Given the vulnerabilities of radio in such a conflict, voice procedures were kept simple and encryption was not always used for tactical communications.
- The systems of the day did not rely on advanced digital communications in order to function. Voice was most important for tactical scenarios, simple telex-style data for higher-level echelons.

The fall of the Berlin Wall in 1989 heralded in a new era and rendered much radio equipment on both sides obsolete overnight. Since that time, the role of western armed forces has changed significantly to cope with the new world situation. Some of the changes include:

- The threat of all-out warfare in Western Europe of the type envisioned during the Cold War has gone. With that, the certainties that the scenario implied have also gone. Future conflicts would be fought on unfamiliar terrain in far flung parts of the world. Some of these places feature radio propagation characteristics very dissimilar to that of Western Europe, as we will find later in the book.
- High-intensity conflict has not been realised at higher than the tactical level in many cases (although some of the infantry exchanges in Iraq and Afghanistan have met and in some cases exceeded the intensity found in World War Two). This means that radio devices beyond personal role level may not require such damage resistance as was necessary before. This can result in a higher probability of damage to a radio; however, this can be offset by the reduced price and availability of replacement units, so long as the necessary logistics are in place.
- Military operations may well have to co-exist with ongoing civilian communications and other systems. Indeed, those forces may be asked to assist in the setting up and maintenance of civilian systems.
- Military operations have been lengthy and semi-static, meaning that it has been possible to set up semi-permanent communications over the theatre of operations.

- In many cases, particularly in Afghanistan, taking the war to the enemy has meant dismounted patrols operating in harsh environments. In some cases, vehicles cannot get to the patrol objectives and have to stand off, ready to provide cover from a distance. This increases the importance of personal and mobile radio for tactical scenarios. Soldiers on the ground also need to be able to communicate with friendly aircraft accurately and in a timely fashion.
- The modern conflict is heavily dependent on reconnaissance. Getting the information from the source to the decision makers in a timely fashion has become crucial. This means that high-bandwidth communications are critical to operational success.
- Asymmetric warfare is now the norm, with traditional warfare less common. Enemy communications are less likely to be of the classic military type. In fact in many cases, civilian radio systems have distinct advantages over older military types. Being able to exploit enemy transmissions is if anything even more important than it has been in the past.

For these reasons, radio communications equipment designed for the Cold War became less useful for the modern scenario. However, many of the original principles remain intact. There is still the need for tactical radio, command and control, a plethora of radar systems and so on. Some of these technologies are described in Table 1.1.

1.2.3 Quasi-Military Type Operations

Quasi-military operators are those that use similar radio communications equipment to that of military forces, but they are not direct combatants. They are present not to support an operation, but to work within it to achieve their diverse aims. There can be a number of them present and functioning in an operational environment. Typical users include:

- border forces, paramilitaries, etc.;
- local emergency services;
- infrastructure services such as road, rail and air services;
- humanitarian operators, working to maintain humanitarian supplies;
- UN or regional monitoring operators, observing local combatants and their behaviour;
- news gathering organizations.

In terms of managing their use of the spectrum, they often need to coordinate their activities with military forces. They also may need short-term access to the spectrum and their use may be dynamic, involving movement as the situation develops. In this way, they are different from purely civilian users who will normally be managed separately.

Table 1.1 Sample military uses of the radio spectrum

Type of system	Typical radio technology	Typical range and comments
Personal Role Radio	Encrypted or clear voice at VHF, UHF or low end of SHF	<1 km. Used to provide contact between members of a patrol
Remote ground sensors	Depends on intended target characteristics	<1 km to a few km. Used to provide detection of tactically important signals
RCIED[a] jammer	Bespoke	<1 km to several km for mobile phone base station jamming. Used to counter radio-controlled jammers
ILS[b] systems	VHF voice	Out to several km. Used for aircraft landing
SHORAD[c] radar	UHF Air defence radar	A few km to tens of km depending on system. Used to protect key points
Point-to-point links	UHF or SHF trunked systems	Tens of km depending on clear line of sight; can be extended by use of relays
High-bandwidth tactical links (terrestrial)	UHF spread spectrum	Tens of km depending on modulation scheme used. Used to provide video and other high-bandwidth data over relatively short ranges
Targeting radar	SHF or EHF	Range depends on system; can be a few km out to the horizon (for surface-air or air-air links, which can be many 10's of km)
Maritime port control	VHF voice	Typically 10's of km
Tactical command and control (VHF)	VHF or UHF voice	50–100 km. UHF normally used for aeronautical links
UAV command and control	UHF or SHF	Varies according to type of system from a few km to worldwide using satellite systems for strategic UAVs
Tactical command and control (HF)	HF voice (groundwave or NVIS[d])	Groundwave out to 50 km depending on conditions; NVIS out to 400 km depending on conditions
Surface search radar	UHF radar	From the horizon out to 10's of km
Air surveillance radar	UHF radar	10's to many hundreds of km, depending on platform (surface or air)
Aircraft navigation systems	VHF or UHF data	Hundreds of km
Command and control (HF)	HF voice	Worldwide dependent on conditions
Command and control (satellite)	Satellite	Worldwide
Navigation systems	GPS	Worldwide

[a]RCIED: Radio controlled improvised explosive device.
[b]ILS: Instrumented landing systems.
[c]SHORAD: Short range air defence system.
[d]NVIS: Near vertical incidence skywave.

Typical equipment they may use includes:

- PMR, e.g. TETRA, TETRAPOL;
- Family Radios;
- satellite phones;
- tactical military communications;
- GSM and other public services;
- semi-fixed infrastructure for backhaul and high-capacity networks.

Some of these may also be used concurrently by enemy forces, particularly irregulars. In some cases, local forces may have some contact with enemy forces, whether officially or not. This makes their management and monitoring particularly important.

1.2.4 Civilian Communications

Civilian communications are a vital part of modern life. For this reason, it is not always possible to suppress civilian communications during an operation and in many ways it is preferable to keep them working to minimise disruption as far as possible. The range of modern civilian communications is very wide and uses large parts of the spectrum. Typical uses include:

- public telephony, internet and messaging services;
- backhaul to support mobile operations and for telephone/internet backhaul;
- broadcast, both digital and analog (TV, radio, etc.);
- navigation services such as GPS;
- aeronautical civil communications;
- maritime civil communications;
- local paging systems, such as for hospitals and other essential services;
- medical equipment (some of which is vulnerable from RF interference);
- TV and radio outside broadcast.

Some of these are common technologies, used by both civil and military users and others have been historically distinct from military systems. However, it is important that military planners understand these technologies as they may cause interference to military systems; it may be necessary to prevent interference into the civil systems and because the opposition may be using them for military purposes it may be necessary to detect, localise and jam them.

1.2.5 Cross-Over Technologies

One trend that has emerged over the last few years is that the development effort into civil radio technologies has grown to eclipse that spent on military systems. Civil systems are also increasingly robust and have higher performance than legacy military systems. This now means that in some cases, military users are taking advantage of civil systems for their own uses. In many cases, civil systems and their ancillaries are cheaper than military alternatives and it can be more cost-effective to use cheaper, disposable handsets and having spares available.

The same is also true in the other directions; who, now, does not use the GPS system at least occasionally?

The simple truth is that there are now an increasing number of cross-over technologies that are used for both military and civil applications. This makes the role of military communicators and spectrum managers far more complex as there are a wider variety of systems available to use for particular circumstances.

We will now look at technical aspects of the basic radio link.

1.3 Factors Constraining Radio Communications

1.3.1 The Basic Radio Link

Radio systems are dependent upon a number of intermediary elements, all of which are necessary to ensure that the transmitted message is received correctly. Figure 1.7 shows some basic blocks in the hardware part of the system. The transmitter is the system that receives or generates the message to be transmitted. It carries out conditioning, conversion and modulation of the signal into an RF electrical signal. For example, a mobile phone picks up voice from its microphone, converts the signal

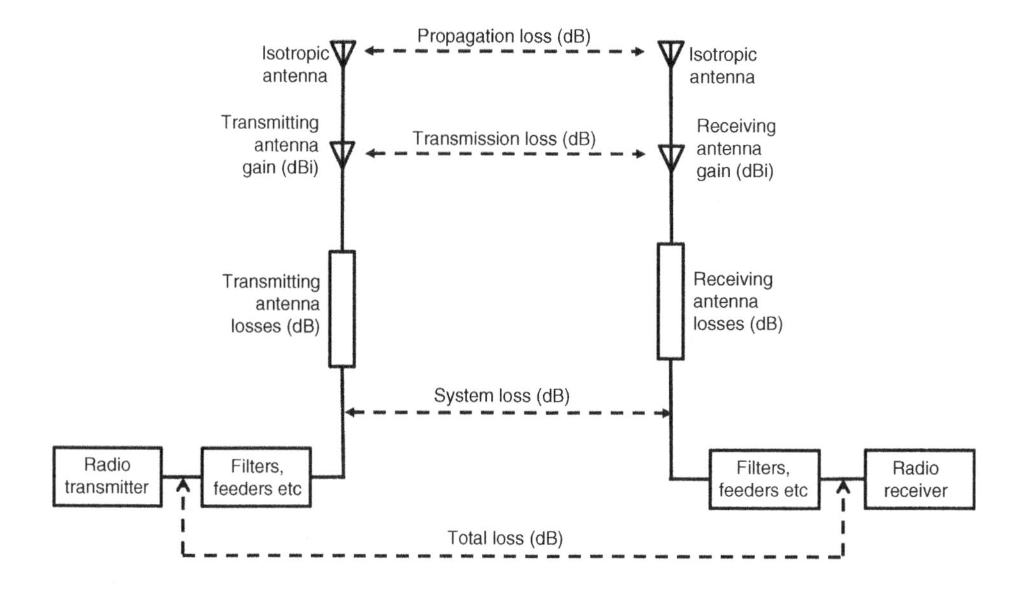

Figure 1.7 Basic loss diagram, derived from ITU-R P.341-1.

into a suitable digital format, modulates the signal with a carrier and transmits the resulting RF signal into the antenna for radio transmission to a base station or another mobile. The diagram is also suitable for the representation of more complex systems such as multiplexed microwave carriers or radar systems.

All of the solid lines in Figure 1.7 represent physical hardware factors, including electrical feeders, connectors, filters and antennas. The hardware configuration of the transmitting and receiving antennas determine their response when compared to an isotropic antenna in this case. It is also possible to use other reference antenna types, of which a perfect dipole antenna is the most common. In this case, the reference is not expressed as dBi, but in terms of dBd (dBs versus a dipole).

The dotted lines represent loss taken between two points in the system, as described in Table 1.2. The key point is that there are a number of potential ways of expressing loss, and it is important to understand which way a particular loss value is expressed.

A typical transmitter might be a self-contained box, rack-mounted or embedded in a larger system. A block diagram of a standard transmitter is shown in Figure 1.8. The input is the message to be transmitted. This is the message that needs to be recovered at the output of the receiver. Ideally every designed aspect of the radio link is optimised to ensure that the recovered message to within an acceptable degree of fidelity. Note that this is different from attempting to recover the *entire* transmitted message, which would be technically far too demanding in many circumstances.

In simple systems, the output of the transmitter can be directly fed into the antenna. However, in many cases there will be additional intermediate electrical entities. These can include filters designed to condition the energy transmitted to minimise interference with other radio links. There may also be amplifiers to increase the power of the

Table 1.2 Loss terms and their meaning

Loss	Description and comments
Propagation loss	Loss between two perfect isotropic antennas. This is dependent on transmission frequency, distance and environmental factors. It does not include any consideration of hardware gains and losses
Transmission loss	Loss between the two antennas actually used. It includes the effects of antenna gain but not associated system losses
System loss	Loss taken between the input of the transmitting antenna and the output of the receiving antenna
Total loss	Loss between a specified point from the output of the transmitter and another specified point from the input to the receiver. Typical points are the physical output of the radio transmitter and the input into the receiver, which may coincide with physical connectors into and out of the two devices. This is often used to reference the maximum loss that can be tolerated to achieve a specific objective; for example 12 dB SINAD, a specified bit error rate, or an acceptable quality of reception

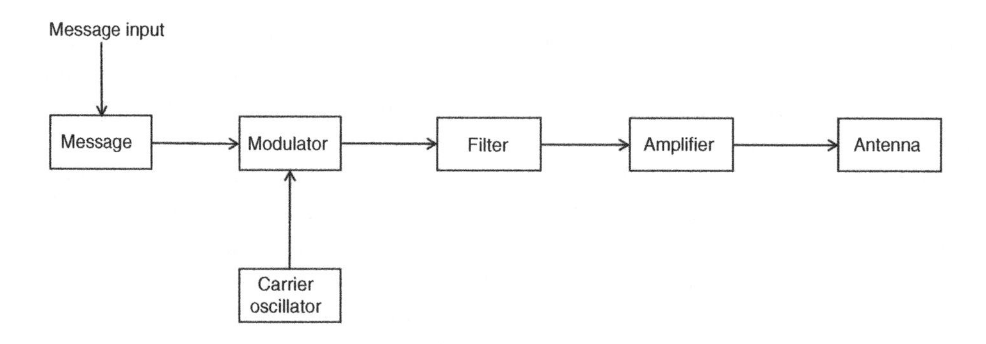

Figure 1.8 Simple transmitter block diagram.

signal output from the transmitter. Even simple electrical entities such as cables used to connect parts of the system (feeders) and RF connectors between sub-systems may affect the power of the signal fed into the antenna itself.

The transmitting antenna may have losses due to impedance mismatching or antenna efficiency, or there may be gain from its directional polar response.

Once the energy entered into the antenna has been launched into the propagation environment, its behaviour will be dependent on many environmental factors, which are discussed in the next section. In general, however, the radiated energy will suffer loss due to distance, reflections, scattering and diffraction and these may also degrade the quality of the received energy.

The normally very small energy available at the receiving antenna will be affected by the receiving antenna efficiency and the gains or losses of the antenna in particular directions. The received energy may be filtered and amplified prior to arriving at the receiver. Like the transmitter, connectors and feeders may also have an effect.

The receiver will take the input signal, de-modulate it and in many systems will process the demodulated signal to produce the output desired.

Each of these elements are important to understanding how a given radio system will perform in practice. Table 1.3 shows a list of typical entities and their normal effects, although there will be exceptions. Note that the entities are listed in alphabetical order only, not in the sequence in which they are used in a system.

The diagram of Figure 1.7 can be re-drawn as a linear process for convenience in analysis. This is useful as a step in understanding the form of the equations associated with radio signal links. This is shown in Figure 1.9.

Often these entities are described in a tabular format as part of a link budget. A simple example of this is shown in Table 1.4. In this table, the unit types in bold have been used for the sample values expressed.

Only some of the elements described previously are included in this simple link budget. However, any other elements that have associated gains or losses are treated in the same manner.

Table 1.3 List of radio entities and their uses and effects

Entity	Use	Effect
Amplifier	Used to boost both transmitted and received electrical energy	Increases range
Antenna	Used to convert electrical energy into radio frequency energy and vice versa	Determines coverage
Combiner	Used to combine two or more radio signals to be transmitted via the same antenna	Improves equipment efficiency
Connector	A connector is required between each discrete component in an electrical system	Used to electrically connect components
Diplexer	A diplexer is a passive device used to combine signals	Convenience and reduction of inter-modulation product
Duplexer	A duplexer is used to combine two or more signals into a common signal to be sent through the communications channel	Improves spectral efficiency
Feeder	A feeder is the name for an RF cable from one part of a radio system to another. Co-axial cables are often used	Allows signal to travel through the system between components
Filter	A filter is used to allow transmission of a band of energy while applying attenuation to other frequencies	Improves spectral efficiency, prevents interference
Receiver Radio	The receiving part of a radio	Necessary to down-convert the baseband signal from the carrier, demodulate the modulated signal and retrieve an acceptable portion of the transmitted message
Splitter	A device to split electrical energy between two or more paths	Used to send electrical energy to different antennas or other devices
Transmitter Radio	The transmitting part of a radio	Used to encode the transmission message using a modulation scheme, and up-convert to a carrier frequency signal

Table 1.4 Generic link budget example

System element	Sample value	Typical units
Transmitter side components		
Transmitter output power	50	W, dBW, **dBm**
Feeder losses	3	dB
Connector losses	1.5	dB
Antenna gain	4	**dBi** or dBd
ERP or ***EIRP***[a]	***49.5***	***dBi****or dBd*
Receiver side components		
Antenna Gain	1.5	**dBi** or dBd
Feeder losses	2	
Connector losses	1.5	
Receiver sensitivity[b]	−100	**dBm**, μV, dBμV
Minimum required input signal[c]	**−98**	**dBi**
Maximum tolerable loss[d]	**147.5**	**dB**

[a] EIRP is calculated in this example based on the output power, losses and antenna gain.
[b] Sensitivity is expressed for a given value of performance for system output, e.g. 12 dB SINAD.
[c] This is the energy that must be present at the receiving antenna in order to supply the receiver input.
[d] This is the maximum transmission loss between the antennas that will allow the minimum required input signal at the receiver. This must include all propagation effects.

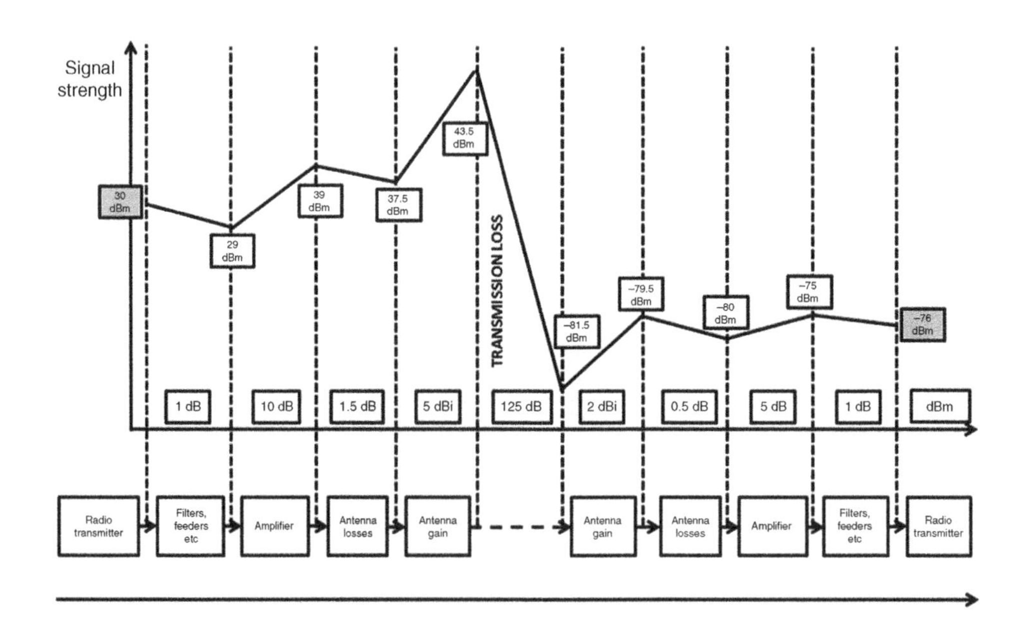

Figure 1.9 Linear diagram of system losses, with an example calculation.

For the communications and electronic warfare operator, most of the hardware parameters are fixed based on the equipment they and the enemy have. In fact, in many cases the only factors under the operator's control will be usually parameters such as:

- Radio output power, which may be selected from pre-set values or may be available in a range between minimum and maximum values.
- Transmission frequency, either from pre-selected values or a range. This will affect propagation loss.
- Radio mode, which may affect certain performance parameters. For example, a system that can transmit voice, encrypted voice, video, etc. will have different characteristics.
- Selection from a range of feeder cables and connectors.
- Selection of sub-systems such as matching tuners or external amplifiers.
- Selection of antenna type.
- Positioning of the antenna main beam in terms of location, height above local ground, direction and tilt.

Some of these selections will be determined by the type of signal being sent and others will depend on a variety of factors such as adherence to spectrum management rules, installation design rules, limitations due to available equipment and environmental factors. These environmental factors are considered next.

1.3.2 Useful Conversion Formulae

Many units are used in radio engineering, and it is often necessary to convert from one type of unit to another. This section lists a number of useful formulae.

1.3.2.1 Power Conversions

These formulae are used to convert between power in W, dBW and dBm.

$$Power(dBW) = 10\log[Power(W)]$$

$$Power(W) = 10^{\left(\frac{Power(dBW)}{10}\right)}$$

$$Power(dBm) = Power(dBW) - 30$$

1.3.2.2 Receiver Sensitivity Conversions

These formulae are used to convert between sensitivity units of dBm, μV and dBμV.

$$Power(W) = \frac{v^2}{R}$$

$$Power(dBW) = 10\log\left(\frac{V^2}{R}\right)$$

$$Sensitivity\ (dBm) = 10\log\left(\frac{Sensitivity(\mu V)^2}{R}\right) + 30$$

where R is impedance, usually 50, 75 or 300 Ω

$$\text{Sen}sitivity(\mu V) = \sqrt{R.10^6.10^{\left(\frac{Sensitivity(dBm)-30}{10}\right)}}$$

$$Sensitivity(dB\mu V) = Sensitivity(dBm) - 107$$

$$Sensitivity(dBm) = Sensitivity(dB\mu V) + 107$$

1.3.2.3 Field Strength Conversions

These formulae are used to convert electrical field strength to and from the power induced in an isotropic antenna.

$$Power(dBm) = -77 + 20Log(E) - 20\log f$$

$$Power(dBm) = -77 + E\left(\frac{dB\mu V}{m}\right) - 20\log f$$

$$Power\left(\frac{dB\mu V}{m}\right) = Power(dBm) + 20\log f + 77$$

$$Power(E) = 10^{\left(\frac{Power(dBm) + 77 + 20\log f}{20}\right)}$$

where

E is electric field strength in μV/m.
f is frequency in MHz.

1.3.2.4 Antenna Conversions

These are the values to use to convert between the relative gains of common reference antennas. This is shown in Table 1.5

Table 1.5 Antenna correction values

Antenna type	Gain (dBi)
Isotropic	0
$\lambda/2$ dipole	2.15
Hertzian dipole	1.75

1.3.3 Environmental Factors

In line engineering, such as for fibre optic systems, it is possible to expect a reasonably steady transmission path between transmitter and receiver in a system. This is however not typically true for radio systems. The energy converted from electrical to RF energy is radiated in a pattern determined by the antenna characteristics at the transmission frequency. This energy is then subject to spreading loss, scattering, reflection and absorption before a small portion of the energy is used to excite the receiving antenna system.

The dominant mechanism may depend on a number of factors including:

- frequency of transmission;
- link length;
- link topology;
- structures near the transmit and receive antennas.

For skywave propagation at HF, conditions vary diurnally due to the influence of the Sun on the Earth's ionosphere. During daytime, extra energy enters the ionosphere and this causes increased ionisation. It is this ionisation that is used to reflect HF skywave transmissions back towards the Earth's surface at far distant locations. When the Sun's energy is occluded by the Earth at night, the ions break down but normally some will remain to provide night time reflections.

This is illustrated in Figure 1.10. The diagram is not to scale. The various layers of the ionosphere are called the D, E and F layers. During the day, the F layer separates into two distinct regions, known as F1 and F2. Of these, the F2 layer is most important for communications.

Of these layers, the D-layer acts to block some frequencies and the others are used for long distance communications. We will see how this affects HF skywave propagation in more depth in Chapter 9. HF groundwave communications depend partly on line of sight and it also travels partly through a shallow region of the Earth's surface, depending on the conductivity of the region.

As well as diurnal variations, HF prediction is also affected by other non-continuous mechanisms, of which the most important is the solar cycle, also known as the Sun-spot cycle. The solar cycle varies over an 11-year period, with minima and maxima

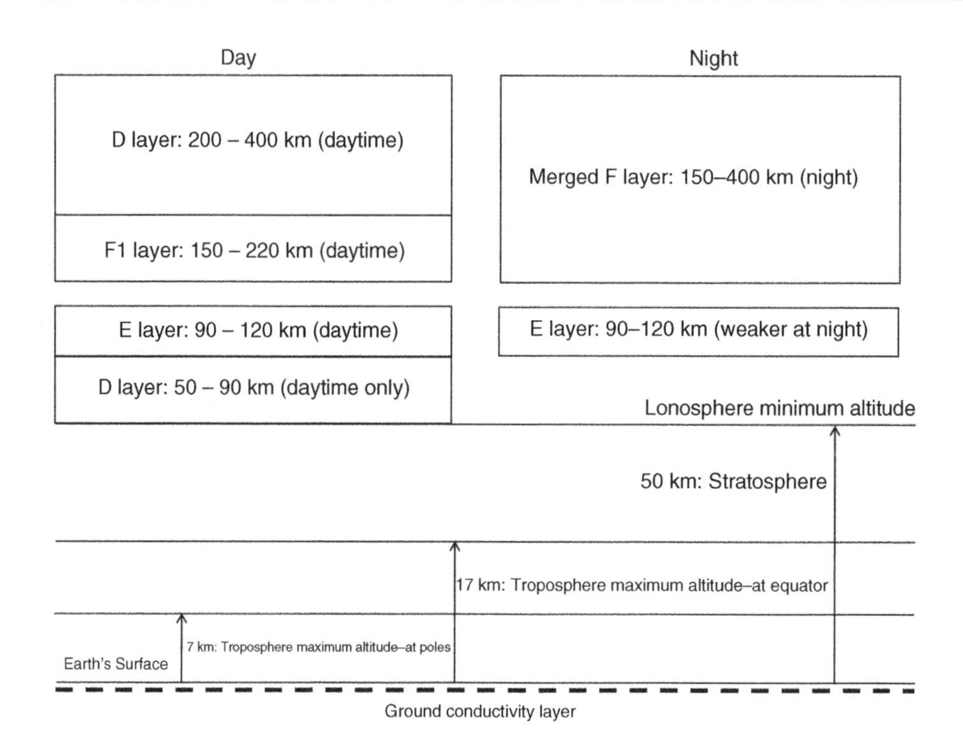

Figure 1.10 The Structure of the ionosphere.

occurring during this period. Other solar events also affect HF significantly; particularly those associated with Coronal Mass Ejections (CMEs). These and other effects are also described in Chapter 9.

At VHF and above, in general transmissions other than those intended for satellite communications will propagate through the troposphere. The troposphere is the lowest layer of the atmosphere and one in which nearly all life on Earth is found. Its maximum altitude varies from the poles, where it is at its lowest altitude of approximately 7 km, to the Equator, where it reaches approximately 17 km.

For terrestrial, aeronautical, maritime and littoral radio links, the main factors involved in radio propagation apart from the effects of link length relate to the topology of the link. This is illustrated in Figure 1.11.

The diagrams in Figure 1.10 show some different link topologies. Diagram (a) shows a link with clear line of sight between transmitter and receiver. Because of the way radiowaves propagate, it is also necessary to have clearance from mid-path obstructions, even if they do not obstruct the direct line of sight. Diagram (b) shows the scenario where such a mid-path obstruction exists; the obstruction will affect the radio signals that arrive at the receiver. In diagram (c), mid-path terrain blocks line of sight

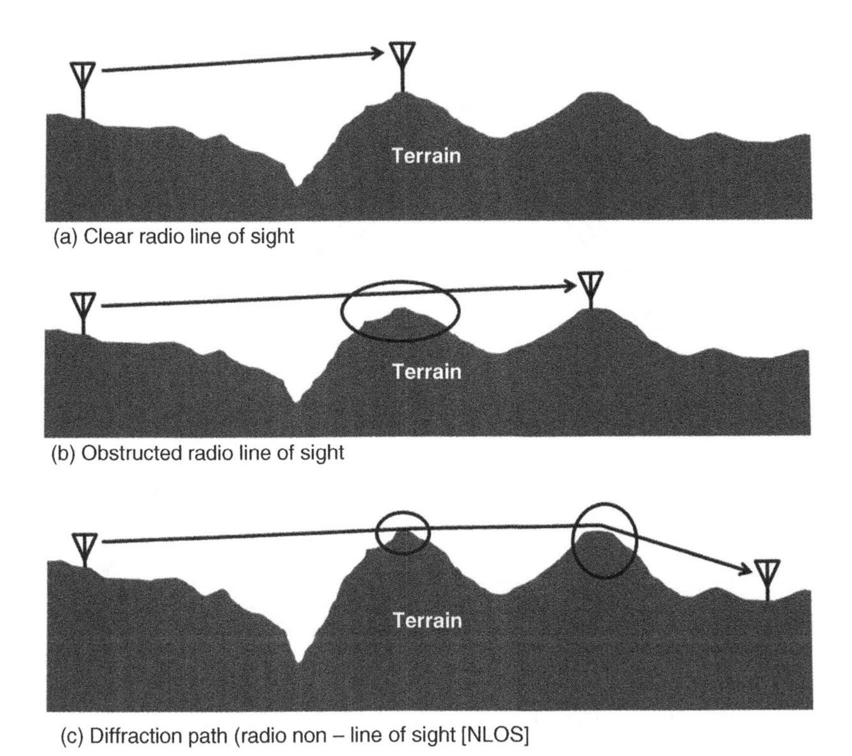

Figure 1.11 Different path topologies.

and the signals arriving at the receiver are diffracted over the obstructions. The received signal will be significantly less than for the line of sight case; however, it still may provide a workable signal level. In Figure 1.12, the effects of Tropospheric scattering are shown. This is normally not the dominant mechanism for propagation except for very long paths where diffracted signals are very small. Some systems do use this method for long-range communications.

In addition to these long term-effects, other shorter duration mechanisms can have an effect. These are illustrated in Figure 1.13. These include ducting at ground level or above it, hydrometeor scatter caused by precipitation and multipath effects. These factors can have severe effects at some times, either providing extremely long paths or inhibiting radio communications at all in adverse conditions.

The performance of radio links is also heavily affected by the environment in the immediate vicinity of the transmitting and receiving antennas. Antennas located within dense urban environments, jungle or forestry, poor ground conductivity or close to sources of radio noise or interference. These will be covered in more depth in Chapter 10 and throughout the rest of the practical part of the book.

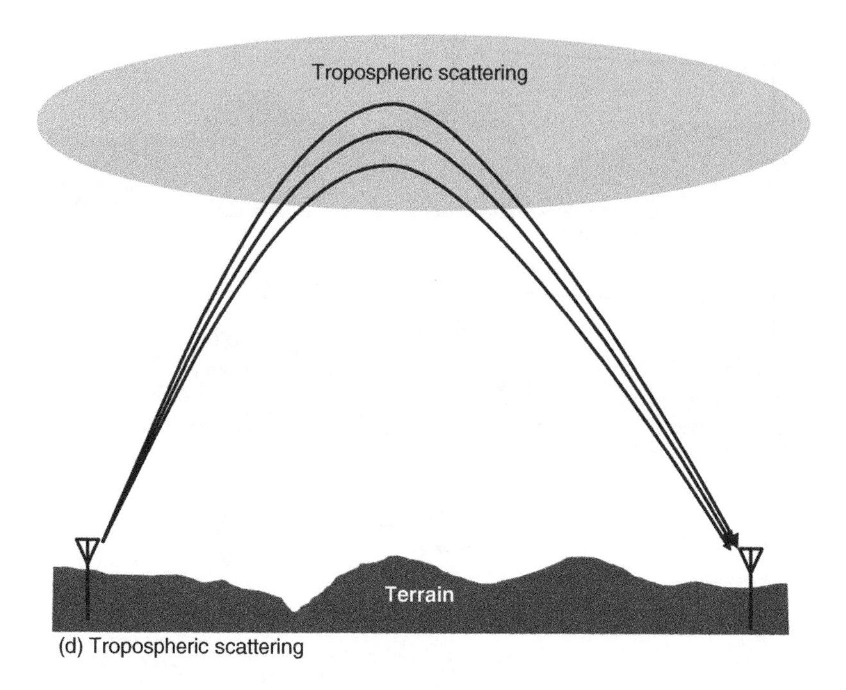

Figure 1.12 Long-term propagation mechanisms.

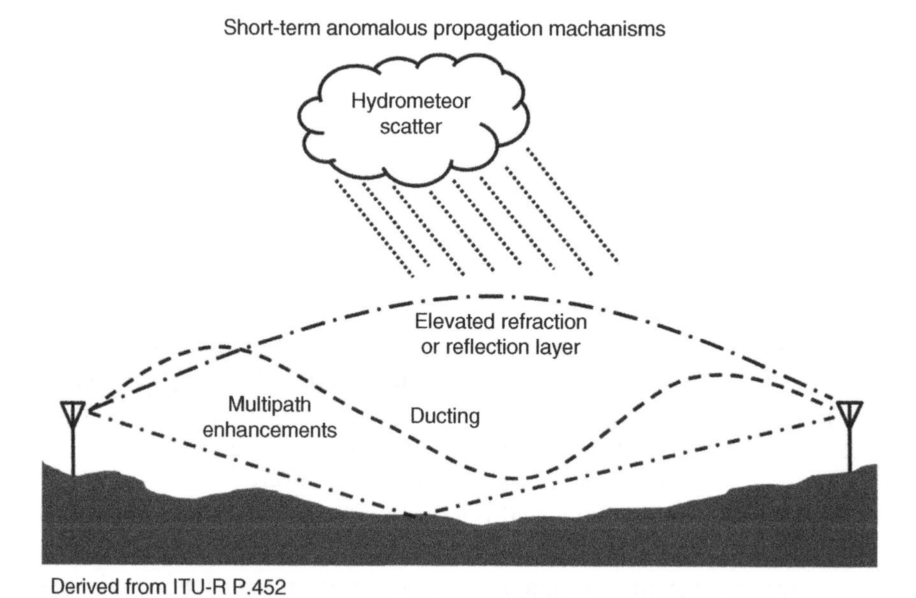

Figure 1.13 Short-term propagation effects. Diagram derived from ITU-R P.452.

For microwave frequencies, it is also necessary to consider a range of other factors that are less important at lower frequencies. This includes precipitation and atmospheric attenuation factors. These are also covered more in Chapter 10.

At all frequencies, a limiting factor will always be noise and potentially interference or deliberate jamming at the receiver.

1.3.4 Fading in Radio Systems

All of the factors we have so far discussed are applicable over relatively large areas and when catered for appropriately be used to predict signal strength variations fairly accurately. However, they do not account for another, very important class of variations so far not considered. This is termed 'fading' and it is used to account for sometimes large signal variations over small areas; in fact, their variations occur over distances from half a wavelength to many wavelengths. Figure 1.14 shows the half-wavelength variations over a short range.

Variations in these ranges are typically not possible to calculate directly, because to model them directly requires a generally unacceptable level of detail of the link environment. Instead, statistical models must be used.

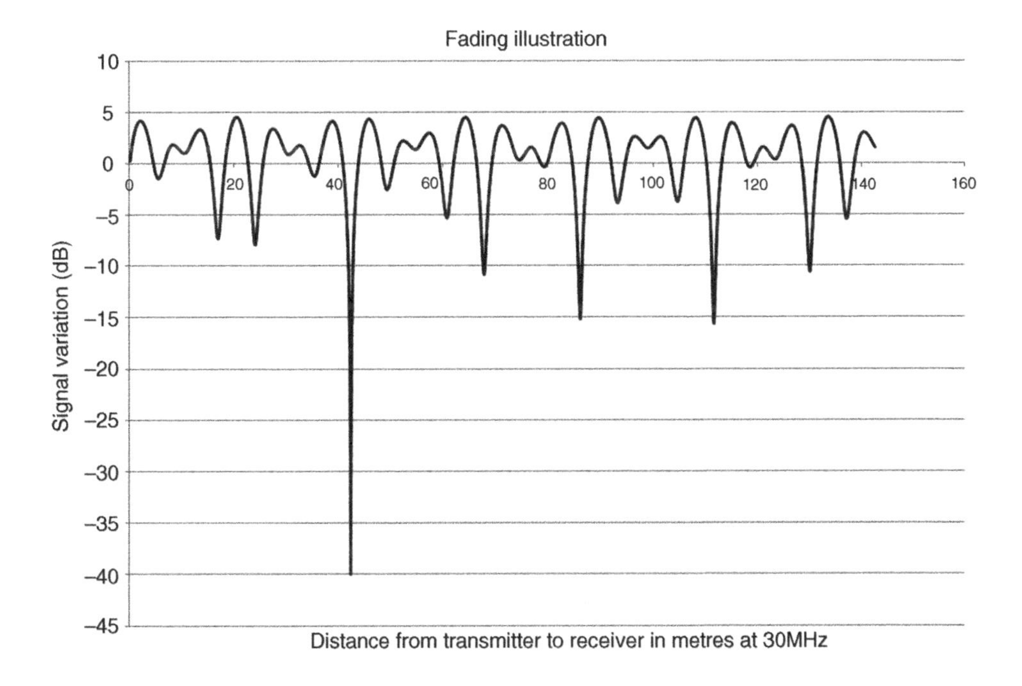

Figure 1.14 Typical variations over a distance of a few wavelengths.

As can be seen from the figure, signals vary substantially over these short distances. Typical fades can easily exceed 40 dB or more. This means that a link that may appear to be usable may not, simply because of the effects of fading. Fortunately, we can use statistical methods to determine the probability of the depth of fades in a system and thus determine the overall Probability of Successful Operation (PSO). We will see later that Rayleigh or Ricean fading characteristics are capable of modelling line of sight links and non-line of sight links equally successfully.

Without going into the detail of modelling fading, it is vitally important that the reader understands that fading is a crucial factor and without considering it, we run the risk of inaccurately assessing any kind of RF link.

References and Further Reading

All internet references correct at time of writing.

Australian Government IPS Radio and Space Services, *Introduction to HF Modelling,* http://www.ips.gov.au/Category/Educational/Other%20Topics/Radio%20Communication/Intro%20to%20HF%20Radio.pdf.

Bradley, P.A.; Damboldt, Th.; Suessmann, P. (2000), *Propagation Models for HF Radio Service Planning,* IEE HF Radio Systems and Techniques No. 474, 2000.

Freeman. L. (2007), *Radio System Design for Telecommunications,* John Wiley & Sons, NJ, USA, ISBN 978-0-471-75713-9.

Graham, A.W.; Kirkman, N.C.; Paul, P.M. (2007), *Mobile Radio Networks Design in the VHF and UHF Bands: A Practical Approach,* John Wiley & Sons ISBN 0-470-02980-3.

Recommendation ITU-R P.341: The Concept of Transmission Loss for Radio Links.

Recommendation ITU-P.452: Microwave Interference Modelling.

Sklar, B. (2001), *Digital Communications: Fundamentals and Applications 2nd Edition,* Prentice Hall, USA, ISBN 978-0130847881.

2

Management of the Radio Spectrum

2.1 Spectrum Management Fundamentals

Spectrum management is a crucial aspect of ensuring that the best use can be made of the radio frequency spectrum. Without such management, there is the risk of unintentional service denial, fratricide and lack of exploitation of opportunities. This is even truer for electronic warfare than it is for communications, radars and navigation systems. Part of this is due to the far more dynamic nature of EW operations, and part of it due to the intentional use of spectrum crippling techniques – often the difficulty is constraining the effect on the target rather than own and collateral systems.

Spectrum management is also important because all communications, non-communications and EW systems that use the spectrum work under the umbrella of a spectrum management system.

In this chapter, we will look at spectrum management as it affects spectrum-dependent systems. First we will look at some methods of spectrum management to minimise interference between authorised users. We will look at differing spectrum management philosophies and compare their potential in a variety of scenarios. We will look at civil and military spectrum management, and why in some ways their requirements are converging, and then we will look at the management of EW within the context of military spectrum management.

2.2 Civil Spectrum Management

Radio signals do not respect national borders. Because of this, it is necessary for there to be an internationally-agreed mechanism for spectrum management and indeed, there is. Management is carried out under the auspices of the International Telecommunications Union (ITU), which is a body of the United Nations (UN). The ITU

Communications, Radar and Electronic Warfare Adrian Graham

has been in existence since the beginning of the use of radio, and it has evolved as new technologies and applications have emerged. Its headquarters are in Geneva, Switzerland. Every four years, there is a Plenipotentiary Conference to determine the strategy and goals for the following four years, and there are World Radiocommunication Conferences (WRCs) to review the Radio Regulations (RRs) that form the basis of further recommendations and standards for the design and use of radio systems. These recommendations and standards are being continually revised and improved to meet emerging needs.

Spectrum is allocated at an international level for specific primary and secondary uses by band. These allocations are made for the three ITU regions. Region 1 covers Europe and Africa, Region 2 covers the Americas and Region 3 covers the rest of the world.

Although allocations are determined within the ITU, each nation has its own national regulator, responsible for all allotment and assignment of services and operators to spectrum. In this context, the terms allocation, allotment and assignment have specific meanings:

- Allocations are made by allotting spectrum to particular services, whether there are any such systems within a country or not.
- Allotments are made within the context of the Table of Allocations within a specific country or group of countries; for example, the allotment of the band 225–400 MHz to NATO UHF air services (although some countries now use the sub-band 280–400 MHz for TETRA systems, by mutual agreement).
- Assignments are made from within an authorised allotment. In the NATO air band example, it might mean assigning a specific air mission with a frequency of 229.750 MHz for the duration of the task.

An illustration of how spectrum management fits together is shown in Figure 2.1. The national regulator is responsible for liaising with the ITU, ITU regional bodies and the regulators of adjacent countries. The national regulator is also responsible for all national spectrum management although, as shown in the figure, parts of the responsibility may be devolved to other organisations, of which the military is typically one. Over all, the national spectrum regulator has the task of providing the best benefit from radio spectrum for their country without causing undue interference to other nations.

There are a number of different ways of managing spectrum. The traditional approach was termed ‘command and control’, where spectrum is managed centrally and rigidly by the regulator. However, in recent years, the benefits of spectrum deregulation and moving management closer to the network operators have become apparent. In this more modern model, management of a subset of the spectrum is devolved from the national regulator to SMOs (Spectrum Management Organisations). An SMO may be responsible for particular parts of the spectrum nationally

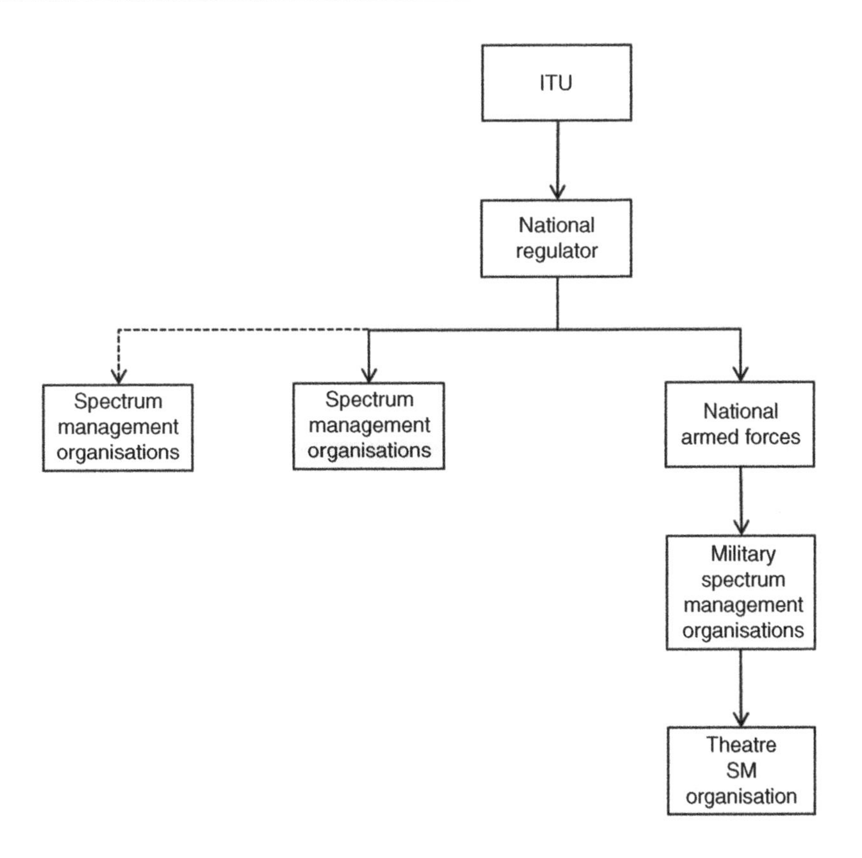

Figure 2.1 Example of national spectrum management structure. The use of Spectrum Management Organisations (SMOs) implies that the system used is not the traditional 'command and control' method.

or regionally. For example, an SMO may be in given responsibility for national TV broadcasts, for managing outside broadcast assignments or for managing taxi communications.

The spectrum management task is to make best use of the spectrum on behalf of the stakeholders. One of the principal tasks is to ensure that authorised spectrum users are protected from interference from other spectrum users. This is achieved via a number of mechanisms:

- Allocations that limit the potential co-existence of different services that may interfere with each other. This is an administrative method that provides spectral separation.
- Allotments designed for specific services or operators. This again is an administrative process providing spectrum separation between dissimilar systems and between services offered by different network operators.

- Assignments for particular up-links and down-links (where they both exist; for example in broadcast there is no uplink). Assignments are normally determined by technical means, based on factors such as:
  - distance from other stations using the same frequency;
  - geographic limits of interfering signals from stations, determined usually by computer predictions;
  - time-based, when some networks do not need to operate continuously;
  - time slot de-confliction, for systems that use Time Division Duplexing (TDD);
  - use of spreading codes (also known as PN codes);
  - use of polarisation in some cases.

It can be seen from this list that spectrum management is partly an administrative process and partly a technical process. Systems that rely more on the administrative methods tend to be conservative and less flexible, as well as being less optimal than those systems based on technical de-confliction. However, technical de-confliction is more demanding to achieve. It requires advanced spectrum management tools, experts to use them and all of the logistics and planning required to achieve and retain the level of knowledge required. However, given that spectrum management in the civil sphere normally changes very slowly and licences tend to be issued to last for a number of years. This is necessary for licence holders to be able to commit to the large investments required to implement large networks and for them to recover their investment. As we will shortly see, this is very different from spectrum management for military operations.

2.3 Military Spectrum Management

2.3.1 Introduction

Civil spectrum management can be characterised in the following way:

- It occurs in a known geographical location.
- The assignment process can be slow, but this is not a problem because the time to acquire a licence is normally known in advance.
- Licences, once assigned, last a long time; usually years.
- Plans can be made years in advance.
- Information on spectrum users is easily obtainable and all important parameters are known explicitly.
- The national regulator has complete control over the spectrum; with the limited exception of pirates, normally in radio broadcast, there will be no unauthorised or third-party spectrum users.

Military spectrum management includes assignments with the same characteristics; for example, air defence networks to protect a nation are unlikely to change rapidly and the area remains the same. However, military operations have very different requirements:

- It can occur virtually anywhere and anytime with little or no notice.
- Plans must be made and maintained rapidly.
- Plans often change at short notice.
- Spectrum demand for forces can greatly exceed demand.
- Spectrum demand changes during an operation and its aftermath.
- In coalition warfare, there will be limitations imposed by the different equipment in use and due to differences in operational procedures.
- For legacy reasons, some equipment parameters are likely to be unknown.
- Information on the location and movements of lower force elements may not be explicitly known.
- There will often be contention with civil spectrum users. While in high-intensity warfare, civilian users can be suppressed if necessary and justified, but for longer operations, co-existence is necessary.
- There may be enemy forces with their own Electronic Attack (EA) and Electronic Support (ES) assets who have their own objectives.

Thus, military planners have to deal with a far more dynamic, less well understood and more hostile scenario. It is possible to break this problem down into a number of well-defined areas and address each in turn.

1. Determine own forces spectrum demands for each phase of the operation.
2. Identify spectrum to be protected.
3. Determine and express operational areas as accurately as possible.
4. Perform frequency assignment and allotment for distinct time phases of the operation.
5. Create the Battlespace Spectrum Management (BSM) plan.
6. Disseminate BSM plan.
7. Manage interference via interference reports.

We will now look at each of these in turn.

2.3.2 Spectrum Demand Assessments

The first thing is to identify the spectrum requirements of the forces involved in the operation. This is far from a straightforward task; however, it can be done with experience and normally by using computer software tools as a decision aid. One way of achieving this process for a given phase of an operation is to use the following approach:

- Determine forces to be deployed during the phase.
- Determine spectrum requirements for deployed forces, based on network requirements.
- Identify locations of forces as far as possible.
- Collate radio equipment holdings for the forces.
- Generate a spectrum demand plan.

As far as possible, spectrum usage should be determined by operational requirements; spectrum management is a subordinate process to military command.

Let us assume that the following hypothetical forces have been identified for a small operation (the example is not meant to be realistic, but rather to illustrate the principles). These are shown in Table 2.1. This would be specified by the military command; it is an input to the spectrum management process, not part of it.

The spectrum requirements are based on the number of channels required rather than the number of individual radios used in each network. Note also that in Table 2.1, the HF requirements are shown as 'circuits' rather than channels. This is because each HF network may require several frequencies to maintain availability over 24-hour periods and perhaps longer. HF prediction methods must be used to determine how many channels are required for each circuit before the total 'spectrum bill' can be produced. An interim step is shown in Figure 2.2. The data has been manipulated to collate spectrum requirements.

Table 2.1 Hypothetical force spectrum requirements

ID	Force element	Spectrum requirements
0001	SPECIAL FORCES DETACHMENT	4 × HF circuits 8 × VHF channels 4 × UHF channels 2 × SHF channels
0002	WARSHIP NONSUCH	6 × HF circuits 12 × VHF channels 12 × UHF channels 6 × SHF channels
0003	WARSHIP NONSUCH AIR GROUP	8 × UHF channels
0004	1 REGIMENT	6 × HF circuits 12 × VHF channels 6 × UHF channels 12 × SHF channels
0005	1A BATTALION	2 × HF circuits 6 × VHF channels 12 × SHF channels
0006	1B BATTALION	2 × HF circuits 6 × VHF channels 12 × SHF channels
0007	TEMPORARY AIRFIELD	2 × HF circuits 5 × VHF channels 4 × UHF channels 6 × SHF channels
0008	LOGISTICS DETACHMENT	2 × HF circuits 6 × VHF channels 6 × SHF channels

	HF	VHF	UHF	SHF
Special forces det	4	8	4	2
Warship nonsuch	6	12	12	6
Warship nonsuch air group	0	0	8	0
1 Regt	6	12	6	12
1A Bn	2	6	0	12
1B Bn	2	6	0	12
Airfield	2	5	4	6
Logs det	2	6	0	6
Requirement	**24**	**55**	**34**	**56**

Figure 2.2 Sample partly-processed spectrum bill.

It is likely that some of the spectrum demands can be expressed more explicitly. For example, the frequency and bandwidth of the airfield primary and secondary radars are likely to be fixed, and some of the SHF channels may be used for specific SATCOM channels. Thus, rather than placing a requirement for a band, it will be possible in some cases to specify individual channels.

We can also look at the deployment of forces. This is shown in Figure 2.3. The warship is at sea, providing helicopter support to land-based forces. The regiment level

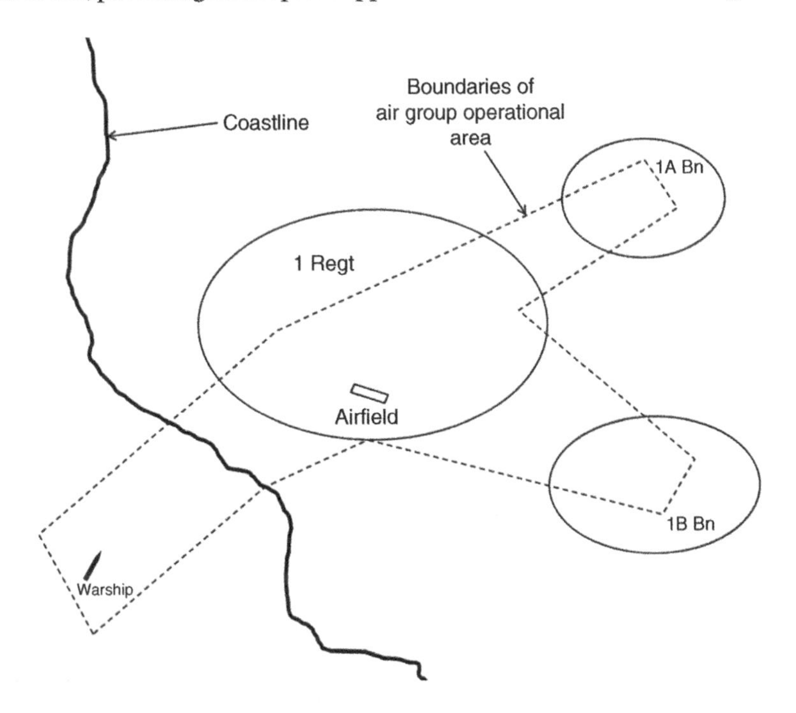

Figure 2.3 Hypothetical force deployment.

communications are primarily within the ellipse shown, and likewise for the two battalions. The logistics detachment may be within any of the three areas shown and the location of Special Forces is unknown.

The spectrum manager can also refine the spectrum demands to determine the actual requirement. This can be based on Standard Operating Procedures (SOPs). For example:

- HF channels are not reused within the operational theatre because they cannot be separated by sufficient distance.
- HF channel requirements are calculated using HF planning tools. The tool shows that in total 96 HF 3 kHz channels are required.
- Where possible, VHF channels are to be re-used. This will depend on further de-confliction analysis.
- UHF channels for PRR will be re-used where possible. Since PRR is a short-range system, it is very likely that channels in each of the three ground areas can be re-used without significant risk.
- The UHF channels not used for PRR are used to communicate with aircraft. Therefore some of the requirement between the air assets and the ground assets are actually common rather than separate.
- Two SATCOM links can be used on a sharing basis by all deployed forces.

From this, we can refine the spectrum bill further as shown in Figure 2.4. The reduction in VHF-SHF requirement is achieved through frequency re-use.

The collated spectrum bill can be presented to the spectrum authority as a spectrum request. In total warfighting, this would be the relevant military authority. Often, however, it will be the country in which the operation is occurring (normally referred to as the 'host country'). The host country will then analyse the requirement, compare it to the spectrum available in the operational areas and then reply with a statement of which requirements can be met. Once that has been received, the planning process can continue.

2.3.3 Spectrum Protection

Before starting any frequency assignment or allotment, it is necessary to identify particular spectrum blocks that need to be protected so that they can be removed

	HF	VHF	UHF	SHF
Requirement	96	40	16	14

Figure 2.4 Refined spectrum bill.

from the list of available frequencies. Spectrum that needs to be protected theatre-wide includes:

- spectrum not allocated to the services to be assigned (by the Radio Regulations);
- emergency and distress channels;
- specific channels and spectrum blocks identified by the host country.

In addition, it is important not to cause interference to adjacent countries so care must be taken when assigning channels that may propagate across the borders.

Also, there may be other parts of the spectrum that may require protection over part or all of the operational theatre. These may include:

- safety of life spectrum;
- channels of interest to intelligence and subject to ES exploitation;
- other channels that require protection for other reasons.

Protection of this spectrum is managed via the Joint Restricted Frequency List (JRFL) process ('JRFL' is often pronounced as a word, as in 'jirfil', for shorthand). Under this process, there are three different categories:

- TABOO: friendly or neutral frequencies that are so important that they must not be jammed or interfered with. This includes distress frequencies and vital command networks.
- GUARDED: enemy frequencies being exploited for intelligence purposes via ES systems, which must not be jammed or interfered with.
- PROTECTED: enemy frequencies being exploited for combat intelligence which should not be jammed until the commander provides permission.

This method allows a complete list of frequencies and bandwidths to be generated. This list then forms an input into the assignment and allotment process.

2.3.4 Assignment and Allotment Process

The assignment and allotment process is designed to allow authorised users to obtain spectrum that meets their needs and to avoid interference that may cause unacceptable degradation of their service. Assignment is the process of directly assigning individual or groups of frequencies to specific emitters and networks. Allotment is the process of providing a group of frequencies or blocks of spectrum to another spectrum manager for them to assign. For both assignment and allotment, it is vital not to interfere with the entries in the JRFL.

One way of processing the spectrum management data is as follows:

- Start by looking at the spectrum requested under the Spectrum Supportability Request (SSR).
- Subtract any limitations imposed by the Host Nation Declaration (HND).
- Subtract all *theatre-wide* JRFL entries.
- The remaining spectrum is available as an input to the assignment and allotment process.
- Collate all relevant location and technical data about each frequency request.
- Perform technical assignment and allotment.

This is illustrated in Figure 2.5. The left hand side of the diagram shows the process and the right hand side illustrates how the available spectrum for assignment and allotment is derived by subtracting the theatre-wide JRFL entries from the Host Nation Response.

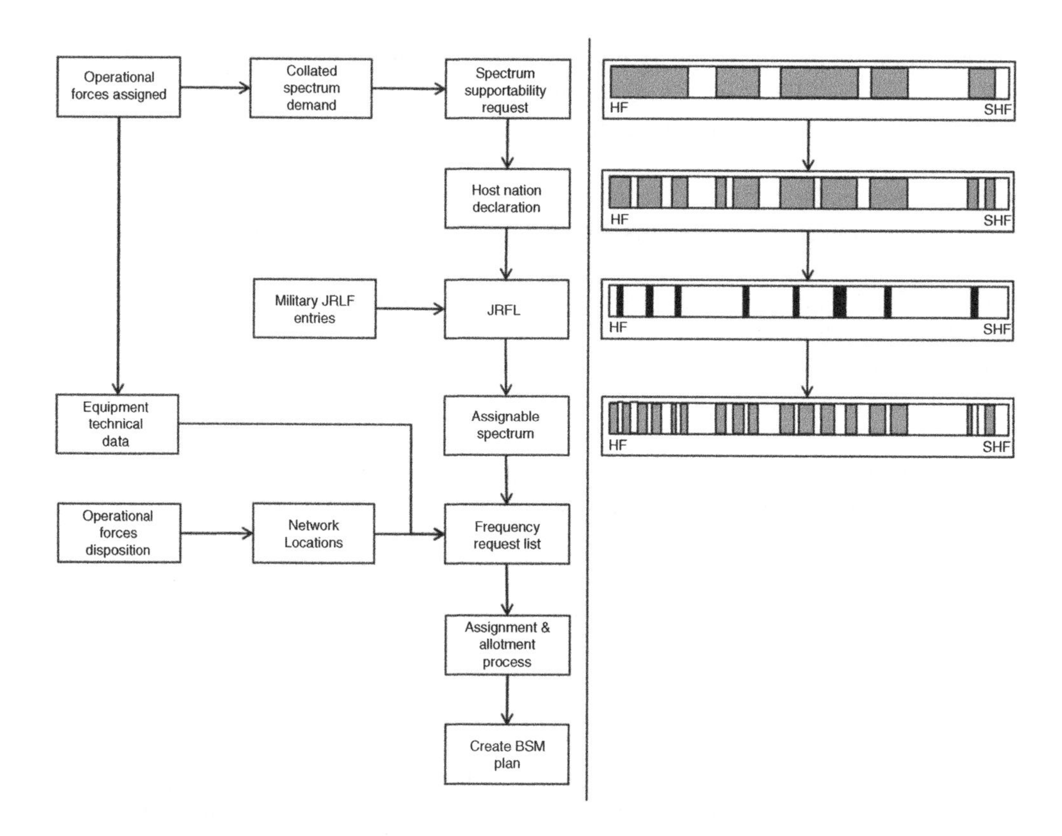

Figure 2.5 Illustration of an approach to deriving the BSM plan and the available spectrum for assignment and allotment.

The frequency request list is the totality of spectrum requirements from all users. This requires having technical data about each emitter in each network and also the Area Of Responsibility (AOR) of each force element.

The technical parameters required for each network include:

- transmit power levels out of the antennas;
- minimum and maximum frequencies the equipment can support;
- bandwidth required;
- tuning properties; whether the system can be tuned continuously, only by specific frequency steps or only at specific frequencies;
- antenna directional characteristics, if the antennas are directional;
- antenna heights above ground and, for directional antennas, their direction and tilt;
- the minimum signal strength that must be available at each receive antenna;
- the ability of the system to reject interfering signals.

The location of the network can be expressed in a number of ways, such as:

- an explicit set of geographic coordinates;
- within a given number of kilometres of a known geographic location;
- within a polygon that describes the force element AOR;
- within multiple AORs;
- undefined within the operational theatre.

The use of polygons to describe network locations is a simple and effective method to describe the scenario where explicit locations are either not known or may change during the lifetime of the assignment. A polygon identifies where the forces using the network will be within without giving an explicit – and unknown or estimated – actual location.

An example is illustrated in Figure 2.6.

The reason why this description of force element excursion boundaries is important is that we can use this as a basis of spectrum de-confliction using a method called 'sterilisation area'. This is based on the principle that if the maximum extent of transmission origination in each area is known, as expressed by the polygon, then we can calculate a region around the polygon that describes the limits that interfering signals will propagate at sufficient strength to cause interference to a specific type of external radio receivers. An example of how a sterilisation area works is shown in Figure 2.7. This shows both the AOR and its corresponding sterilisation area.

We will cover interference in later chapters but for now, it is enough to know that the sterilisation area is not set by the ability of the radios within the AOR to tolerate interference, but rather by the ability of other networks that may be operating near this AOR to tolerate interference from the transmitters within the AOR. This will change

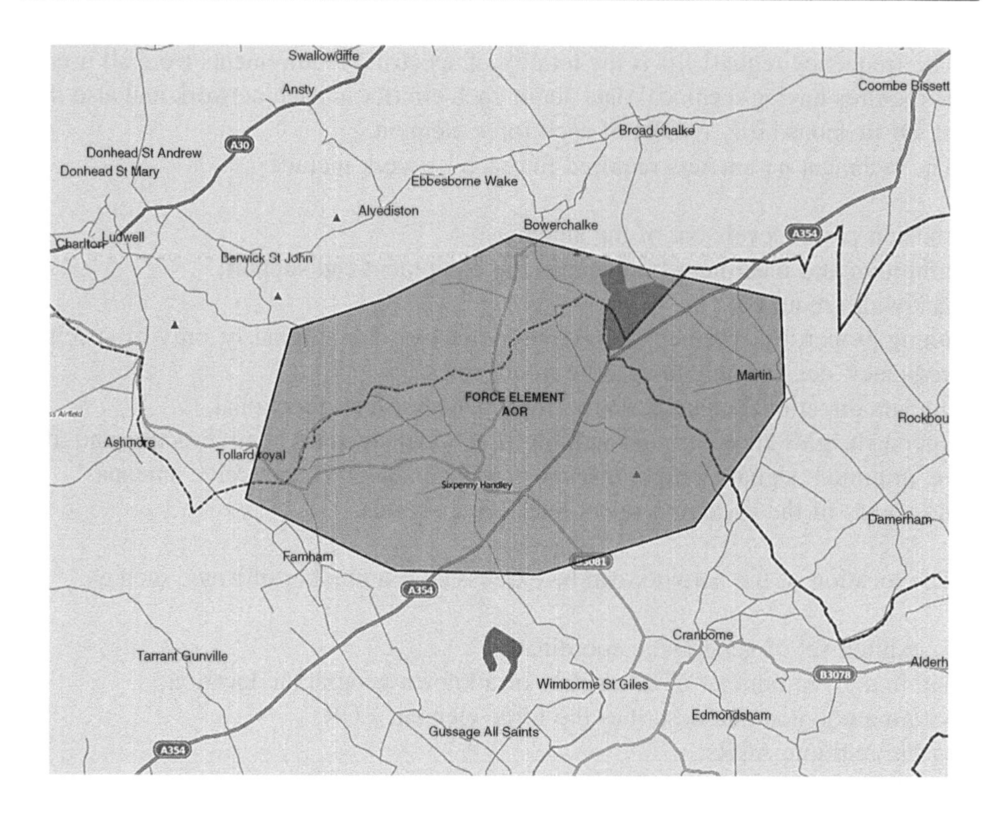

Figure 2.6 A hypothetical Area of Operation (AOR) for a force element during an exercise. All we know if that the members of the force element will be somewhere within the AOR during a specific timescale. However, if we know that the force element will not leave this area, then it acts as a method of describing where radio transmissions from the force will originate from. This helps with the frequency assignment and allotment process.

between different types of networks, so if there are 40 different networks that may suffer interference and they have different interference tolerance, then 40 different sterilisations need to be calculated. This is however easy computationally.

Using this approach, the assignment and allotment process becomes a task of determining potential overlaps of many AORs and their associated force elements. This approach provides four potential outcomes between two sets of AORs and their associated sterilisation area:

- The AORs overlap. This makes frequency re-use very difficult without creating interference problems.
- The AORs do not overlap, but one or both the sterilisation areas overlap on or both of the AORs. Again, this makes it difficult the same frequency without problem, although the likelihood of actual interference is less than when AORs overlap.

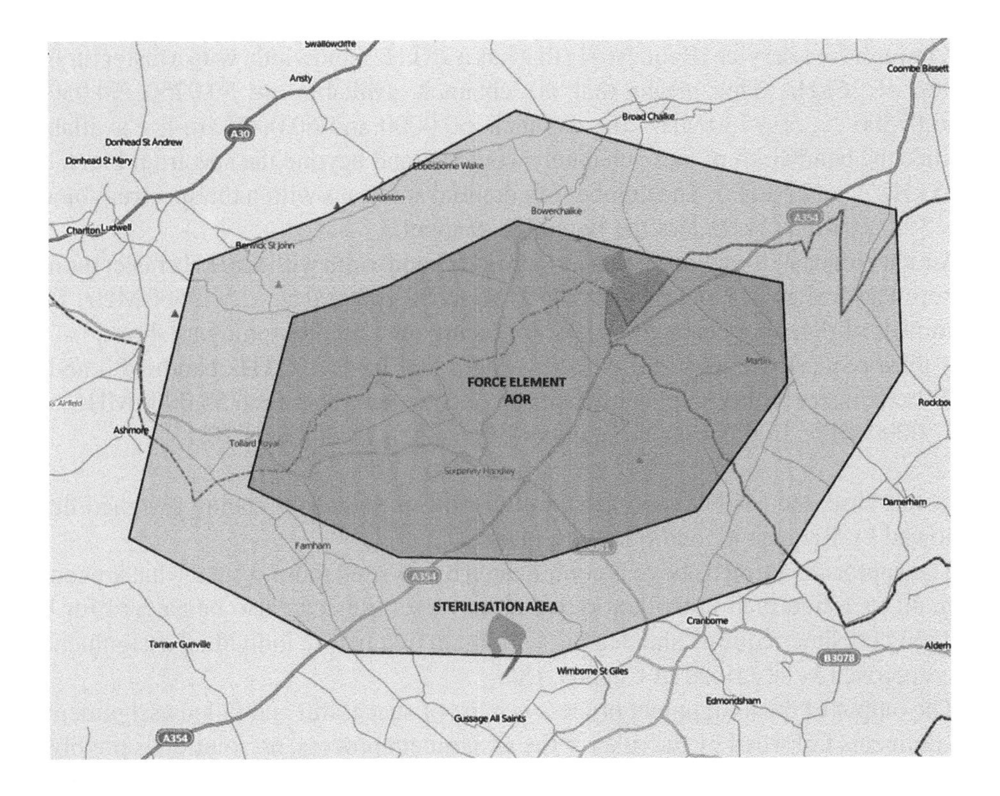

Figure 2.7 Based on the example of Figure 2.5, this figure shows a sterilisation area around the AOR. The sterilisation area is a number of kilometres around the AOR. The size of the sterilisation area does not depend on the radios used by the force element within the AOR, but rather the limits of interfering signal strength that a nearby (other) network(s) can tolerate.

- The AORs do not overlap, but the sterilisation areas overlap. Depending on how accurate the sterilisation areas are, there may still be potential interference, but the probability is less than the previous two scenarios.
- Neither the AORs nor the sterilisation areas overlap. In this case, interference should not occur and the same frequencies can be re-used safely.

It is also possible to use mathematical algorithms to optimise the assignment and allotment process. Some of these are highly complex and require substantial information about the systems to be assigned. One such set of approaches are based on the concept of Most Constrained First Assigned (MCFA). In this case, frequency requests are sorted on the basis of the number of potential solutions there are given the available spectrum. For example, consider the following.

There is a block of spectrum available from 50–60 MHz. This is not assigned or allotted and can be used for new assignments. There are potentially three types of assignment that need to be made in the spectrum block. These are:

- A request for a service (Request #1) that has a 25 kHz bandwidth, with a raster tuning step of 26 kHz. This means that the channels available are 50.0250, 50.0500, 50.07500 ... 59.9750 MHz (the channels 50.0000 and 60.0000 are not available since the bandwidth of these channels would extend beyond the spectrum block by 25 kHz/2 = 12.5 kHz). The number of potential solutions within the spectrum block is 398 (10 MHz/2 kHz less the two edge channels).
- Another request (Request #2) that has a 50 kHz bandwidth with a 50 kHz raster tuning step. The available channels are 50.0500 MHz, 50.1000 MHz ... 59.9500 MHz. The number of potential channels is 198, following the same reasoning as above.
- A third request (Request #3) is for a system that has a 100 kHz bandwidth and a 200 kHz raster tuning step, but which can only be tuned between 55.0000 MHz and 57.0000 MHz. In this case the number of channels available is 10.

In this case, the MCFA approach would result in Request #3 being assigned first, followed by Request #2 and finally Request #1.

The approach outlined above is complicated by co-sited radio systems that are more difficult to assign than individual requests; however, this can also be catered for by collating co-sited requests and dealing with them first before individual assignments. The approach is described in Chapter 18.

The output of the assignment process is a list of successful and failed assignments. Assignments fail when by the rules of the assignment process, no solution is possible. If this occurs, the assignment run can be repeated with some randomisation or seeking algorithms to attempt to improve on the original assignment outcome. This can be repeated a large number of times and the best solution chosen, even if it is not perfect. Management of imperfect assignment is the job of the spectrum manager.

However good the solution is, the outcome of the assignment and allotment process is a list of assignments and allotments. These are then incorporated into the BSM plan.

2.3.5 The BSM Plan

The NATO standard BSM plan, used as an example of the type of plan which any armed forces may produce, consists of a number of sections:

1. References: Documentation that underpins the operation and its execution.
2. Commander's Spectrum Strategy: Concept of operations, scope and mission.
3. Spectrum Management Control Process: Authority and command relationships.
4. Spectrum Stakeholders: List of stakeholders and points of contact.
5. Supported Systems: List of equipment approved for the operation.
   Force Spectrum Requirements, as annexes:
   a. Deployed forces.
   b. Map of theatre of operations.
   c. Supported radio systems.

d. Assignments and allotments.
e. Spectrum request format.
f. Interference reporting and resolution process.
g. JRFL and request format.

The assignments and allotments list is included as Annex D as shown.

The BSM plan is a living document that will change during the operation as it develops.

Once the BSM plan for a given phase has been completed and authorised, it can then be disseminated as appropriate.

2.3.6 BSM Plan Dissemination

The BSM plan has to be disseminated to all of the relevant stakeholders. To achieve this, some especially sensitive EW and intelligence information must be 'sanitised' to prevent readers from receiving information that they do not need to know. Plan management must also be carried out by recipients to ensure they are using the correct version and not a superseded one. Plans must also be sent out well before they are implemented and by sufficiently robust means that each recipient gets a copy. If some do not, there is the risk that the plan will fall apart as frequency changes to avoid interference are not implemented. There is also the possible risk that if this occurs, attempting to fix the problem over radio channels will be impossible because of the interference problems.

2.3.7 Interference Management

A major part of the military spectrum management plan involves managing and resolving interference issues. For best efficiency, this process should be carried out according to the operating procedures included in the BSM plan.

If an authorised user experiences interference, they will generate a formatted interference report. This will be sent to the spectrum manager, who will attempt to identify the interferer and determine actions to be taken. The spectrum manager should be able to identify the culprit from the interference location and the channel(s) affected. If the interferer does not appear in the BSM plan, then the interference is caused by either another party or by an authorised user not adhering to their assignments. If it is another party, it may be deliberate or unintentional enemy action or interference from civilian or other users in the area. Whatever the cause, the spectrum manager will determine the action to be taken, such as re-assigning a new frequency for the interference victim. The spectrum manager will then notify the victim of the action to be taken, or will explain how the issue has been resolved in another way.

2.4 Management of EW Activities

Electronic warfare activities need to be carried out within the context of the BSM plan, with particular reference to the JRFL. Failure to do so is likely to lead to fratricide (jamming own force communications and systems) in the worst case. This means that spectrum management and electronic warfare are intertwined activities.

In this section, we will look at EW management, and how fratricide and other problems are avoided. First, we will look at the types of EW that can be carried out. These are shown in Figure 2.8.

Jamming and directed energy weapons pose the greatest threat to radio systems. Jamming should be managed within the BSM and JRFL processes; otherwise, there is a great risk of fratricide. Directed energy weapons are still in their infancy, but in any case, they normally rely on highly-directional antennas that will minimise the risk to systems other than the intended target. The hardkill option is arguably not an EW activity, but attack planning must be carefully carried out to prevent the missiles hitting the wrong target.

Protection of EW assets and their associated communications from interference can be achieved through the spectrum management process, EMCON (Emission CONtrol) policy and by using emission shielding. EMCON is an administrative process that manages rules whereby particular systems can emit or when they are barred from doing so. Emission shielding can be achieved by trying to ensure that when RF energy is radiated, so that it is not available to the enemy. An example would be placing a communications centre and its associated antennas on the slope of a hill facing away from the enemy; the hill will block transmissions in the direction of the enemy.

EW activities are normally managed by a dedicated cell, usually referred to as the Electronic Warfare Coordination Cell (EWCC). The EWCC is the authority that liaises with the theatre spectrum manager to coordinate their activities with the wider RF activities.

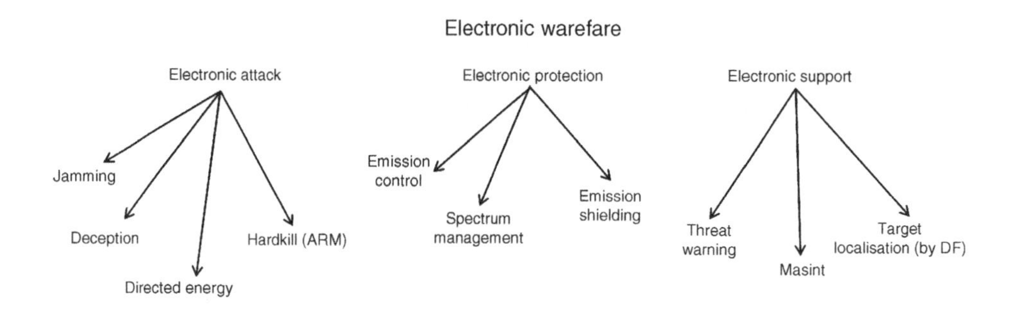

Figure 2.8 Typical electronic warfare-related activities. ARM stands for Anti-Radiation Missile.

References and Further Reading

All internet references correct at time of writing.

The following are a small sample of the type of information available freely on the internet (except for the Radio Regulations and Cave et al). However, the reader is recommended to use their internet search engines to obtain the latest copies and documents available.

Cave, M.; Doyle, C.; Webb, W. (2007), *Essentials of Modern Spectrum Management*, Cambridge University Press, USA, ISBN 0 521 876 699.

ITU (2008), Radio Regulations, http://www.itu.int/publ/R-REG-RR-2008/en.

Recommendation ITU-SM.1047 NATIONAL SPECTRUM MANAGEMENT.

Recommendation ITU-SM.1370-0 DESIGN GUIDELINES FOR DEVELOPING ADVANCED AUTOMATED SPECTRUM MANAGEMENT SYSTEMS (ASMS).

US Army (2007), US Army Concept Capability Plan for Electromagnetic Spectrum Operations for the Future Mobile Force 2015-2024, http://www.tradoc.army.mil/tpubs/pams/p525-7-16.pdf.

US DoD (1997), Army Management of the Radio Spectrum, http://www.army.mil/usapa/epubs/pdf/r5_12.pdf.

US DoD (2007), Joint Publication 3-13.1 Electronic Warfare, http://www.fas.org/irp/doddir/dod/jp3-13-1.pdf.

References and Further Reading

3

The Radio Channel

3.1 Frequency Aspects of the Radio Channel

In our brief introduction to radio propagation in Chapter 1, it was identified that the 'radio channel' is in fact a complex medium that is dependent on a wide range of mechanisms that can affect some radio transmissions heavily and some others less so. This gives the concept of radio bands; portions of the radio spectrum that behave in largely the same way and differently to those outside of that band. Figure 3.1 illustrates the terminology given to different bands by those working in different aspects of radio.

These are the bands often referred to for convenience. However, in terms of understanding how radio prediction works, it is often worth thinking not about frequency but rather in terms of wavelength. The wavelength of a radio signal is given by the following simple formula:

$$\lambda = \frac{c}{f}$$

where
λ is wavelength in metres.
c is the speed of light; approximately 3×10^8 metres/second.
f is frequency in Hz.

If the frequency is expressed in Megahertz (MHz), then this formula can be simplified to

$$\lambda = \frac{300}{f(\text{MHz})}$$

The reason why wavelength is a more useful unit to work in than frequency when considering propagation effects is that in general, objects need to be typically 10's of wavelengths in size to have significant effect. Thus, from Figure 3.2 at 3000 MHz

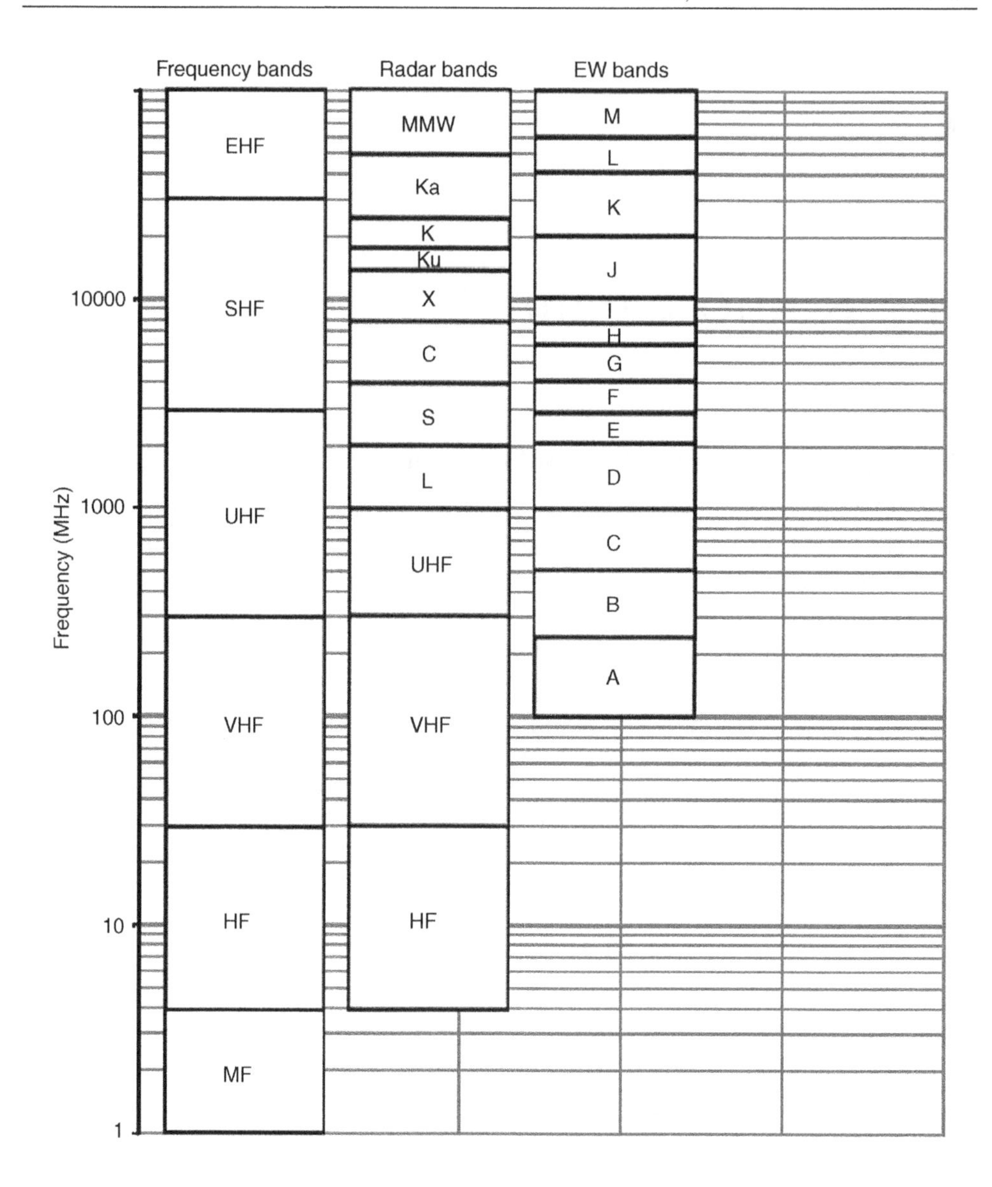

Figure 3.1 Frequency bands and terminology used to describe them.

(3 GHz), objects only a few centimetres across may be relevant. At 30 MHz, however, the object must be many 10's of metres to have an effect. This is the size of a building, ship or large aircraft, but more often, hills and other terrain features are likely to be the dominant mechanisms. At lower frequencies, obstructions need to be even larger to have an effect.

The graph shown, like many others in the book, can be used by the reader to directly read off the required values. This is why these figures are reproduced at as large a scale

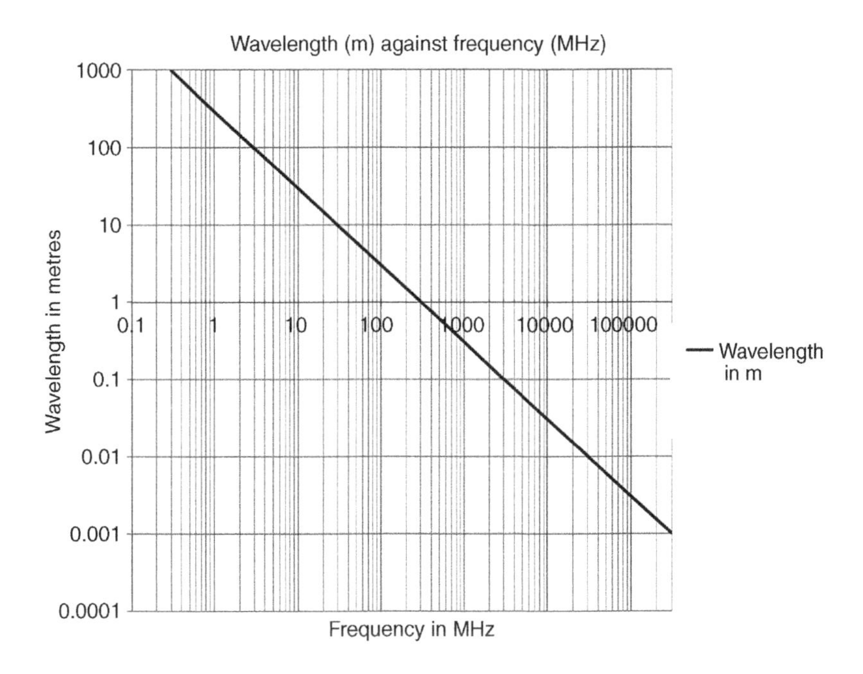

Figure 3.2 Relationship between frequency and wavelength.

as possible. Alternatively, such graphs can be reproduced by the user using Microsoft™ Excel™, MATLAB or similar packages.

Later in the book, we will be looking at something called the 'short sector', which is approximately 40–50 wavelengths. Figure 3.3 shows the short sector as 50 wavelengths, for easy reference. For now, it is enough to know that the short sector is the boundary at which radio engineers switch from using radio prediction techniques to determine the received signal strength and instead start to use statistical methods. This is described in greater depth later in this chapter.

The short sector is important because over its length, we can expect nominal signal strength to remain much the same, but we will be able to isolate and treat independently fast fading mechanisms.

From the last two figures, it should be clear that wavelength is a crucial aspect of radio propagation. This is indeed true and in fact, it is an important factor in any equation to represent radio propagation. This can be seen by considering free space loss against distance for a range of different frequencies. It can be seen that for a decade difference in frequency, there is a 20 dB difference in transmission loss. Free space loss is the term used to describe the loss due to spreading of the signal as it radiates into a non-ionising medium.

The free space loss equation can be derived from simple principles. Figure 3.4 shows the basic scenario. In this diagram, there is an 'isotropic radiator' at the centre of a

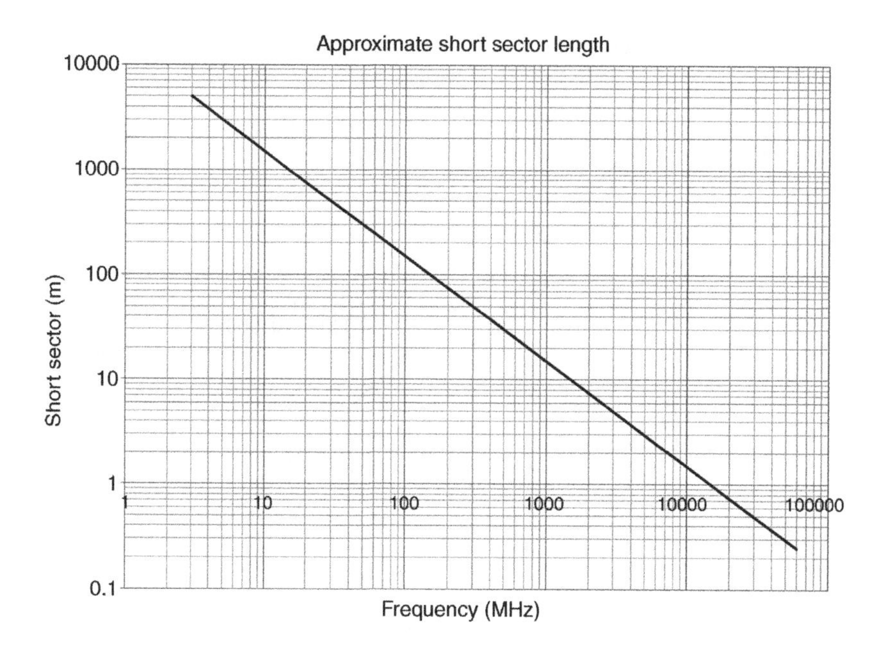

Figure 3.3 The length of the short sector against frequency, based on the assumption that the short sector is fifty times the wavelength of the transmission frequency.

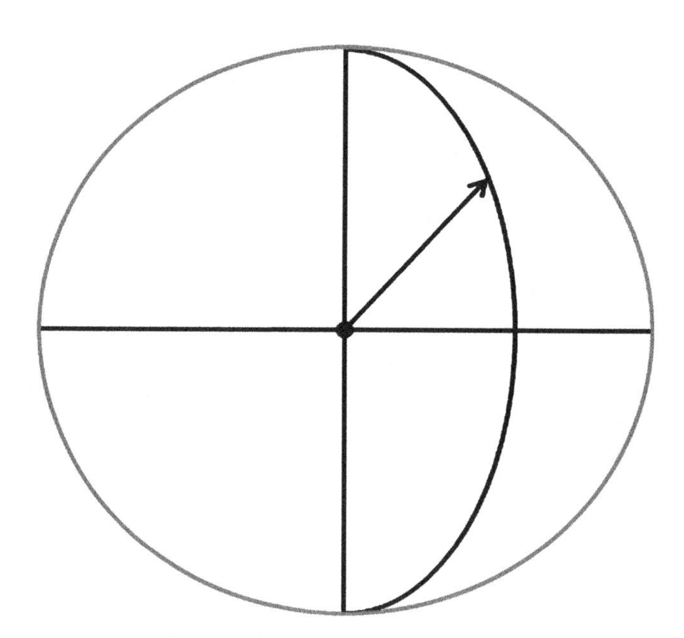

Figure 3.4 The figure shows an isotropic radiator at the centre of a sphere. Because an isotropic radiator radiates in all directions equally, the energy produced is reduced as a function of range as the advancing wavefronts cover a larger surface area.

sphere. An isotropic antenna is a theoretical antenna that radiates equally in all directions.

The sphere represents a given distance from the radiator (d). As the distance increases, the surface area increases accordingly and, since the energy produced by the radiator does not increase, the energy per unit of surface area decreases accordingly. This can be described by:

$$L = \frac{(4\pi)^2 d^2}{\lambda^2}$$

where
L is the loss ratio.
d is distance in metres.
λ is wavelength in metres.

Since we have an equation linking wavelength to frequency, we can re-write this formula as:

$$L = \frac{(4\pi)^2 d^2 f^2}{c^2}$$

Normally, radio engineers work in decibels (dB). For those who need it, there is a refresher on decibels in Appendix A. Converting the above formula into its logarithmic equivalent, we get:

$$L = 20\log(4\pi) + 20\log(d) + 20\log(f) - 20\log(c)$$

The term $20\log(4\pi)$ is a constant and is equal to 22.0. The same is true for $20\log(c)$, which is 169.5. Thus, the equation becomes:

$$L = 20\log(d) + 20\log(f) - 147.5$$

The units for distance are in metres and the units for frequency are in Hz. We can convert the formula into a more useful form by converting the distance units into kilometres by adding 20log(1,000), which is 60, and converting the distance units into MHz by adding 20log(1,000,000), which is 120. Because we are working in dB, this is entirely acceptable. The constant term thus becomes $180 - 147.5 = 32.5$, giving us the form most often quoted:

$$L = 32.5 + 20\log(d) + 20\log(f)$$

Using this equation for different distances and frequencies, we can calculate and graph free space loss as shown in Figure 3.5. We will see later that although free space

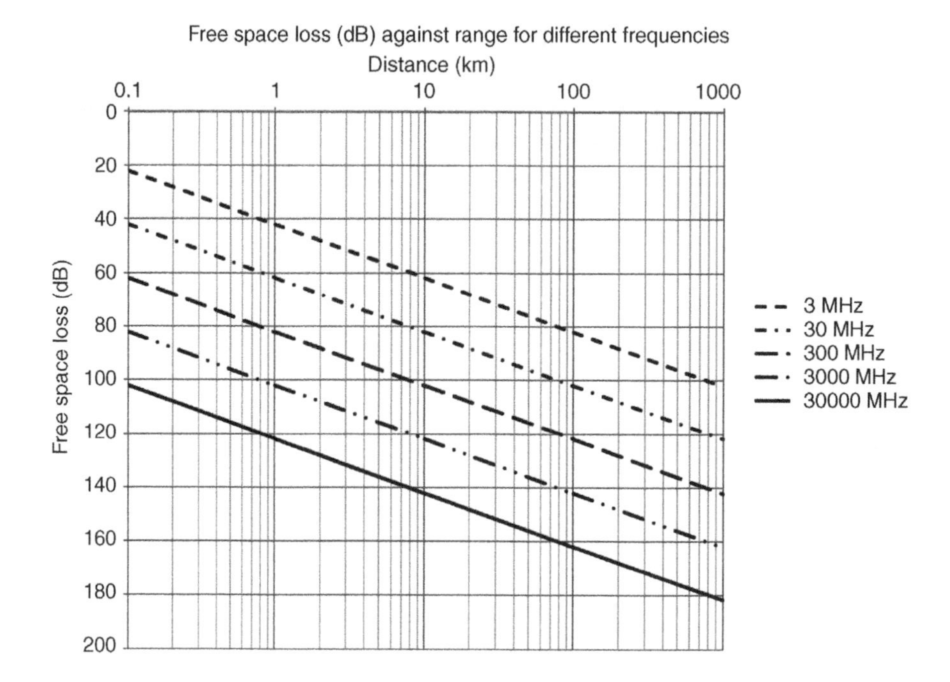

Figure 3.5 Free space loss calculated for different distances and frequencies. Note that although the concept of free space loss is important in radio engineering, it is seldom possible to use this equation for practical radio links.

loss is a very useful concept, it can rarely be used to predict real radio signal strength received at an antenna. However, it gives us a good understanding of the way frequency affects radio propagation as shown in Figure 3.5.

The graph shows that the higher the frequency, the higher the loss; note the vertical axis in the diagram. It increases as towards the bottom of the graph. The reason for showing it in this manner is that loss is often expressed as a negative value and it is worthwhile for the user to be familiar with the different ways that this is expressed in different books and other sources.

A simple thought experiment demonstrates the importance of the basic relationship between loss and frequency. Consider the situation where a system will function as long as there is no more than 125 dB of loss between the two antennas. This is illustrated in Figure 3.6.

The implication of this is that frequency selected will have an impact on system range and this is in fact true. In general, the lower the frequency, the longer the potential link given the same maximum loss value. However, this is only one aspect of link design. It is also important to consider the type of message to be carried, and the modulated scheme to be used. If the bandwidth of the signal to be transmitted is relatively small, such as for voice and low-data rate systems, then there is little

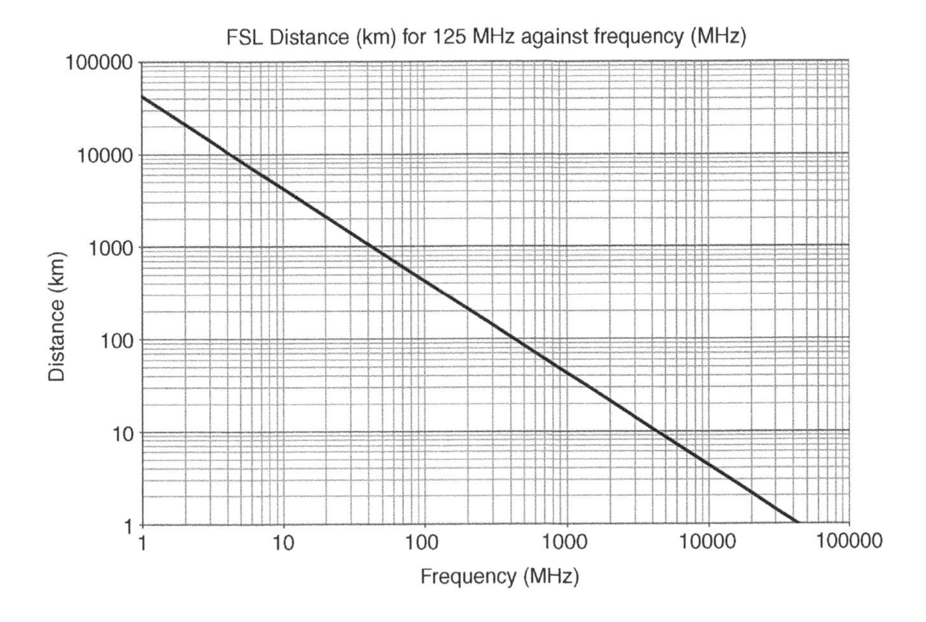

Figure 3.6 Graph showing frequency (MHz) versus distance (km) for a fixed loss of 125 dB.

variation in propagation effects over the band. This is referred to as being 'narrowband'. If the signal bandwidth is larger, then variations must be considered over the band of transmission. This is referred to as 'wideband'. It is typical for high data rate systems, such as video links and microwave systems. For the sake of clarity we will also differentiate frequency hopping systems, which cover wide frequencies but which may have the characteristics of broadband or narrowband systems.

We will next consider narrowband signal and their characteristics before looking at other types of transmission. In this discussion, we will introduce the concept of fast fading and its effects on radio system performance.

3.2 Narrowband Signals

'Narrowband' signals are those that have a narrow bandwidth compared to their transmission frequency. Because of this, their propagation is considered in a single calculation that assumes that propagation is consistent across the bandwidth and also that fading characteristics are correlated. Historically, most transmissions for radio communications have been narrowband in nature and their bandwidth has been set by the modulation scheme selected or, in digital systems, according to the gross bit rate. By contrast, wideband signals have been more common in modern systems and in some of these the transmission bandwidth is wider than that necessary for the transmission of the base information. It is instead combined with a higher rate code modulating the base band signal into a wider bandwidth that can offer benefits in terms of robustness and stealth.

Why do we consider narrowband and wideband signals to be different in nature? Narrowband signals have the following characteristics:

- The bandwidth of the transmission is small compared to the frequency of transmission.
- We can generally assume that interferers have, or will approximate to, a flat response across the bandwidth of transmission. This is referred to as AWGN (Additive White Gaussian Noise).
- There is no capability to reject interference by clever use of a wider bandwidth to counter fast fading.

We will see more of the difference when we compare wideband signals in Section 3.4.

For both analog and digital systems, voice communications are typically narrowband, with transmission bandwidths between 3 kHz (for HF) and 8.3–50 kHz for tactical VHF and UHF voice communications.

In Section 1.3.3, we saw that signals vary over short distances due to the receiver and transmitter antenna environments. We will now look at the statistical methods of accounting for narrowband signal fast fading.

Firstly, we must consider why fast fading occurs at all.

Consider a receiver operating in a built-up environment. In this case, it is unlikely that there is a direct line of sight between a transmitter and receiver. Buildings are very likely to get in the way, and there may be other obstructions such as groups of trees. All of these will affect the received signal due to:

- Signal attenuation (absorption) as the signal passes through, is diffracted over or around obstructions.
- Reflections, which are more likely to be destructive in nature than they are to be constructive. These will therefore reduce the level of usable signal.
- Scattering; this is where signals bounce off an obstruction and scatter in many directions. An analogy is the situation where light scatters from sunlight allowing objects to be seen from any direction. Thus, if you were in a dark room with a torch, you can point a torch at a mirror and see the reflection on the wall behind you, but you see the mirror and other objects in the room due to scattering. Without scattering, we would not be able to see objects clearly. Scattering in radio terms allows walls along a street to ‘fill in’ coverage within the street, even if it is not in direct line of sight of the transmitter or a distinct reflection from the wall. The key difference is that scattering radiates in many directions, not just one. This can be due to relatively small variations in the wall surface structure for example.

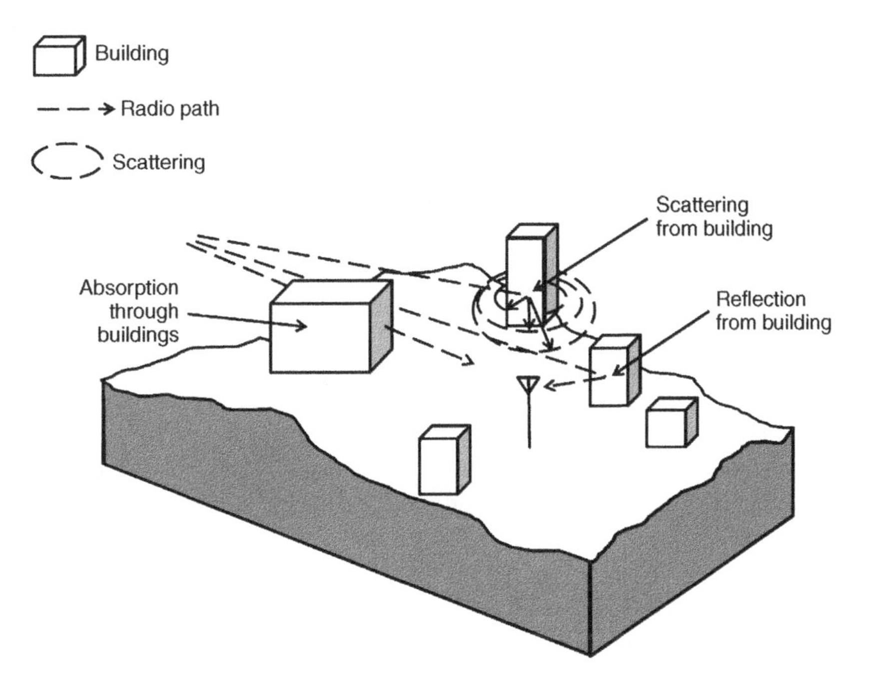

Figure 3.7 Absorption, reflection and scattering near a radio receiver antenna.

This is illustrated in Figure 3.7.

As previously noted, we cannot expect to work anywhere in the world and still be able to predict how signals will vary at small scales. This is due to two main factors; inability to model the scenario in sufficient depth and also that the scenario will be dynamic – only a few vehicles and moving people will render a high-fidelity model less than useless. The solution is statistics. The mathematics may be fairly complex, but fortunately, its application is somewhat simpler in practice. To provide the reader with the background, we will look at this mathematics and then show how it can be used practically using statistical tools that do not require a detailed use of the complex equations.

There are many mathematical models that can be used to model fast fading, but we will restrict ourselves to the analysis of two common and robust models; those of Rayleigh and Ricean fading. The other models approximate to these and for communications and electronic warfare applications, the use of both these models is normally sufficient.

3.2.1 Rayleigh Fading

So-called Rayleigh fading is used for the situation where there is generally no direct line of sight between transmitter and receiver. In this case, the energy arriving at the receiving antenna is comprised of a number of components, with none being dominant over the others. This is typical for most mobile radio links, including tactical VHF, mobile phones and many other systems. It is also typical for electronic warfare links with the exception of aeronautical systems, which are often better described using Ricean fading. Rayleigh fading is most applicable in situations such as:

- When the receiver is in a built-up environment such as encountered in urban warfare and the links are not line of sight. It is applicable therefore to RCIED scenarios in many cases.
- For tactical air to ground links, where the airborne platform is at a low altitude. This is often the case for helicopter to ground or tactical close support applications.
- Where the receiving antenna and transmitting antenna are close to the ground and effectively embedded in the radio clutter (where radio clutter is the term used to describe objects that influence radio propagation). This is the case in many short-range tactical scenarios.

Now for the mathematics; the Rayleigh fading distribution characteristics are:
The most probable value is

$$\sigma$$

The median value is

$$\sigma\sqrt{2\ln 2}$$

The mean value is

$$\sigma\sqrt{\frac{\pi}{2}}$$

The root mean squared value is

$$\sigma\sqrt{2}$$

The standard deviation is

$$\sigma\sqrt{2-\frac{\pi}{2}}$$

The Probability Density Function (PDF) is

$$P(x) = \frac{x}{\sigma^2}\exp\left(-\frac{x^2}{2\sigma^2}\right)$$

The Cumulative Density Function (CDF) is

$$F(x) = 1 - \exp\left(-\frac{x^2}{2\sigma^2}\right)$$

The CDF and PDF are illustrated in Figure 3.8 for a standard deviation of 1.0. The CDF is shown in Figure 3.9.

Of these, the CDF is of most importance because it identifies the likelihood of a given value to be exceeded. The standard deviation for radio signal variation will vary according to circumstances, but a value of approximately 10 dB is roughly typical. With this figure, it is possible to relate the probability of exceeding a value directly to variations in signal strength in dB.

If we use the 0.5 probability value as our normalised result, we can correct the value in dB to determine variations from this value. An example is shown in Table 3.1.

For radio signal fading, the normalised values give us metrics known as 'availability'. Availability gives the probability of receiving a working signal based on the threshold set at the 0.5 normalised value. The number of dB above this normalised value is known as the 'margin'. Thus, if the minimum wanted signal value is, say, −100 dBm and the actual value calculated by a propagation model for the short sector is −90 dBm, then the margin is 10 dB and the availability is just over 90% for that location.

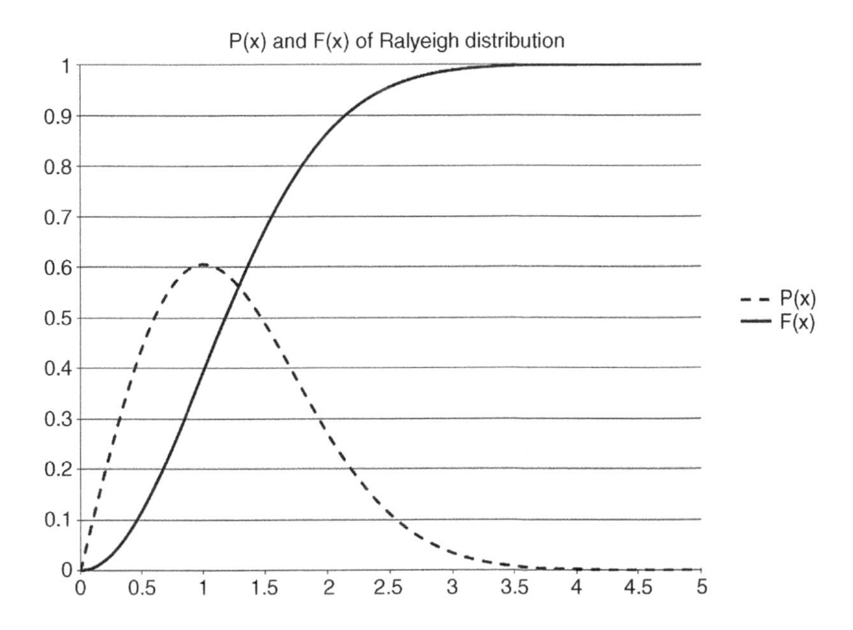

Figure 3.8 PDF (P(x) and CDF (F(x) for the Rayleigh distribution with $\sigma = 1.0$.

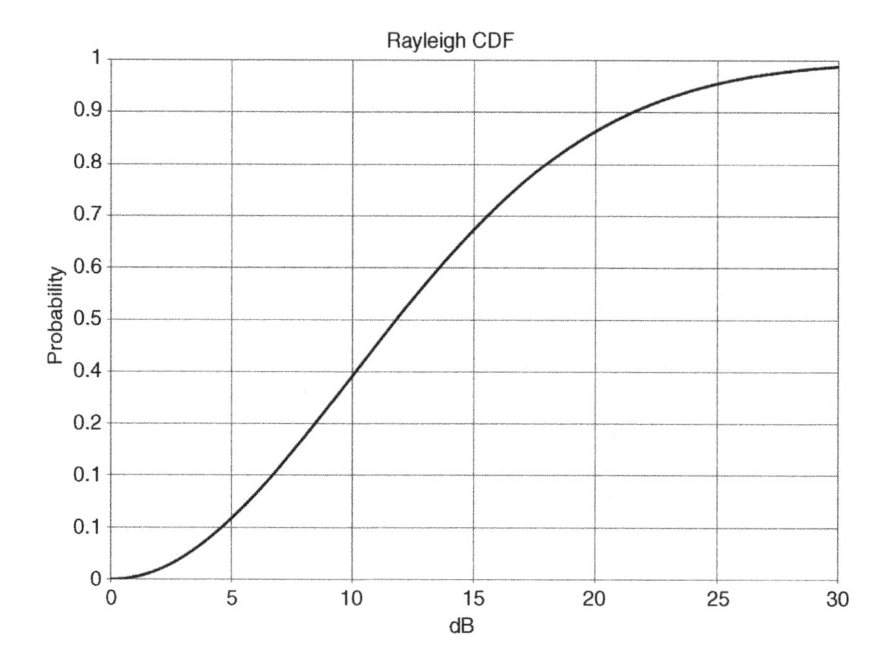

Figure 3.9 Rayleigh CDF with a standard deviation of 10 dB.

3.2.2 Ricean Fading

Ricean fading offers a useful way of describing a wider variety of scenarios than Rayleigh fading. By selection of the properties used in the Ricean fading calculations, it can be made to give identical results to Rayleigh fading but with other properties, it can also model the scenario where there is a dominant main path, perhaps caused by a direct line of sight path, and many subsidiary paths that provide smaller components at the receiver. In terms of practical scenarios, it is applicable to many fixed links, for open ground links and for aeronautical radio links.

The mathematics for the Ricean distribution can be expressed in a number of ways. We will focus on the form most applicable to radio.

In this case, we want to be able to look at the steady part of the signal, most likely caused by direct line of sight between transmitter and receiver, and the varying part

Table 3.1 Rayleigh CDF values and normalised values against $F(x) = 0.5$

F(x)	Uncorrected value in dB	Corrected against median value in dB
0.5	11.8	0
0.9	21.5	9.7
0.99	30.4	18.6
0.999	37.2	25.4

caused by reflections and scattering. We can express this using the dimensionless parameter k where:

$$k = \frac{P_c}{P_r}$$

where
P_c is the power of the constant part.
P_r is the power of the varying part.
The PDF is

$$P(x) = \frac{x}{\sigma^2} \exp\left(-\frac{x^2}{2\sigma^2}\right) \exp(-k) I_0 \left(\frac{x\sqrt{2k}}{\sigma}\right)$$

where I_0 is the Bessel function of the first kind and zero-th order.

The PDF is illustrated in Figure 3.10 for a range of values of k.

Note that with $k = 0$, the Ricean distribution is the same as the Rayleigh distribution.

The CDF is derived from the integral of the PDF. Rather than try to solve this mathematically, it can be approximated numerically as shown in Figure 3.11 below.

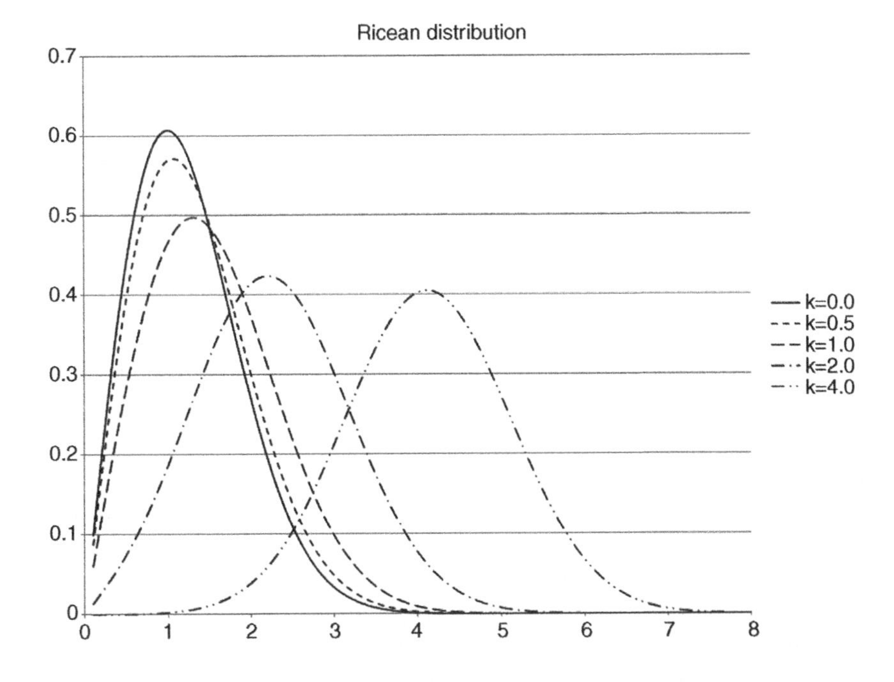

Figure 3.10 The Ricean distribution for given values of k.

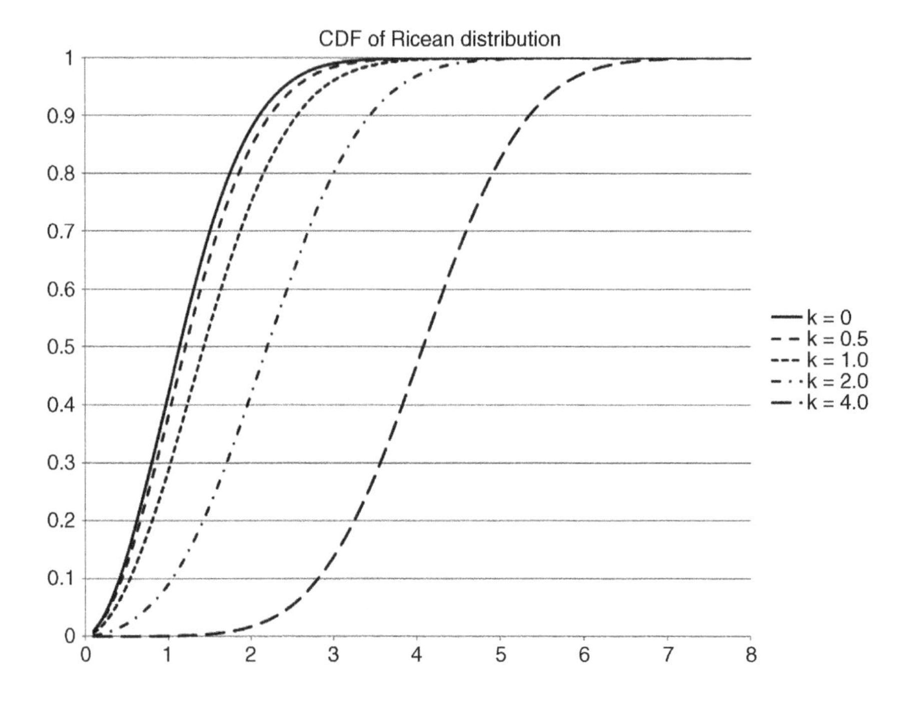

Figure 3.11 CDF of the Ricean distribution.

We will see how these are applied to practical applications in the second part of the book.

3.3 Frequency Hopping Signals

As the name suggests, frequency hopping systems move dynamically across a range of spectrum over time. This is illustrated in Figure 3.12. For the duration of a given timeslot, the signal is static on a particular channel. At the end of that time, the signal is removed from that channel and transmitted onto a new frequency. This is carried out repeatedly until the transmission is over. The sequence of channels used is termed the 'hopset'.

Hopsets have to be carefully designed so that multiple transceivers can work together without causing interference. Additionally, the rise and fall time of each time slot transmission must be carefully managed because sharp transmissions lead to wideband interference. Because of these factors, system design and implementation of frequency hopping systems is far more complex than fixed frequency systems. However, in terms of the physics of radio propagation, there is no difference other than the necessity to model not just one potential link, but many. There are two key factors that must be considered; the gain response of the antennas used may vary across the frequency range as shown in Figure 3.13.

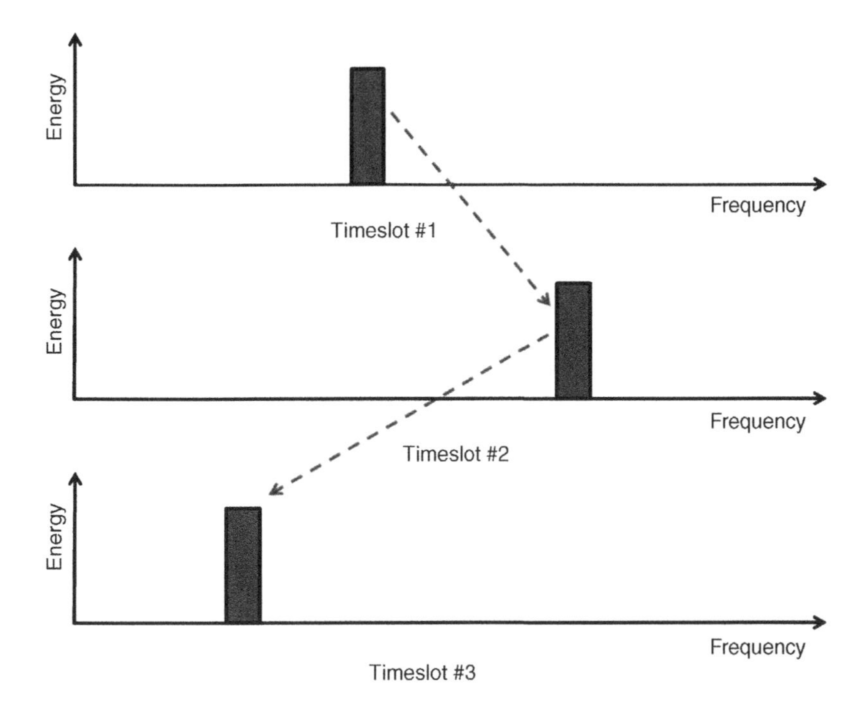

Figure 3.12 Frequency hopping system, showing how the signal uses time slots at different frequencies.

In Figure 3.13, we can see the effects of the antenna gain response across frequency, the propagation loss, which in general increases with frequency and the received signal level. In this case, the minimum working signal level is shown and it can be seen that the combined effects of antenna response and propagation loss mean that only the bottom part of the band can be used effectively for this link.

In terms of assessing the performance of individual time slots, we can apply the statistics relevant to the transmission at that moment in time. If, as shown in the figure, the individual time slot transmissions are narrowband in nature then the statistics described above can be used. But frequency hopping systems can also be used to hop wideband signals. In this case, we would have to use the statistics described in the next section.

3.4 Wideband Signals

Wideband signals differ from narrowband ones in that the response across frequency varies. This is caused by the effective cancellation of the signal strength due to strong reflections. Figure 3.14 shows an example at just above 400 MHz with two combined signals arriving at the receiver with the same signal strength. This would be the case if

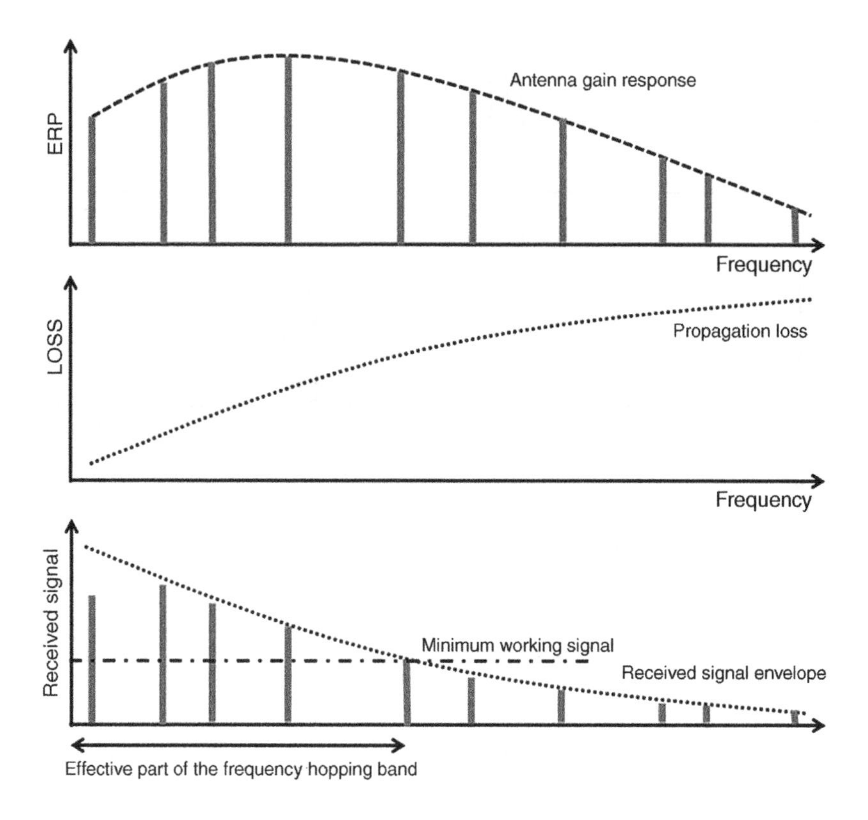

Figure 3.13 Over a wide frequency range, the antenna gain response may vary. Propagation losses generally increase in frequency, meaning that the combined effects can affect the range of frequencies that can be used for longer links.

the direct signal and a single reflected signal at the same power arrive at the receiver. In this case, the summation of power is expressed by:

$$y = a\sqrt{2(1 + \cos(\omega\Delta t + \Delta\theta))}$$

where
ω 2πf.
Δt path delay in seconds.
$\Delta\theta$ phase delay.

When converted to logarithmic terms, the result is as shown in Figure 3.14.

This is known as the channel response and must be taken into account when considering the system receiver characteristics. This is usually expressed by the system sensitivity, which includes an analysis of the channel response as described by a channel model. The channel model is derived from a summation of delay responses with

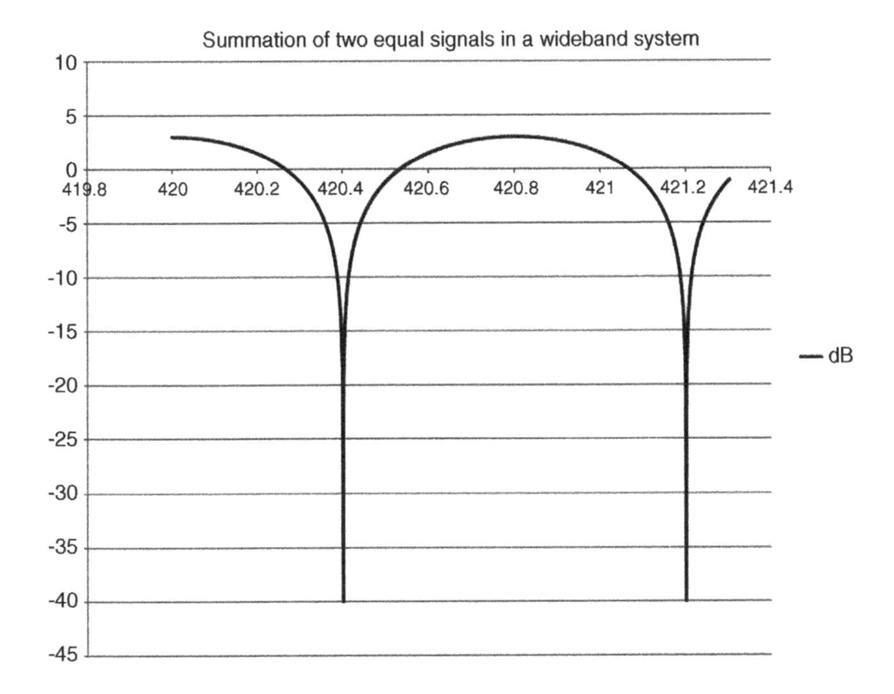

Figure 3.14 Summation of two equal strength signals arriving with a phase difference at a receiver. Note that the signals cancel at just over 1 MHz. Information that includes components at or near the cancellation points will be degraded due to the channel response.

expected reflection coefficients. Note the deep fades at 420.4 MHz and 421.2 MHz. These are caused by phase cancellation of the reflected signal. If the bandwidth of the signal is wider than the nulls shown, then the nulls will be included in the received signal, and thus some information may be lost. However, the case shown is for a perfect single reflector. In practice, there are likely to be multiple reflectors, and their amplitudes will also vary, so the situation is likely to be more complex. This can affect the statistics of fading, which are often accounted for by system design and are likely to be included in the receiver sensitivity figures for the radio system.

Note that for a moving receiver, the reflection characteristics will vary substantially. System design can also counter short-term deep fades by the 'fly-wheel' effect, in which error correction allows for poor instantaneous reception by ignoring their effect.

3.5 The Effect of Movement on the Radio Channel

We have discussed the effects of fading on both narrowband and wideband signals. If there are fixed transmitter and fixed receiver antennas then if the immediate

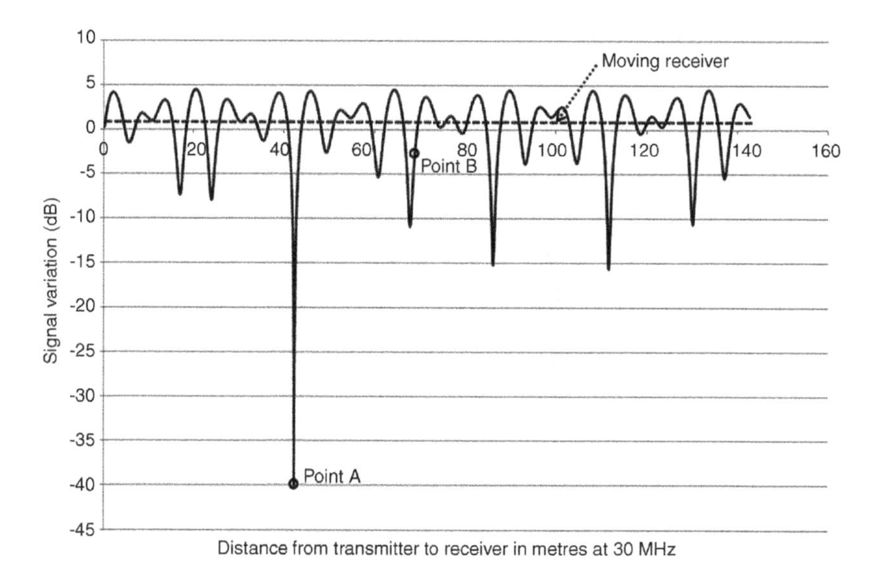

Figure 3.15 A moving receiver crosses fades but receives a decent signal level on average. The location of a stationary mobile system on the other hand could be at any point on this type of graph. If it were at Point B, then the fade is not too deep. If however, it happens to be at a location like Point A, it would be far less likely to be able communicate effectively.

environment characteristics do not change, the signal between the two is effectively a standing wave that does not change in nature over time in propagation terms. However, if one or both of the antennas are moving then the environment does change substantially over time and the effects of deep fades can more easily be countered by error correction to overcome short-term fades. This is illustrated in Figure 3.15.

A stationary receiver may be at any point along the line. If it happens that the receiver is at point A, then it finds itself in a deep fade and is likely to be unable to communicate. If it happens to be at Point BG, then the situation is different; however, there is no way to determine what the fade depth will be. The dotted line shows a receiver moving with respect to the transmitter (either towards it or away from it). The average power received is shown; it is at the non-faded level on average. Using a fly-wheel within the receiver, which allows the system to work despite short fades, the communications link can be maintained.

Point-to-point links also suffer from fades but generally, more work is carried out to ensure that the received signal is not in a deep fade in normal conditions. Fades may occur temporarily due to changes in atmospheric conditions and if this is likely for a given link, then diversity antennas may to used to counter this.

References and Further Reading

Graham, A.W.; Kirkman, N.C.; Paul, P.M. (2007), *Mobile Radio Networks Design in the VHF and UHF Bands: A Practical Approach*, John Wiley & Sons ISBN 0-470-02980-3.
Recommendation ITU-R P.525: Calculation of Free Space Attenuation.
Recommendation ITU-R P.1057: Probability Distributions Relevant to Radiowave Propagation Modelling.
Recommendation ITU-R P.1406: Propagation Effects Relating to Terrestrial Land Mobile Services in the VHU and UHF Bands.

4

Radio Links in the Presence of Noise

4.1 Sources of Radio Noise

Radio links cannot be created over infinite distances and will be limited in range by background noise even if there are no interferers. Such noise is always present and it is more prevalent at lower frequencies than at higher ones.

We will look at the limiting effects of noise in this section before going on to look at interference, which is a different situation caused by the presence of intentional signals being transmitted for other radio services.

Noise sources can be split into different categories. These include sources beyond the atmosphere, atmospheric noise and unintended noise generated by man-made sources. In this chapter, we are not considering energy derived from other intentional radio systems – these are covered in Chapter 5.

Celestial noise sources are generated by stars such as the Sun and from other stars, which are most prevalent from the angle of the Galactic plane. The angle of the Galactic plane can be seen on very clear nights in dark locations. Directional antennas pointed directly at the sun will be subject to noise generated from it.

Atmospheric noise is caused by atmospheric gases and hydrometeors. This varies according to time of day and season and due to random events.

Man-made noise is caused by machinery that produces radio frequency energy as an unintended by-product. For example, at VHF, car ignition impulses are a major source of noise. For urban environments, man-made noise is often the dominant factor.

Whatever the dominant noise source, it can be the limiting factor for radio link performance. However, we also have to consider the effects of noise internal to the receiver before we can compare external noise to determine which is the most important. We will now look at receiver internal noise before looking in more depth at the effects of external noise.

Communications, Radar and Electronic Warfare Adrian Graham

4.2 Effects of Noise

In the absence of intentional or unintentional energy, the fundamental limit on radio communications is caused in the receiver itself by the thermal noise floor. This is caused by random fluctuations of electrons in the radio system, and is described by the expression:

$$N = kTB$$

where
N is the noise floor.
k is the Boltzmann constant $= 1.38 \times 10^{-23}$ (J/K).
T is the noise temperature of the receiver in Kelvin.
B is the bandwidth in Hz.

The noise floor is thus dependent on the receiver temperature and the bandwidth of the signal received. Note that there is no dependence on the frequency of the signal.

Often, the concept of reference noise power is used, with T = 290 K for terrestrial receivers. For a 1 Hz bandwidth, this is equal to 4×10^{-21} W, which is -204 dBW (in a 1 Hz bandwidth). It should be noted that technically, the bandwidth is not the same as the 3 dB limit normally quoted on equipment data sheets, but rather is the area under the power transfer curve. However, in practice, the two values are normally typically fairly similar, so it is normally reasonable to use the 3 dB bandwidth figure.

From the noise floor equation, it is clear that the thermal noise floor is predicated on the temperature of the receiving system and also on the bandwidth of the system. This means that highly sensitive systems should be cooled, and that the system should be limited in bandwidth if possible. Space borne receivers will normally be very cold, unless they are heated by sunlight.

In practice, physically realisable systems suffer from additional noise above this fundamental limit. To account for this, the concept of noise figure (F_r) is used. This is described as an additional noise component that adds to the thermal noise floor to produce the same response actually measured from a receiver. The noise figure is expressed in decibels and it varies according to the design of the radio. To obtain the total noise in the system, the following expression is then useful.

$$P_n = F_r - 204 + B$$

where
P_n is the receiver noise floor in dBW.
B is 10 log (bandwidth in Hz).

Any practical signal must exhibit wanted characteristics with respect to the receiver noise floor. In non-CDMA systems, the signal level must be at a higher level than the noise floor (known as the Signal to Noise Ratio or SNR) in order for the receiver to

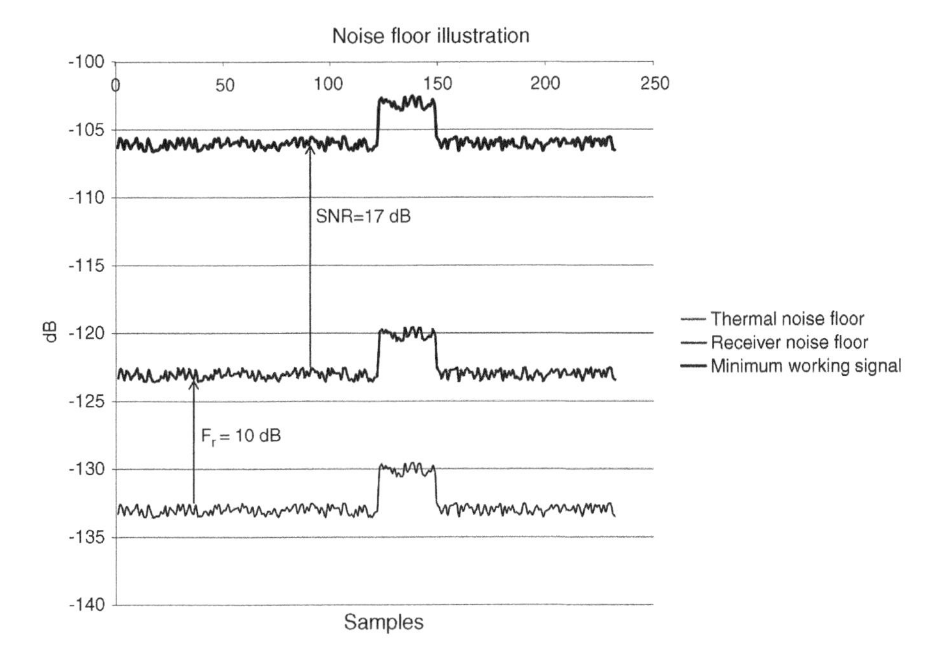

Figure 4.1 Illustration of thermal noise floor and receiver noise floor given a bandwidth of 12.5 kHz and a noise figure F_r of 10 dB. Note that when the noise floor is increased by a sudden burst of noise, the receiver noise floor also lifts by the same amount. This leads to reduced receiver sensitivity for the duration of the noise burst. The minimum working signal level is given by adding the required SNR to the receiver noise floor figure.

achieve a given level of performance such as a given level of SINAD or BER. This is illustrated in Figure 4.1. The noise floor varies due to small fluctuations around a constant value. Figure 4.2 shows the basic form of a receiver.

The thermal noise floor of a receiver is shown at the bottom, with the receiver noise floor shown F_r dB (10 dB in this case) above it, and the minimum level of signal

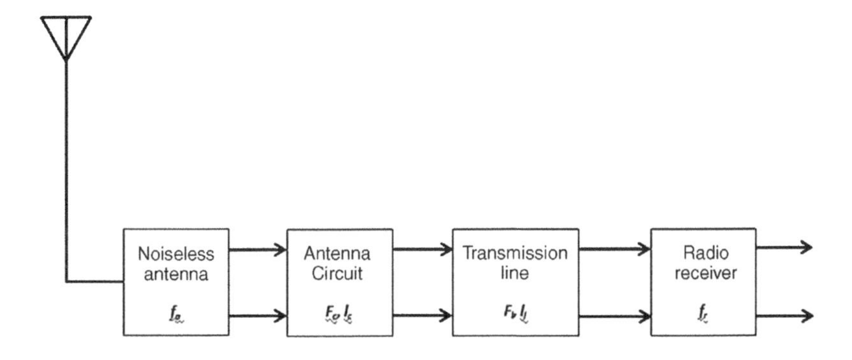

Figure 4.2 The elements of the front end of a receiver.

required to achieve a given degree of performance is shown as the required SNR dB above that (17 dB in this case). In practice, the noise floor value used may need to take account of the probabilistic distribution it obeys, and a correction based on a wanted ordinate value may be needed in order to perform this, but for the moment, we will assume white Gaussian noise and ignore greater complexities. This is because in practice, the mobile radio network engineer will use appropriate SNR figures provided by equipment manufacturers or technology specifications that will already have accounted for some of these additional factors.

To illustrate this basic approach numerically with a practical example, assume that a receiver has a bandwidth of 25 kHz, a noise figure of 4 dB and an SNR value required of 10 dB, thus

$$B = 10\,\mathrm{Log}\,(25{,}000) = 44\,\mathrm{dB}\ \text{(approximately)}$$
$$N = -204 + B = -160\,\mathrm{dBW}$$
$$P_n = N + F_r = -160 + 4 = -156\,\mathrm{dBW}$$
$$\text{Receiver sensitivity} = P_n + 10 = -146\,\mathrm{dBW} = -116\,\mathrm{dBm}$$

The radio receiver sensitivity is the minimum signal level at which the desired performance will be achieved. This value can then be used in the link budget along with all the aspects of radio prediction and fading required, as normal.

This is a slightly simplified diagram from ITU-R P.372. From this simplified diagram, the system noise can be calculated by:

$$f = f_a + (f_c - 1) + l_c(f_l - 1) + l_c l_l(f_r - 1)$$

where
f_a is the external noise factor.
f_c is the noise factor associated with the antenna circuit.
l_c is the loss figure associated with the antenna circuit.
f_l is the noise factor associated with the transmission line.
l_l is the loss figure associated with the transmission line.
f_r is the noise factor of the receiver, which is the linear version of F_r.

The external noise factor is defined as:

$$f_a = \frac{p_n}{k\, t_0\, b}$$

where
p_n is the available noise from a lossless antenna.
t_0 is the reference temperature.
b is bandwidth in Hz.

An alternative definition often used for satellite communications is expressed as a temperature:

$$f_a = \frac{t_a}{t_0}$$

where t_a is the effective antenna noise. The noise figure can be converted to its equivalent noise temperature by:

$$T_{RX} = T_R\left(10^{\left(\frac{F_r}{10}\right)} - 1\right)$$

where
T_{RX} is the noise temperature of the receiver.
T_R is the reference temperature.
F_r is the receiver noise figure in dB.

Thus, for a noise figure of 8 dB at a reference temperature of 290 K, the equivalent noise figure is:

$$T_{RX} = 290\left(10^{\frac{8}{10}} - 1\right)$$

This gives the equivalent noise temperature as 1540 K. The converse formula is:

$$F_r = 10\log\left(\frac{T_{RX} + T_R}{T_R}\right)$$

Apart from the external noise factor, all of the other factors are dependent on the radio design itself. The antenna noise factor is dependent on the environment in the vicinity of the receiver. This will depend on the dominant mechanism, which is often caused by manmade noise for mobile radio systems in the VHF and UHF bands.

A graph of the median value of radio noise estimated from measurements taken in the 1970s by the ITU and more recently by Mass Consultants Ltd is shown in Figure 4.3. This shows the median value, around which there will be variations.

We can use this graph to determine the noise power for a given environment. For example, assume the radio is working at 400 MHz in a city centre environment. From the graph, the approximate noise figure is approximately 18 dB.

Thus,

Man-made noise at the receiver = 18 dB, so $f_a = 10^{(\text{noise}/10)} = 63.1$
Omni-directional antenna, with loss factor = 0 dB, so $f_c = \exp(0/10) = 1$
0 dB antenna circuit loss factor = 0 dB, so $l_c = \exp(0/10) = 1$
0 dB noise or loss in the transmission cable, so $f_l = l_l = 1$
$F_r = 10$ dB, so $f_r = 10^{(10/10)} = 10$

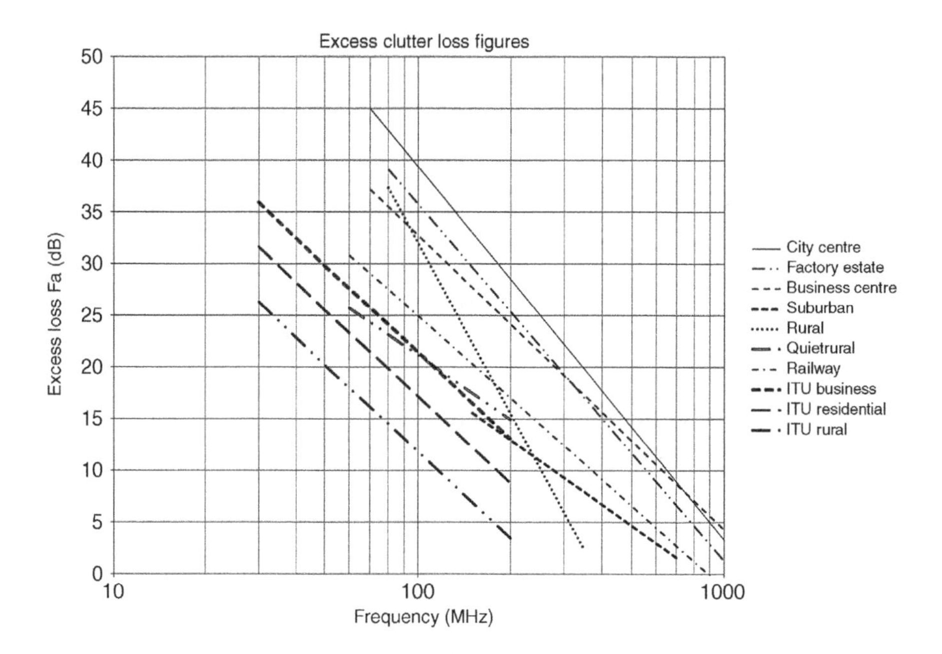

Figure 4.3 Excess noise caused by the receiver environment. The ITU values extend to lower frequencies, but were gathered some years ago. The non-ITU values were gathered in the UK in recent years by Mass Consultants working for the UK regulator, Ofcom.

Using the formula for system noise:

$$f = f_a + (f_c - 1) + l_c(f_l - 1) + l_c l_l(f_r - 1)$$

$$f = 63.1 + (1 - 1) + 1 \times (1 - 1) + 1 \times 1 \times (10 - 1) = 72.1$$

$$\text{System noise } F = 10\log(72.1) = 18.6\,\text{dB}$$

From this the noise power for a system with a bandwidth of 25 kHz can be calculated from:

$$P_n = F - 204 + 10\log(25000) = 18.6 - 204 + 44 = -141.4\,\text{dBW} = -111.4\,\text{dBm}$$

If no sources of interfering radio energy are present and the system is limited by the noise power as shown, then the system is known as 'noise-limited'. Conversely, if the system is limited by interference from another radio system, it is known as 'interference limited'. An example of a system becoming interference limited is shown in Figure 4.4. The effect of the interferer is to raise the noise floor above the thermal limit. Because the noise figure and the required SNR do not change, the effect is to raise the level of the minimum required signal at the receiver. This has the effect of de-sensitising the receiver and reducing the effective range.

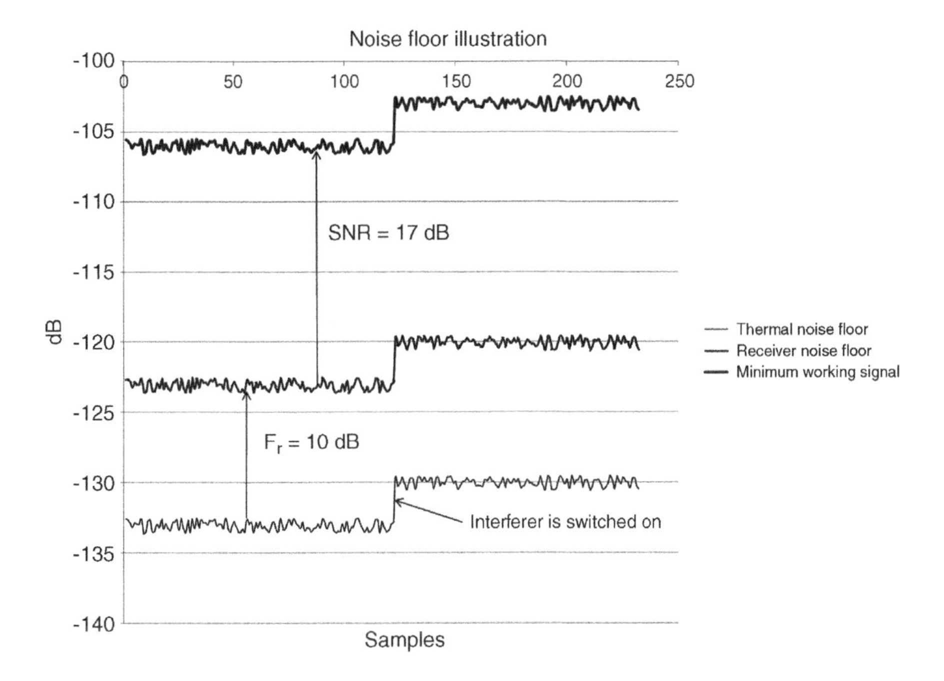

Figure 4.4 The de-sensitising effect on a receiver of an interfering signal. The effect is to raise the minimum working signal required and by doing so, the maximum range of the interfered system is reduced. The effect of intentional interference – jamming – is exactly the same for noise jammers.

Many communications and radar jammers use noise to jam their targets. In this case, the difference between an unintentional interferer and a jammer are technically indistinguishable.

4.3 The Radio Receiver

In section 1.3, we looked at a block diagram of a typical transmitter. In this section, we will look at receivers and receiver performance for different levels of signal to noise and interference.

A simple generic receiver is shown in block diagram form in Figure 4.5.

The RF amplifier shown boosts the signal level prior to the signal being mixed with a local oscillator that down-converts the signal to a fixed intermediate frequency. This signal is then conditioned by the filter(s) and further amplified. The demodulator carries out the inverse process of the transmitter's modulator, leaving the recovered signal, which is hopefully of sufficient quality to allow successful recovery of the original signal. Of course, the diagram is very simple and there may be many other features such as Automatic Gain Control (AGC), error-correction and further signal processing. However, the diagram shows some important features of radio receivers.

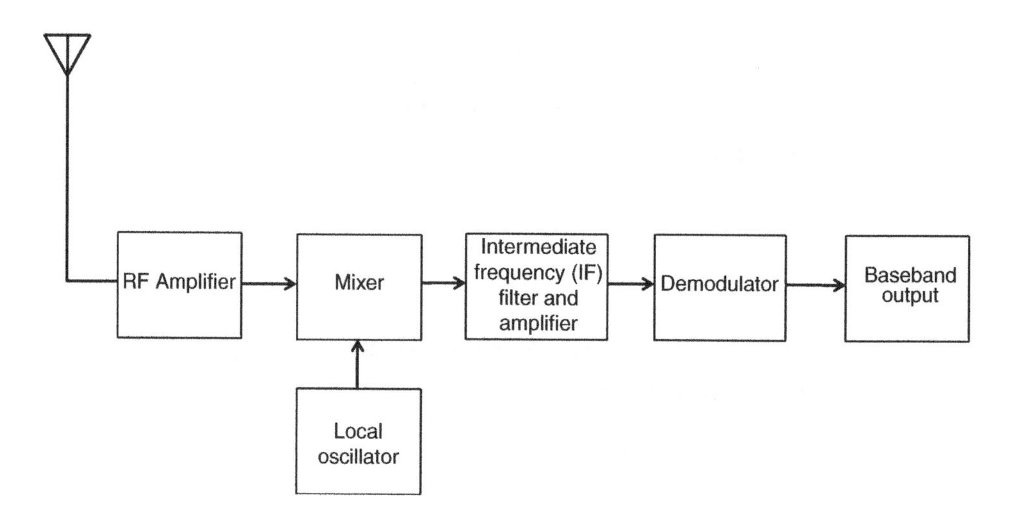

Figure 4.5 Simple generic radio receiver block diagram.

The output of the antenna is not always immediately filtered. This means that any signal within the passband of the antenna is fed into the remaining electronics. If the system features an automatic gain control or a signal higher than can be dealt with by the electronics then a high level interferer outside of the tuned frequency can still prevent successful reception. In the case of automatic gain control, the interfering signal can push the wanted signal down into the noise floor. This is illustrated in Figure 4.6. On the left hand side is the normal condition with no interferer. The automatic gain control amplifies the signal to the wanted level for further processing. The right hand side shows an interfering signal that adds RF energy into the AGC. Since the AGC cannot distinguish between the wanted signal and the interferer, all of the incoming energy is summed and when the signal is amplified to the desired level the wanted signal is at far lower signal strength than wanted. It can in fact be pushed down into the noise, but even if not, the signal to noise ratio will be significantly less and therefore the error rate will be substantially higher than wanted.

Even if there is no AGC, the power from an out-of-band receiver can still add energy into the receiver if it is not filtered out effectively. This can cause problems if the input power level is too high because it can push electronic components into non-linearity, which happens when the input power is too high. If this occurs, the non-linearity introduces additional spectrum components that may cause components downstream to fail to operate correctly.

If the input power from interferers is very much higher than the receiver circuitry can cope with, then the system components can fail permanently. This is the aim for directed energy weapons, which seek to destroy radio receivers by introducing unacceptable power into the electronics. In electronic warfare terms, this would be classified as a hard-kill. Indeed, if the power is so large, the energy may not enter only

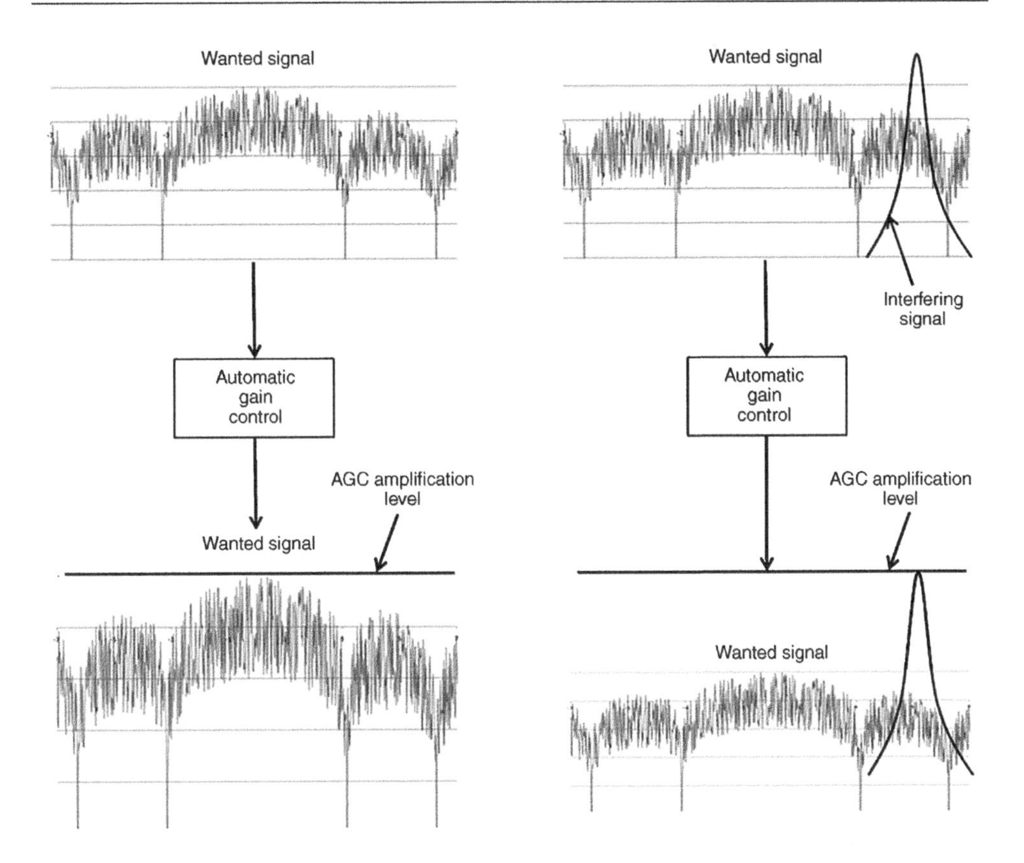

Figure 4.6 The effect of an interferer on a receiver using automatic gain control. The interfering energy adds to the wanted signal energy. Because the AGC is intended to set the input to a certain level for further processing, anything that affects that directly influences the radio system performance. In this case, the wanted signal level is less than wanted due to the extra energy of the interferer – or jammer. This occurs even though the interferer is not within the wanted band if the AGC is upstream of filters to reduce out-of-band responses. When used for jamming, this is known as 'out-of-band' jamming.

via the antenna but directly into the electronics of the receiver system, causing saturation throughout the entire receiver. This can occur not only due to directed energy weapons but also through the phenomenon of Electro-Magnetic Pulse (EMP). EMP is caused by the detonation of nuclear weapons at high altitude and also by EMP weapons. These are a new category of system in which the EMP effects of a nuclear weapon are induced by the near-instantaneous collapse of a high energy electro-magnetic field. This can be generated by a metal coil excited by RF energy and exploding high explosives within the coil. The destruction of the coil and the resultant collapse of the EM field cause a high level, broadband burst of energy. If this is close enough to the target, then the receiver can be destroyed.

We will consider the more normal case of wanted versus interfering signals in Chapter 5.

4.4 Radio Link Budgets in the Presence of Noise

Link performance in terms of Bit Error Rate (BER) can be estimated for different modulation schemes based on the signal energy to noise over the duration of a bit. This is referred to as E_b/N_o; the energy of the signal per bit over noise in the same time interval. For multiple state systems such as 4QAM and higher order QAM systems, the factor is E_s/N_o; energy per symbol over noise over the duration of the transmission of the symbol. Assuming that one bit is transmitted in the same time as one symbol, the two become the same. If transmitted power does not vary but noise is increased, this clearly has an effect on this parameter. In simple terms, if the noise increases, so does the BER.

The BER for digital systems can be determined by considering the probability of error. In this analysis, the Gaussian complimentary error function is important:

$$Q(x) = \frac{1}{2} erfc\left(\frac{x}{\sqrt{2}}\right)$$

For even numbered QAM modulation schemes, the probability of a bit or symbol error (depending on the relationship between bits and symbols) can be approximated by the following for $E_b/N_o \gg 1$ and for even k:

$$P_e = \frac{4}{k}\left(1 - \frac{1}{\sqrt{M}}\right) Q\left(\sqrt{\frac{3k}{M-1}\frac{E_b}{N_o}}\right)$$

where
k is the number of bits per symbol.
M is the order of modulation.

For 16QAM, for example, $k = 4$ and $M = 16$. Note that this is a linear equation, and the E_b/N_o ratio must be converted from dB to its linear equivalent.

For odd-k QAM it is more difficult to determine the exact BER, but it is possible to determine the upper bound by:

$$P_e \leq 4Q\left(\sqrt{\frac{3kE_b}{(M-1)N_o}}\right)$$

Figure 4.7 shows a graph of BER curves for a number of digital modulation schemes, including Binary Phase Shift Keying (BPSK), Quadrature Phase Shift Keying (QPSK), 8-, 16- and 64-Quadrature Amplitude Modulation (QAM). Note that the more complex modulation schemes require a higher value of E_b/N_o to achieve the same BER as lower modulation schemes. However, they offer the capability to transmit information at a higher rate.

One other factor is important in this analysis; noise jammers increase noise, and higher digital modulations are more vulnerable to this as well as noise increases.

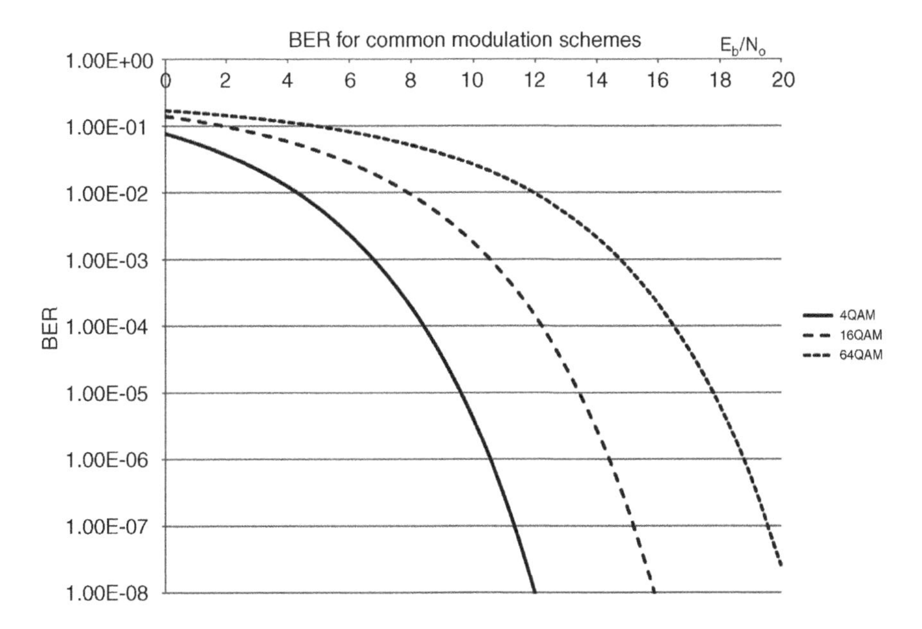

Figure 4.7 Bit error rates for common modulation schemes. The BER for BPSK and QPSK is the same BER as 4QAM.

We can also work with the minimum acceptable figures for different types of radio system. Some typical examples are shown in Table 4.1. The figures for AM and FM are typical at 10 dB and 12 dB respectively. The figures for specific radios were obtained from their technical data sheets. Note that the figures for the MBITR radio for AM and FM correspond with the typical values, which we would expect. We can also see that the signal level for UMTS is a negative value, since it uses a spread spectrum approach. In all cases, the quoted values are for a given system performance, deemed to be acceptable to users. The equivalent for the graphs shown in Figure 4.7 would be the selection of a BER of, say, 1% BER (10^{-2}) being acceptable for a given application. In

Table 4.1 Typical values of signal compared to noise levels and some values derived from specific radio technical data sheets

System	Value (dB)	Description
Analog AM	10	SINAD
Analog FM	12	SINAD
TETRA	19	Eb/No
GSM	9	Eb/No
UMTS	−9	Ec/Io
JTRS MBITR (FM)	12	SINAD
JTRS MBITR (AM)	10	SINAD

this case, the E_b/N_o of approximately 4 dB for 4-QAM, 8 dB for 16-QAM and 12 dB for 64-QAM (as read from the graph).

Let us look at an example of the effect caused by an increase in noise floor at the receiver location. Assume that the system under consideration is a duplex (two-way) radio system with an uplink (mobile-base station) frequency of 380 MHz. The mobile has output power of 1W and in a quiet radio environment, the base station has a working receive sensitivity of − 86 dBm, giving a range in an urban environment of approximately 8 km. If however the noise at the base station is higher than this by 8 dB, the effect is to reduce the nominal range down to approximately 5 km. If the base station antenna is omni-directional, then this range reduction occurs in all directions equally. The effect is illustrated in Figure 4.8. Note that in this case it is the uplink that it limited; the range of the downlink will be determined by the noise present at each mobile location.

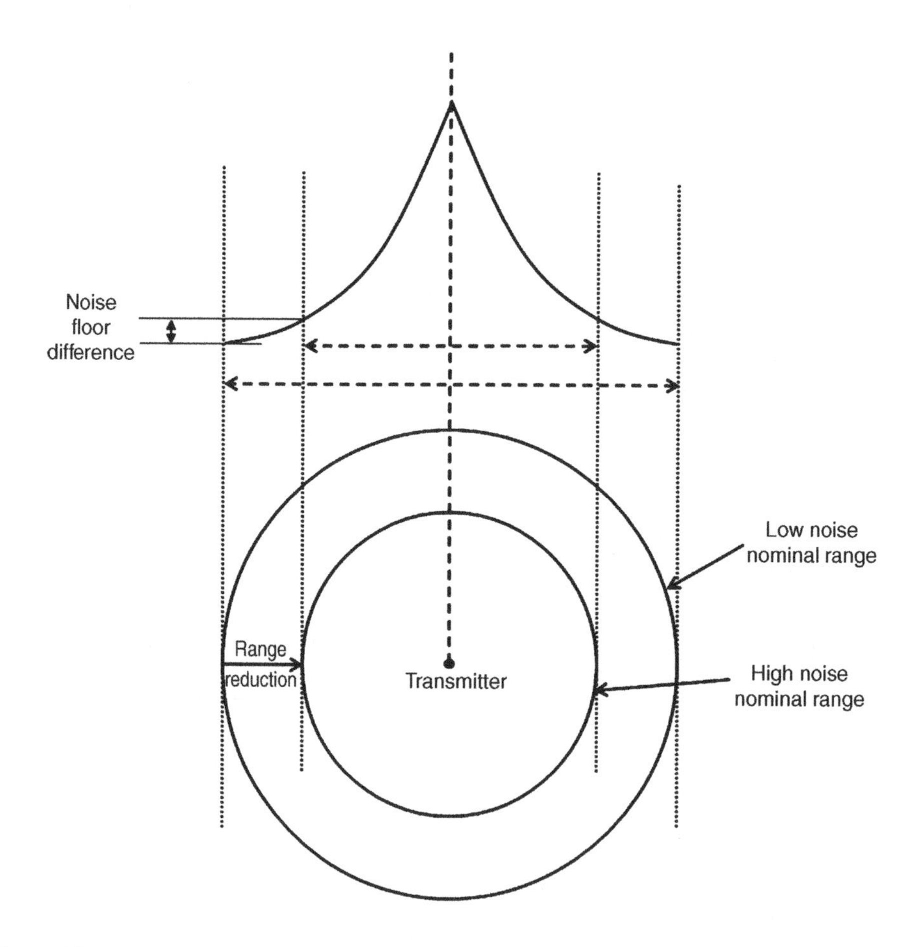

Figure 4.8 Diagram showing how nominal range is reduced by high noise at a base station location.

References and Further Reading

All internet references correct at time of writing.

Graham, A.W.; Kirkman, N.C.; Paul, P.M (2007), *Mobile Radio Networks Design in the VHF and UHF Bands: A Practical Approach*, John Wiley & Sons ISBN 0-470-02980-3.

Recommendation ITU-R P.372: Radio Noise.

Wagstaff, A.; Merricks, N. (2003), *Man Made Noise Measurement Program*, Mass Communications Ltd on behalf of Ofcom, UK, http://www.ofcom.org.uk/static/archive/ra/topics/research/topics/man-made-noise_finalreport.pdf.

5

Radio Links in the Presence of Interference

5.1 Sources of Radio Interference

One of the primary tasks of the network planning engineer will be to manage interference issues in the design. This does not necessarily mean that at the end of the design process, there will be no interference anywhere in the network, but rather that the prevalence of interference is minimised and, if unavoidable, is placed where it will do the least harm (for those technologies that allow this to be done). There are a number of ways to do this, of which the most effective is frequency assignment, which is covered in the next chapter. Others will be discussed towards the end of this chapter, but first we will examine the issues of intra-net interference – in other words interference caused by other transmitters in the same network. We will look at the modelling of co-channel interference, where both the wanted system and the interferer are tuned to the same channel, and then extend this to look at interference from interferers tuned to other channels, and also at the composite effects of multiple interferers. For most mobile networks, we will be examining uplink and downlink interference separately.

Interference is defined as unwanted contributions from other intended radio systems. This is distinct from noise, which is regarded as contributions from unintended radio frequency sources. We can split interference sources into different categories:

- Intra-network interference, where interference is caused by other transmitters within the same network. In this case, we can assume that the interfering signal has the same properties as the victim system.
- Inter-network interference from similar radio networks. This might be the case, for example, where the coverage from two adjacent TETRA networks overlaps.

Communications, Radar and Electronic Warfare Adrian Graham

- Inter-network interference from dissimilar radio systems. In this case, we cannot assume that the interfering signal has the same characteristics as the victim system.

We split potential interferers in this way because we will use different techniques to assess the level of likely interference, particularly for the third case.

5.2 Interference in the Spectral Domain

In this section, we will look at the following interference scenarios:

- On-channel interference (also known as co-channel interference).
- Adjacent channel interference.
- Convolution of interferer and receiver characteristics.
- Off-channel interference.
- Narrowband interference against broadband receiver.
- Multiple interferers.

5.2.1 Co-Channel Interference

The simplest case of interference is when both the wanted and the interferer are tuned to the same frequency, occupying the same frequency band with identical spectral characteristics. This can either be another transmitter in the same network or a different one. An illustration of the scenario for a point-to-point network is shown in Figure 5.1.

There are two things we need to establish in order to be able to determine the effect of the interferer on the wanted link. The first is to establish the level of interference compared to the strength of the wanted signal that the system can tolerate. Secondly, we need to determine the received signal strengths from both the wanted transmitter and the interferer so we can compare their relative strengths to the interference characteristics.

For example, the co-channel rejection level is 19 dB. This means that to function according to the specification, the wanted signal must be 19 dB above a co-channel interferer. In Figure 5.2, therefore for a TETRA receiver, the signal to interferer difference must be 19 dB or greater.

In the case of a system like TETRA or GSM, a mobile receiver or the transmission to a base station from a mobile embedded in clutter will typically be described by Rayleigh fading. We need to take this into account when considering the level of input from both wanted and interfering signal. If the interferer and wanted signal are not co-located, then the fading characteristics are most likely to be de-correlated, meaning that the variations at the receiver due to fading are entirely unrelated. When performing analysis, we may want to consider different levels of fade between wanted and interfering signal. A typical value may be 95% availability for the wanted link and 50% for the interferer. Thus, if the nominal (50%) signal strength at a receiver is −70 dBm,

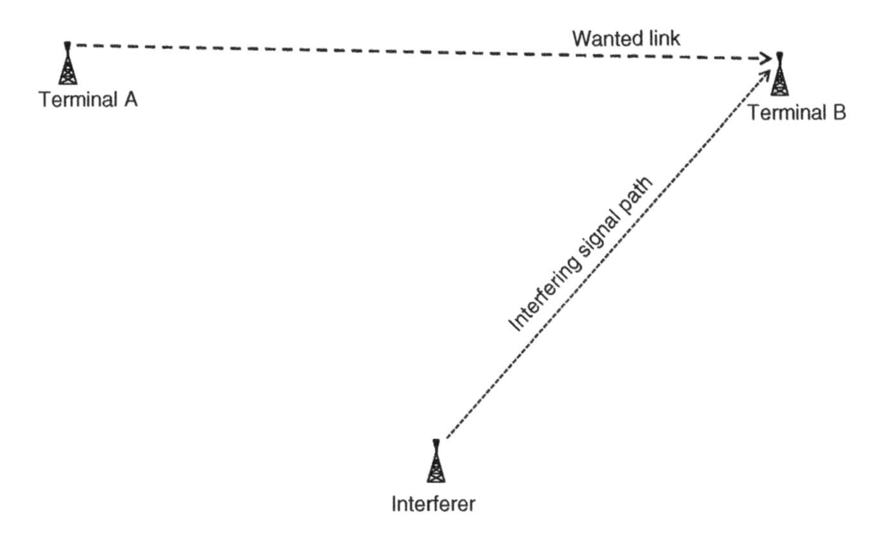

Figure 5.1 Interference scenario for a point-to-point link.

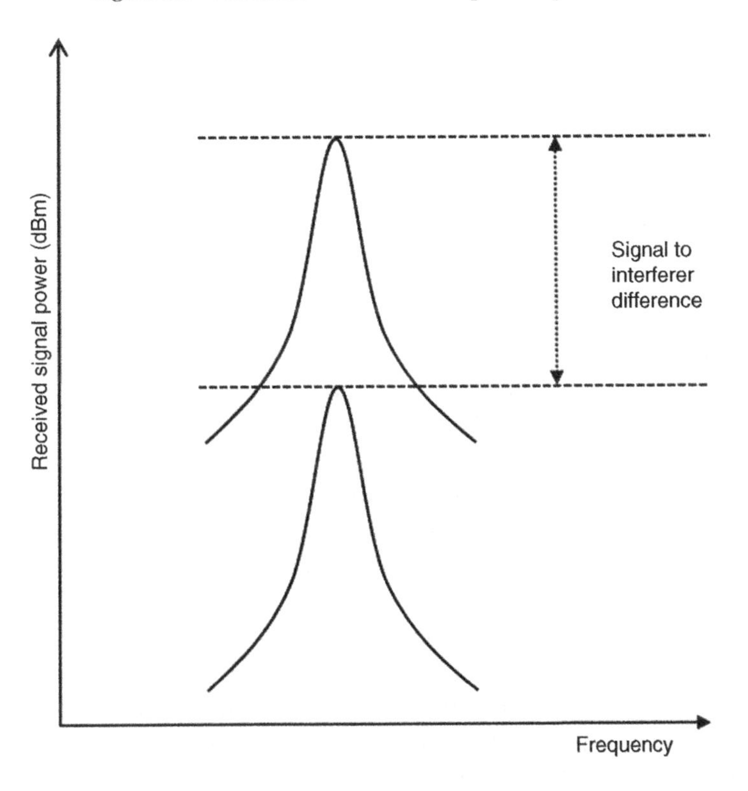

Figure 5.2 Co-channel signal to interferer level. For the example of TETRA, if the interferer is less than 19 dB lower than the wanted signal, then the wanted link will not function at the level prescribed by the interferer.

then using Rayleigh fading the signal exceed for 95% of locations within the short sector is 12.7 dB lower than this, which is −82.5 dBm. For an interferer with a nominal signal strength at the receiver of −90 dBm, the signal level exceeded for 50% of the time will be −90 dBm. The difference between the two is only 7.5 dB, significantly less than the required 19 dB and therefore the interference criteria cannot be met for this level of availability.

In general, it is highly undesirable to have re-used frequencies within potential interference range and normally the system designers will use frequency re-use to prevent this occurring in practice.

In this example, we focussed on interference into a base station and not the mobile subscribers to the network. Interference from base stations to mobiles is unlikely to happen because normally the uplink and downlink frequencies will be separated in frequency. For example, in the UK Airwave network, the difference between paired uplinks and downlinks is 10 MHz, and there is at least 5 MHz difference between the closest uplink and downlink channels. Interference between mobiles is less likely since each mobile is less powerful and is likely to be embedded in clutter. However, the same process as described above can be carried out for mobile to mobile interference.

5.2.2 Adjacent and Other Channel Offset Interference

So far, we have considered interferers on the same channel as the victim. However, this is not the only time that interferers can cause problems. Interferers on other channels can cause problems due to energy spilling over their channel into others. This is worst for the immediate channel above and below the victim frequency. These channels are known as ‘adjacent channels’. To determine potential interference, we need to know the relative strength of energy in the adjacent channels and then proceed to determine the effect of that energy.

The adjacent channel situation is shown in Figure 5.3. If we consider channel n as the victim and channel $n + 1$ as the interferer, it can be seen that although most of the energy is within its own channel, there is still a skirt of energy overflowing into the adjacent channel and beyond. It is this energy that causes adjacent channel interference. Although the level of energy spilled into other bands is relatively low compared to the main channel, it must be remembered that the power transmitted is significantly higher than the receive signal level and can still cause problems. This is particularly the case where transmitters and receivers are positioned in close proximity to one another.

Most systems will give figures for the relative strength of interference into adjacent channels. For example, if the amount of energy radiated into the adjacent channel is 50 dB down on the peak power in the wanted channel, then this can be accounted for in the adjacent channel interference analysis.

For example, assume that the received signal strength for 50% availability is −50 dBm in the interferer’s intended channel. This is a strong signal and it would be likely to occur close to the transmitter. The power in the adjacent channel

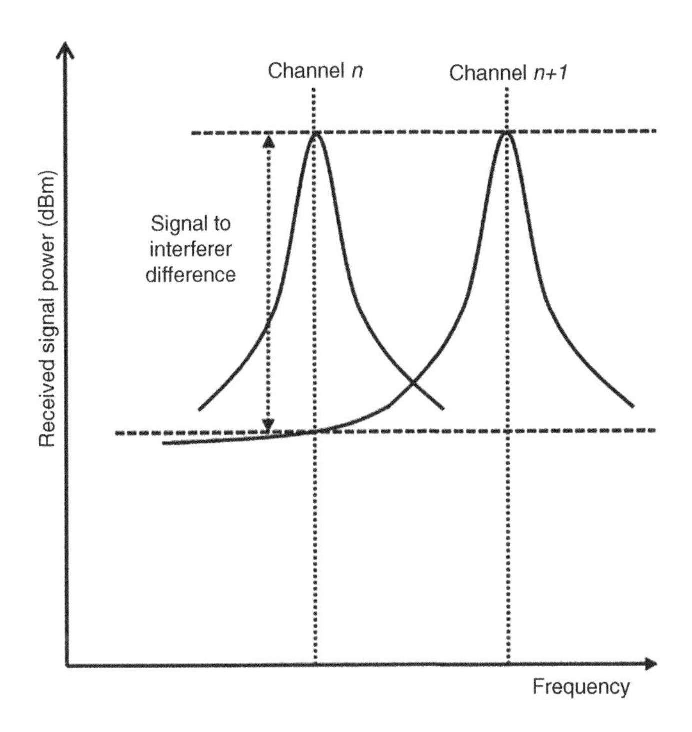

Figure 5.3 Diagram showing energy from an interferer (tuned to channel $n + 1$) spreading into a victim receiver tuned to channel n. Energy from a transmission, even if filtered at the transmitter, will still be present outside of the tuned band. This can cause interference to the adjacent channel as shown or even receivers many channels away.

would be −100 dBm, assuming the 50-dB adjacent channel figure quoted above. If the wanted signal strength for another receiver operating in the adjacent channel is −90 dBm, then the difference between the wanted and interfering signal is only 10 dB below the wanted signal. We would need to compare this to the co-channel interference rejection figure to determine whether interference would be present in this scenario. We can also apply the same corrections to account for fading of both the wanted and interfering signal to determine its performance at the required levels of availability.

Interference does not stop at the adjacent channels, but may continue over a wide range. An illustration of wider transmit spectral power is shown in Figure 5.4.

The level of energy received in these other bands can be described in the manner shown on Table 5.1.

Table 5.1 can be interpreted in the following manner. For co-channel interference, the wanted power must be 12 dB higher than any interferer. For the adjacent channels above and below the wanted channel, the signal can tolerate an interferer as powerful as the wanted channel. For interferers four channels away, the interference power at its own tuned frequency can be 30 dB higher than the wanted signal.

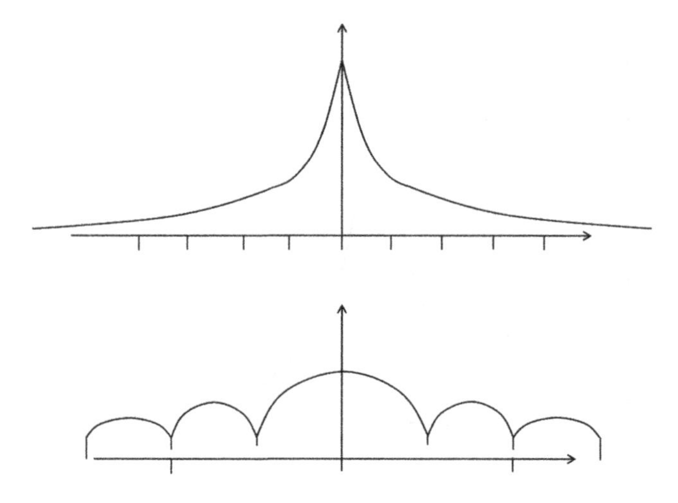

Figure 5.4 Illustration of energy radiated from a narrowband transmission (top) and a wideband transmission (bottom). These images are indicative only, but they show how energy is present outside of the tuned frequency and can extend a wide range beyond the nominal bandwidth. The energy level transmitted is low compared to the received power in the tuned frequency but at close range, this energy can cause interference to receivers tuned many channels away. The diagrams do not include any spurious features that may also be present in some cases at discrete frequencies such as harmonics, images and inter-modulation products.

Specifications for radio transmitters for a particular service can include a 'spectrum mask'. This is not an exact description of the actual transmitted power spectral density, but is a definition of maximum values that cannot be exceeded. An example is shown in Figure 5.5 for TETRA and for the TETRA Advanced Packet Service (TAPS). The frequency offset values are in MHz and the lines for spurious levels in the TETRA bandwidth are shown in dB. Note that the frequency offset is applicable both above and below the tuned frequency.

In the absence of information about the actual performance of specific radios, such spectrum masks allow designers to work using nominal figures.

Table 5.1 Hypothetical C/I figures for a radio system. The channel offset figures are valid for channels above and below the tuned frequency, which is channel offset = 0

Channel offset (No. of channels)	C/I (dB)
0	12
1	0
2	−6
3	−12
4	−30
5	−40

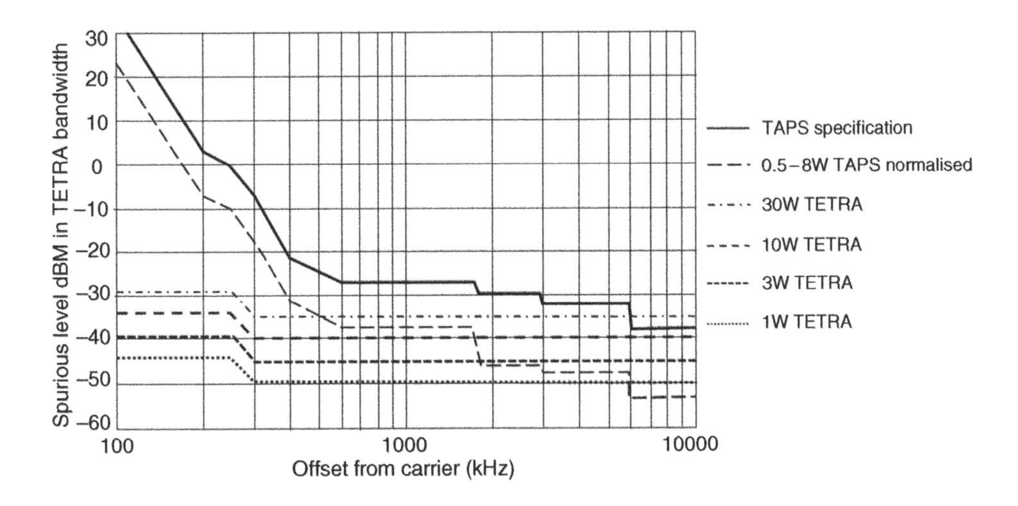

Figure 5.5 TAPS and TETRA spectrum masks.

5.2.3 Overlapping Bandwidth Calculations

All of the calculations carried out so far assume that the wanted and interferer have the same bandwidth and are spaced equal numbers of channels apart. This is not always the case; sometimes where spectrum is scarce, a frequency plan with less than full channel spacing will be used, and also (and increasingly), there is the possibility that different services will share the same spectrum. In this case, we need to consider the degree of spectral overlap in order to determine potential interference effects. This is illustrated in Figure 5.6.

Figure 5.6 shows two carriers with similar bandwidth, which are tuned $^1/_2$ of their bandwidth apart. In this case half the unwanted channel's bandwidth is overlapped with the wanted channel and hence half the power (3 dB down on full power) and which may cause interference. This is a slight simplification of the real situation, since neither signal will have the bandwidth response shown but in practice it is a good approximation. The following formula can be used to calculate the unwanted interference contribution in to the wanted signal in this situation.

$$P_{ei} = P_I - 10\,\mathrm{LOG}_{10} BW_{ol}$$

where

P_{ei} is the effective level of the interfering signal.

P_I is the power of the interferer.

$$BW_{ol} = ((B_w + B_u)/2 - |F_2 - F_1|)/B_w$$

where B_u overlaps B_w, $BW_{ol} \leq B_u$ and $BW_{ol} \leq B_w$.

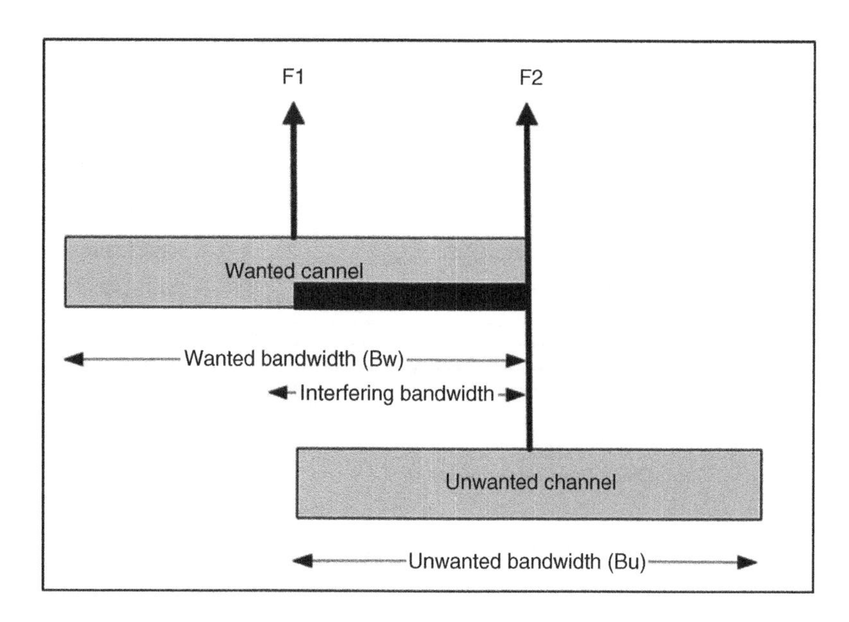

Figure 5.6 An example of an interferer partly in the wanted channel. This can occur when spectrum demand exceeds supply and the channel plan includes overlapping channels.

If B_u is contained within B_w no correction is needed.

The bandwidth correction for different percentage of coverage, with the correction values shown in dB is shown in Figure 5.7.

5.2.4 Interference between Dissimilar Systems

Management of interference between dissimilar systems is complicated by the differences between the spectral characteristics of the interfering transmitter and

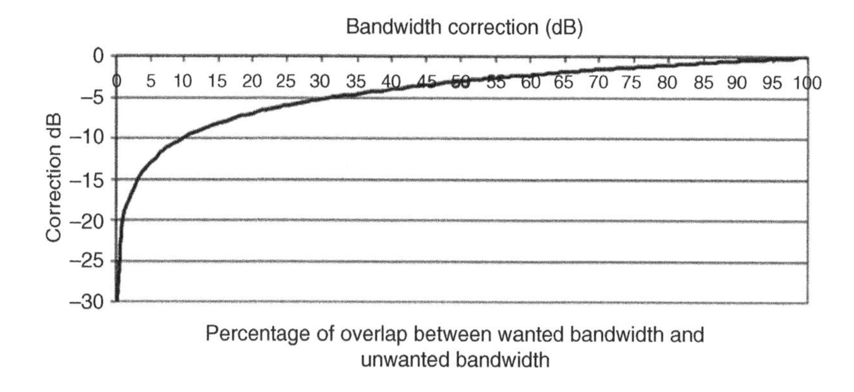

Figure 5.7 Bandwidth correction in dB compared to the percentage overlap.

the type of signal the receiver is designed to capture. We need a generic method of determining how these different systems interact. One possible way of achieving this is to perform actual measurements on the test bench using test gear that as far as possible reproduces the transmit characteristics, the propagation channel and the receive interference rejection characteristics against the interfering signal type. If this is done correctly, then very useful metrics can be derived for use in system design. However, this process is time-consuming and can be expensive and even if it is carried out, the results may not be made publicly available. This is of course particularly true for military systems, where security classification may preclude release of the data.

Another method is to calculate the RF power injected into the receiver from an interfering transmitter. This is a complicated process that needs to consider the energy not just arriving at the victim antenna but also the energy passed through the receiver filters into the input of the discriminator. This is illustrated in Figure 5.8.

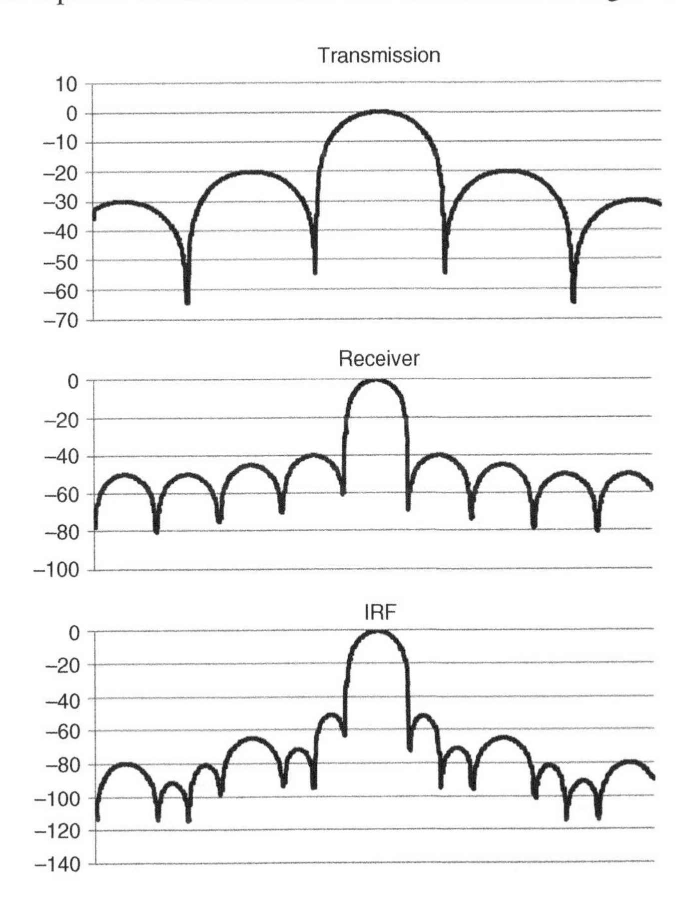

Figure 5.8 The energy transmitted by a co-channel interferer is affected by the receiver response, including filters designed to reject interference. The Interference Rejection Factor is the convolution of the two.

The top diagram shows the transmitted energy from an interferer. The receiver input rejection filter response. The Interference Rejection Factor (IRF) is shown in the bottom diagram. This is the convolution of the input energy and receiver response. The response shown is for a co-channel interferer. This shows the importance of both transmitter and receiver filters to reduce the potential effects of interference. The sum of the energy input into the receiver discrimination circuits determines how much the interferer will disrupt the victim receiver. The lower the unwanted energy, the better the receiver will work.

This process also works for interferers that are not co-channel as illustrated in Figure 5.9. The rejection response is again shown at the bottom. In this case, it is more complex in nature. The important aspect is the sum of energy injected into the receiver after filtering.

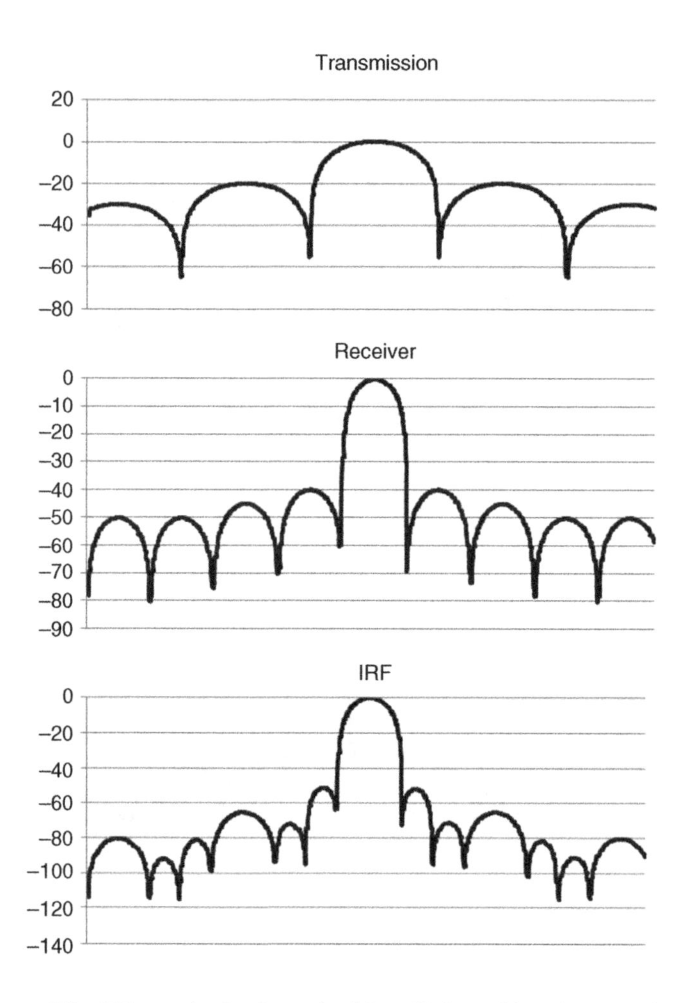

Figure 5.9 IRF can also be determined for off-channel interferers as shown.

The IRF response shown in Figures 5.8 and 5.9 can be resolved into a numeric value for interference rejection for a given type of transmission and receiver, and for a given frequency offset. The overall response is likely to be along the lines of the response shown in Figure 5.5 for the TAPS/TETRA case.

5.2.5 Multiple Interferers

So far in this section we have only considered a single interferer. However, in some cases there may be many interferers, each providing a component of energy to the total interference. We will look at two methods of assessing interference from multiple interferers; the power sum method and the Simplified Multiplication Method (SMM).

5.2.5.1 Power Sum Method

The power sum method can be used for multiple interferers when they are of the same type as the victim signal. The formula for the power sum method is:

$$P_i = 10\,log_{10}\left(\sum 10^{\left(\frac{1}{10}\right)}\right)$$

where

P_i is the calculated total interference.

I is the interfering power level of each interferer (dBm).

We can illustrate the use of this by examples. First, consider two interferers each with a power level in the receiver channel of −100 dBm. Thus, the equation becomes:

$$P_i = 10\,log_{10}\left(10\frac{-100}{10} + 10^{\frac{-100}{10}}\right)$$

This evaluates to −96.9897 dB, or approximately −97 dBm. Consideration of this will show that in the case of two de-correlated interference sources of the same power, the total power is 3 dB above the level of each. Since 3 dB equates to double the power, this is logical. However, the process also works for any combination of interferers with any individual power contribution.

As another example, consider the four interferers:

I_1 = −103 dBm.

I_2 = −97 dBm.

I_3 = −105 dBm.

I_4 = −120 dBm.

Applying this to the formula gives:

$$P_i = 10\,log_{10}(10^{-10.3} + 10^{-9.7} + 10^{-10.5} + 10^{-12.0})$$

Evaluating this gives a result of P_i = −95.49 dBm.

5.2.5.2 Simplified Multiplication Method

The Simplified Multiplication Method (SMM) is fully described in report CCIR-945, from which this section has been derived. The method is based on the following assumptions:

- There is no correlation between the interferers.
- There is one dominant interferer.
- Time dispersion can be ignored.
- Noise is negligible compared to the interferers.

The SMM is based on using the CDF of the normal distribution. The method is described by an example shown below. The process is iterative.

Assume there are four interferers as in the example for the power sum method:
I_1 = −103 dBm.
I_2 = −97 dBm.
I_3 = −105 dBm.
I_4 = −120 dBm.

The SMM process starts by choosing a seed value to begin the calculation. This is chosen as 6 dB above the highest interferer, which in this case is −97 + 6 = −91 dBm. Each of the interferers are compared to this value, giving a difference value Z_i.

After this, the normal distribution is normalised for each interferer by the expression:

$$X_i = \frac{Z_i}{S_d\sqrt{2}}$$

where S_d is the standard deviation due to fading, which at VHF is approximately 8.3 dB, or approximately 9.5 dB for UHF. In this case, we will use the VHF value. For each interferer we determine the normal distribution CDF for X_i. We then multiply the CDF values for each interferer together and finally produce the delta value that we will use to modify the seed value for the next step. The delta value is calculated according to:

$$\Delta = \frac{(0.5 - CDF\ product)}{0.05}$$

where the value of 0.05 has been shown to be the best correction to allow the sequence to converge quickly.

The delta value is added to the seed value to give an improved value, which in this case is −91 – 0.3335 = −91.3335 dB.

The step 2 seed value is modified by the delta value to give a value = −91.37 dB. The process can be continued to refine the value, but improvements will be negligible.

Table 5.2a Step one in the SMM method with seed value = −91 dBm

Interferer	Level	Z_i	X_i	CDF
I1	−103	12	1.0223	0.8467
I2	−97	6	0.5112	0.6954
I3	−105	14	1.1927	0.8835
I4	−120	29	2.4706	0.9933
Product of CDF values				0.5167
Delta				−0.3335 dB

Table 5.2b Step two in the SMM method with seed value = −91.3335 dB

Interferer	Level	Z_i	X_i	CDF
I1	−103	11.6665	0.9939	0.8399
I2	−97	6.6665	0.4828	0.6854
I3	−105	13.6665	1.1643	0.8778
I4	−120	28.6665	2.4422	0.9927
Product of CDF values				0.5016
Delta				−0.0323

The SMM is only one method of assessing interference from multiple interferers; there are others based on modifying the SMM or via other routes. Often these have been designed to deal with specific circumstances. The processes involved in this process are illustrated in Tables 5.2a and 5.2b.

5.3 Interference in the Time Domain

5.3.1 Time Slots, Frequency Hopping Systems and Activity Ratios

Interference is not just dependent on spectrum considerations, but it can also have time-varying characteristics due to a number of factors. Some of these are technology driven and others due to variable atmospheric effects over time. Technologies based on Time Division Duplexing (TDD) use time slots to multiplex several radio channels onto a single channel as shown in Figure 5.10. These time slots help to prevent interference and they also mean that one individual channel is working only 1/n of the time, where n is the number of slots. TETRA, for example, uses four time slots in a single carrier. Timing is vital for the receiver to function properly.

The relative time that a signal is active is called the 'activity ratio'. A continuously-transmitting signal has an activity ratio of 1.0, and a signal that is active only 1% of the time has an activity ratio of 0.01. When considering interference effects of signals with intermittent transmission, this must be included in interference calculations. The precise method of evaluating interference depends on the statistics describing the probability of the wanted and unwanted signals being present at the same instant. This

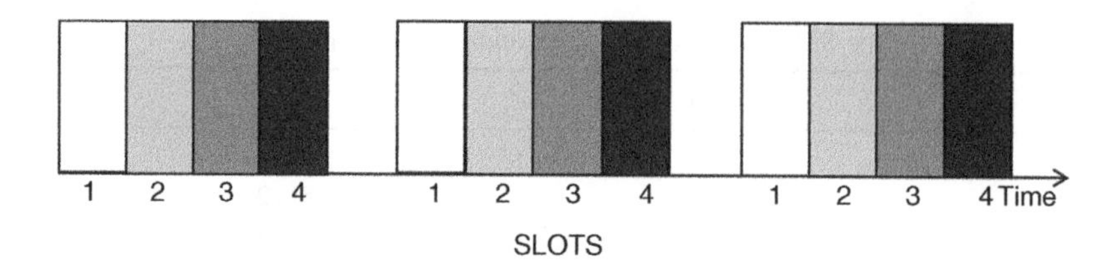

Figure 5.10 Time slots in a TDD system, with a gap for synchronisation and system management.

is described in Figure 5.11, which shows wanted and unwanted randomly distributed transmissions occurring at an activity level of 0.2. The relative locations of the two transmission sources are not considered at present, but we will be looking at this aspect shortly. We are also assuming that the receiver for the wanted transmitter is looking for a signal during the same time slot.

The clashes can be seen from the figure. These are where both systems are using the same part of spectrum at the same time, and thus where interference can occur.

If the two transmissions are truly random, then classic Erlang-B calculations can be used to determine the probability of interference occurring based on common use of the same channel at the same time.

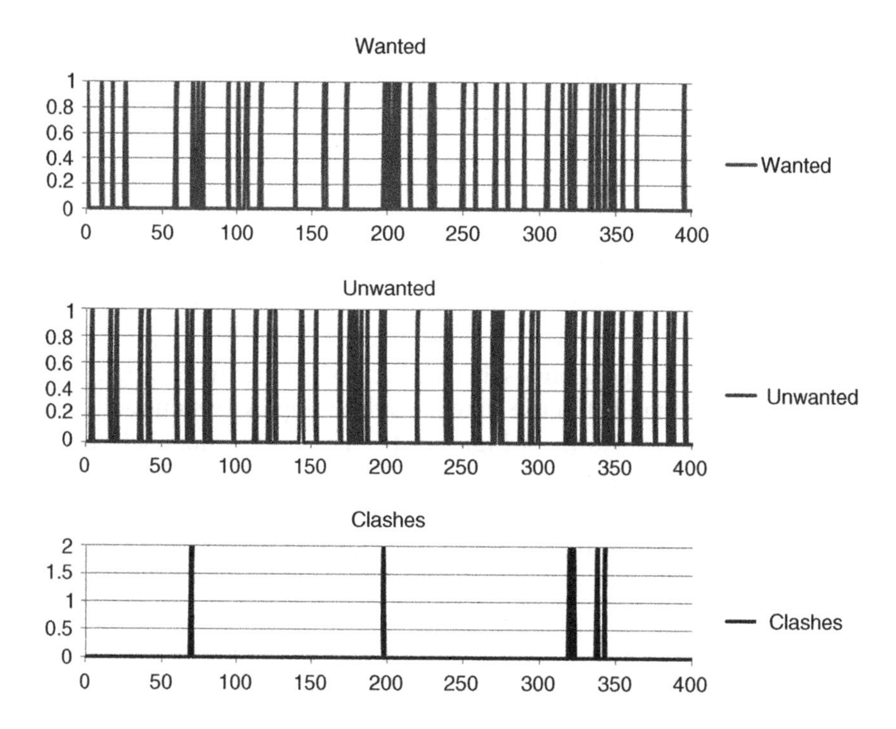

Figure 5.11 Example of two systems operating independently with occasional time clashes.

However, it pays to be careful about applying Erlang-B without considering the scenario to be modelled. Erlang-B requires randomness for the statistics to be valid (technically, this is called a *poisson process*), but there can be many conditions that prohibit this from being the case. These include:

- Systems where one transmission will trigger another one. This might be the relay of orders from higher to lower echelons, a sensor network where communications are used to relay information about intercepts or other tactical information, or systems based on timing (where, for example, transmissions are made at the same time each day). In this case, the transmissions are not truly random. Although the triggering event may be random, once the process has started it follows non-random behaviour.
- Systems where the use of spectrum is correlated between wanted and victim. This may include two frequency hopping systems using the same or similar hop sets. Thus the two systems are linked and not therefore random.
- A tracking system, such as a follower jammer, which will attempt to jam the portion of the link present during a time slot period. Again, the two systems are linked and not truly random.

These cases, which are not poisson processes, cannot be solved using Erlang's equations. Without calculating the mathematics from scratch, one possible approach is to simulate the network performance to gather the metrics necessary to determine interference probability.

5.3.2 Non-Continuous Interference

Interference is not always a constant effect. In particular, signals that propagate a long way are subject to changes in atmospheric conditions. This includes variation in the k-factor due to changes in the vertical atmospheric column, ducting and increases in troposcatter. These can be modelled over long periods to produce special propagation prediction models that account for differences in propagation over time. Such models normally include results for different percentages of time; 50%, 10%, 5% and 1% are commonly quoted in such models.

5.4 Interference Mitigation Techniques

There are several ways of militating against interference. These include:

- power management of the interferer;
- antenna height of both interferer and victim;
- antenna tilt of both interferer and victim;

- sectored antennas or null steering for both interferer and victim;
- frequency re-assignment for either interferer or victim.

The selection of which mitigation method is applied is usually based on the amount of interference and the implications of the potential changes. Some typical considerations include:

- Power management: one approach is to reduce the power of the interferer, to benefit from the change in relative signal strength. However, reducing the power of the interferer will also reduce its service area and this will have an effect on network coverage and capacity in the region around the interfering base station.
- Antenna height: the antenna heights of either the interferer or victim can be reduced (or both). This will have the effect of reducing the range of the interferer if its antenna height is reduced (again this may have a knock-on effect to the overall network coverage and local capacity). The victim antenna height can also be reduced. This can help in the situation where the reduction in height will result in benefits from terrain or clutter shielding. Again, of course, the coverage area of the victim will be reduced in this case.
- Antenna tilting: for antennas that have vertical directivity, it is possible to orient antennas used for paths within the horizon such that the energy radiated towards the horizon is reduced. This will have the effect of maintaining the coverage in the wanted area but reducing interference outside of it.
- Sectored antennas and null steering: antennas with directional patterns in the horizontal plane can be used to minimise interference in the direction of the victim or interferer (as required). Antennas that have a null (very low response in a particular direction) can be used to spatially filter out particular interferers. This approach can only be used when the direction of the interfering energy is known, and thus it is applicable to the condition of base station to base station interference.
- Frequency Assignment: either the victim or interfering system can have their frequencies changed to prevent interference between them. This of course may cause interference with other spectrum users and thus should only be done with care, and with checks to identify any potential problems that may be caused by the change.

References and Further Reading

All internet references correct at time of writing.

CEPT (2002), *ADJACENT BAND COMPATIBILITY BETWEEN TETRA TAPS MOBILE SERVICES AT 870 MHz*, http://www.erodocdb.dk/Docs/doc98/official/pdf/ECCREP014.PDF

CCIR Report 945: *Method for the Assessment of Multiple Interference.*

CCLRC RAL, dB Spectrum, Transfinite Systems, University of York (2007), *Modelling Interference from Multiple Sources*, http://www.ofcom.org.uk/research/technology/research/prop/multinter/report.pdf.

Saunders, S.; Aragon-Savala, A. (2007), *Antennas and Propagation for Wireless Communications Systems*, Wiley-Blackwell, ISBN 978-0470848791.

6

Radio Links and Deliberate Jamming

6.1 The Purpose of Jamming

Radio jamming is an old practice, dating back theoretically to the dawn of radio communications. It is something that has become recognised in the zeitgeist of the Cold War and various conflicts around the world. However, in being recognised by the public some basic mistakes have become entrenched. Principal among these is the concept of 'jamming the transmission'. This is possible in theory but in most cases, it is the receiver that must be jammed. This is the point in the system where the signal is weakest and thus most vulnerable. The second major misconception is that jamming is absolute; once you jam, you prevent reception in any case. Again, this is far from true in practice. In both communications and in non-communications jamming we will come across the concept of 'burn-through'. This is the link robustness needed to overcome the jamming effect. If a link is sufficiently strong, it will become impossible for a jammer to prevent communications (or for radar systems, detection).

So what is the practical purpose of jamming? The purpose of jamming is to prevent the enemy from using radio links freely in a tactical environment. In this case, a radio link can be the link from target to receiver in a radar system, between two communications elements or between navigation transmitters and aircraft trying to use them, for example. In practical terms, the aim is to limit the use of the radio spectrum for one or more target systems until it becomes tactically useless. This is not the same as completely jamming the link. If the effective range can be brought down from 35 km, a tactically useful distance, to a range of a few hundred metres – tactically insufficient – then the jamming is successful in its intent.

This section looks at jamming fundamentals for communications jammers. Chapter 7 looks at non-communications jammers, which have very different modes of operation.

6.2 How Jamming Works

Jamming is all about getting sufficient energy into the victim receiver at the right time and in the right place. This is illustrated in Figure 6.1. The basic scenario shown is appropriate to all forms of jamming, including radar jamming. The idea is that the power received from the enemy transmitter at the victim receiver is overwhelmed by the jamming power. The ratio of jamming power to signal power required for effective jamming is known as the Jamming-to-Signal ratio (J/S) and is normally expressed in dB.

As we will see, the J/S ratio need not always be positive in order to disrupt enemy links.

Any jammer will be limited in the power it can transmit and the bandwidth over which it can be applied. Therefore, jammers need to be configured appropriately to be able to function well.

However powerful the jammer, unless the power received from the jammer at the receiver is at or above its maximum receiver power, a suitably strong link can burn through the jamming. This happens if the enemy link is short enough, or the transmitter power is high enough to overcome the jamming power. This is shown in Figure 6.2.

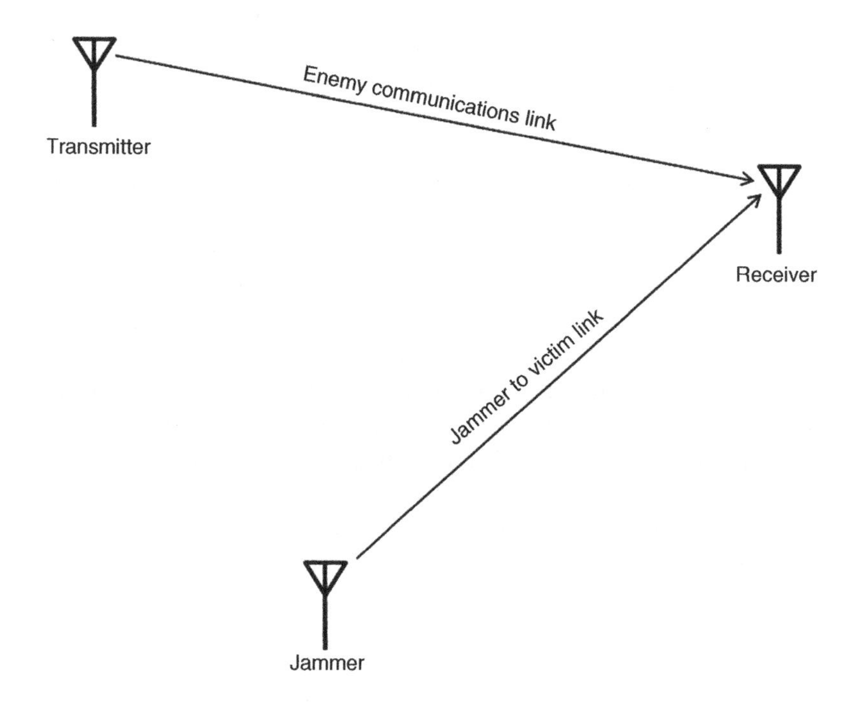

Figure 6.1 Jamming system geometry.

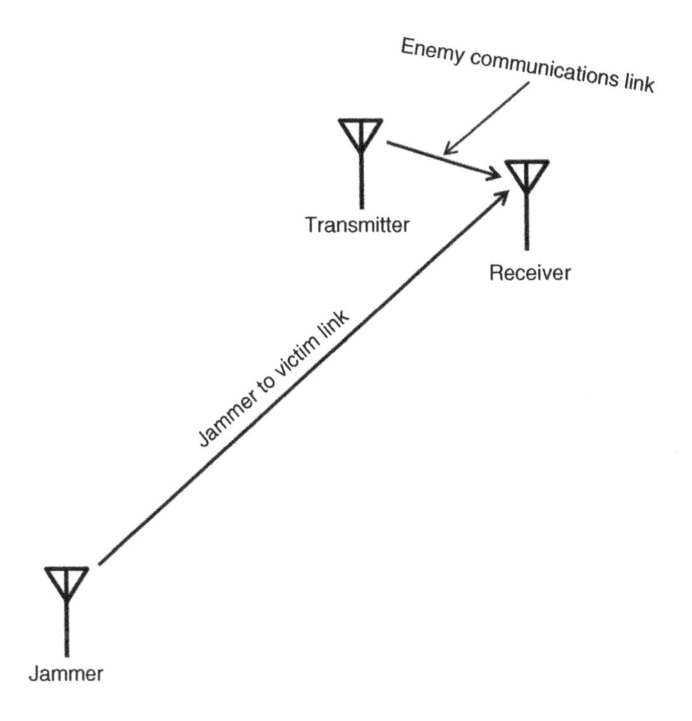

Figure 6.2 Burn-through of jamming.

6.3 Types of Communications Jammers

The limitations in jammer power mean that it is necessary to have different methods to apply the jamming signal to different applications.

The first type is the simplest. This is called 'jam on tune'. In this case, an enemy transmission is detected on a particular frequency and within a measured bandwidth. The jammer operator tunes the jammer to the same frequency and sets the same bandwidth. Then the jammer is switched on. Periodically, the jammer will be switched off for a short time to check whether the target signal is still present. This is illustrated in Figure 6.3.

If the jam on tune jammer can be tuned rapidly, it can be used to jam against frequency hopping radios. This specific type is known as a follower jammer. The frequency hopper will not dwell on a single frequency for long, but as long as the jammer can jam at least one third of the signal then the jam will be effective.

When the target signal is changing frequency or there are multiple signals to jam, a swept jammer can be used. This is illustrated in Figure 6.4. Note that in this figure, the vertical axis is frequency rather than power. The saw shape shown is typical, but any form of sweep may be used.

In this case, enemy signals are not jammed all of the time. For example, assume the jammer bandwidth is 0.5 MHz and that it sweeps through a target band of 30–32 MHz

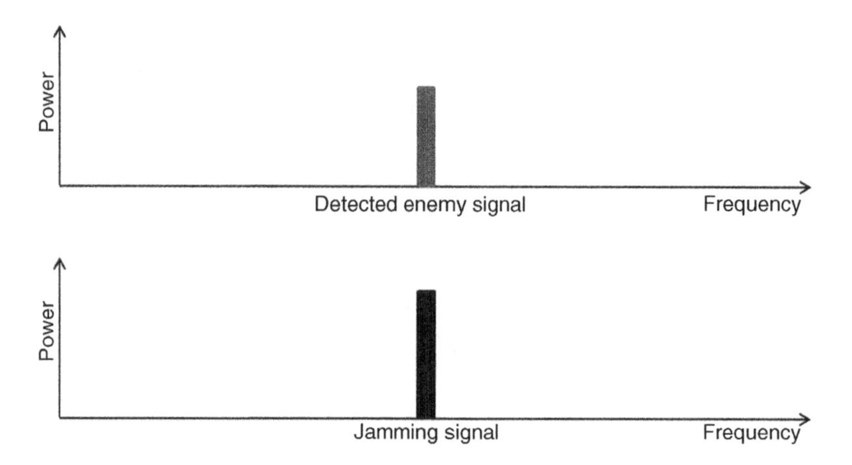

Figure 6.3 A simple jam on tune jammer.

at a rate of 1 MHz per second. It therefore takes 2 seconds to pass through the same frequency twice. If a target signal has a frequency within the band, maximum jamming efficiency would be present for one quarter of the time. This is based on the assumption that the target bandwidth is substantially narrower than the jamming bandwidth. If the target signal is carrying voice, then one quarter of the signal will be lost, but not only that, the audio produced would be very difficult for listeners to interpret.

A different method of jamming a wide bandwidth is to use a barrage jammer. This simultaneously jams a broad band of spectrum. This is illustrated in Figure 6.5.

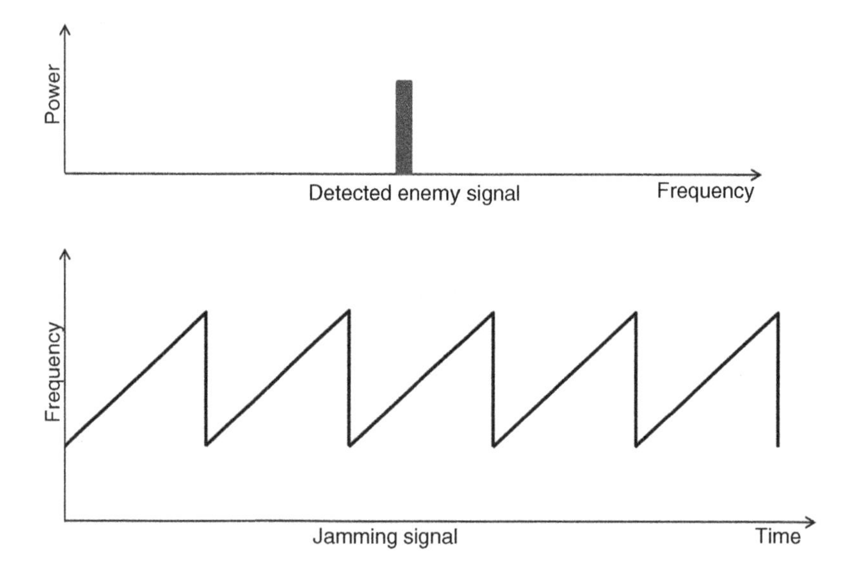

Figure 6.4 A swept jammer. The jamming power is swept through the target band repeatedly.

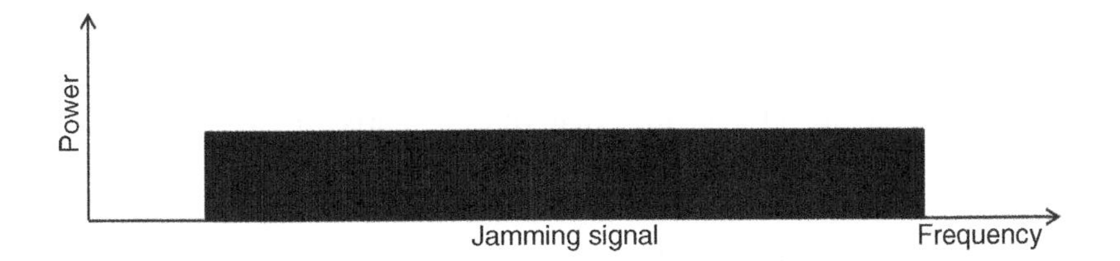

Figure 6.5 A barrage jammer.

Barrage jammers require very high power, because the jamming energy is spread over a wide bandwidth. Barrage jammers can also be used to jam frequency hopping signals. In this case, it is not necessary to jam the entire frequencies used by the frequency hopper. Similarly to the follower jammer, if at least one third of the frequencies used by the hopper are jammed, then the jam will be effective.

If more than one frequency needs to be jammed, an adaptive jammer can be used. This uses the power available to the jammer to jam several signals simultaneously by splitting the available power into the required number of channels. This is illustrated in Figure 6.6. Since the power available to the jammer is limited, the addition of each new channel to be jammed reduces the power available for each channel. Thus, if two channels are to be jammed using the same power, the power in each channel is reduced by a factor of 2, and so on.

If the technology of the enemy radio system is known, it may be possible to use 'smart jamming' techniques. For example, for systems that use a control channel to manage access, the jammer can be applied against this channel alone. This will prevent users from achieving connectivity. Or, the jammer can prevent packet acknowledgements

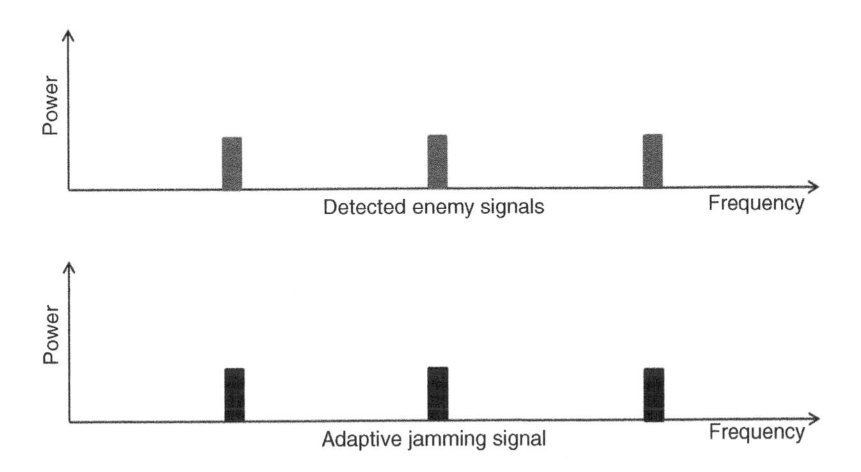

Figure 6.6 Illustration of an adaptive jammer. The available jamming power is split into the number of channels to be jammed, at the expense of the power available in each jammed channel.

being successfully received. This forces the network nodes to continuously re-transmit packages, preventing messages from being received and filling the network to capacity. Another method is to attack the message routing parts of the signal, preventing messages from being sent to the right parts of the network. There are a wide variety of other techniques under development. All of these systems rely on detailed knowledge of network operation and identification of vulnerable parts to be exploited by the jammer. Smart jamming has particular application to countered Radio Controlled Improvised Explosive Devices (RCIEDs), where terrorists use commercial communications equipment to mount attacks.

6.4 Jamming Mitigation Techniques

There are many methods that can be used to protect communications against jamming attack. This starts at the design phase and goes all the way through to operational techniques. Design improvements include the use of spread spectrum systems, which offer three main advantages. The first is that the signal is far harder to find. The second is that because the signal is spread over a wider bandwidth than a non-spread system, the jammer has to jam over a wider band, which reduces its power. The third is that the spreading system used provides anti-jam capability in its own right by rejecting noise signals through its processing gain. This is the decibel ratio of the spread signal bandwidth over the baseband signal bandwidth. Spread spectrum signal can however still be jammed by sufficiently high powered jammers.

Other systems can be protected by operational tactics. Methods include:

- power management to minimise the probability of detection;
- use of terrain and clutter to limit the ability of the enemy to detect and jam signals;
- frequent changes of frequency.
- intermittent use of systems and minimising transmissions;
- use of antennas with 'nulls' (low energy output) in the direction of enemy jammers;
- use of highly directional antennas and orientating them so that neither points in the direction of the enemy. This means attempting to make important links parallel to the threat axis of the enemy.

Other methods are described in Part Two.

References and Further Reading

Ademy, D. (2001), *EW101: A First Course in Electronic Warfare*, Artech House, MA, USA, ISBN 1-58053-169-5.
Frater, M.; Ryan, M. (2001), *Electronic Warfare for the Digitized Battlefield*, Artech House, MA, USA, ISBN 1-58053-271-3.
Poisel, R. (2002), *Introduction to Communication Electronic Warfare Systems*, Artech House, MA, USA, ISBN 1-58053-344-2.

7

Radar and Radar Jamming

7.1 Introduction to Radars

Radars are active radio frequency devices used to determine the location of targets of interest within the radar coverage area. Although they were originally designed to identify aircraft, their use was soon adapted to cover maritime and then terrestrial environments. These days, radars are used for a wide variety of applications:

- early warning (ground- or air-based);
- Ground Control Intercept (GCI);
- ground surveillance (ground- or air-based, maritime or terrestrial);
- surface search (maritime);
- coastal surveillance (usually based on shore);
- tracking systems for missile and gun systems (air, terrestrial and maritime);
- fire control, intercept and target illumination (air, terrestrial and maritime);
- mortar, artillery and sniper tracking systems (usually terrestrial);
- mapping radar to measure ground and clutter characteristics;
- weather sensing (terrestrial and airborne);
- navigation (air, terrestrial and maritime);
- air traffic control (air, terrestrial and maritime);
- range tracking and space tracking (air, terrestrial and maritime).

To meet the varying demands of these types of system, different techniques and frequencies are used. Types of radar are covered in Section 7.3.

A list of frequency bands with the nomenclature used by radar practitioners is shown in Figure 7.1.

Different types of radar operate within the HF band up into the EHF band. In general, the lower frequencies have longer range but lower resolution. Higher frequencies are

Communications, Radar and Electronic Warfare Adrian Graham

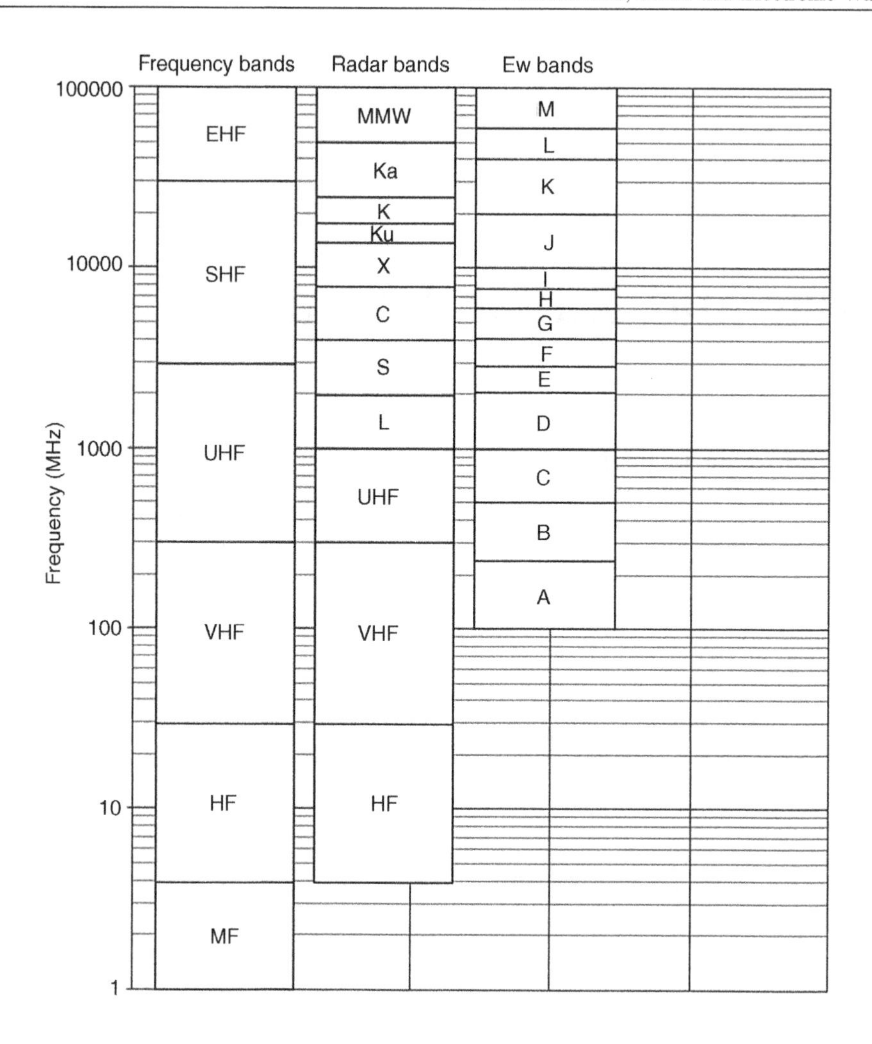

Figure 7.1 Radar and EW frequency band terminology.

used for shorter-range applications such as missile terminal guidance and high-resolution mapping.

Radar systems operate in the basic manner shown in Figure 7.2.

Some systems operate in so-called 'bi-static' mode, in which the transmitter and receiver antennas are physically separated as shown in Figure 7.3. In this diagram, the transmitter and receiver antennas are located in different geographic locations. The target aircraft is at an altitude within range of both antennas.

Although the concept of radar is very simple, its implementation is often far from simple and the 'arms race' of radar technology has led to many innovations that add complexity in return for performance. We will be looking at some of these in this chapter and later in the book.

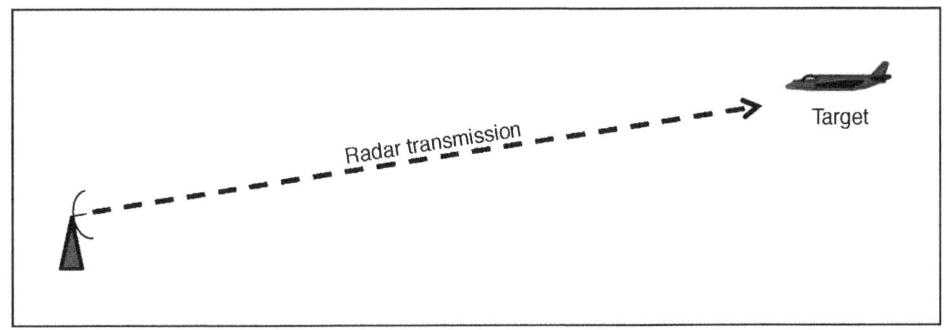

Radar transmit phase

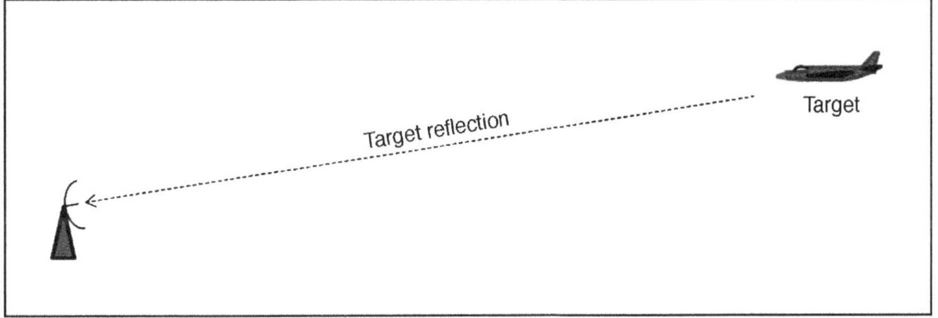

Radar receive phase

Figure 7.2 Mode of operation of generic simple radar system.

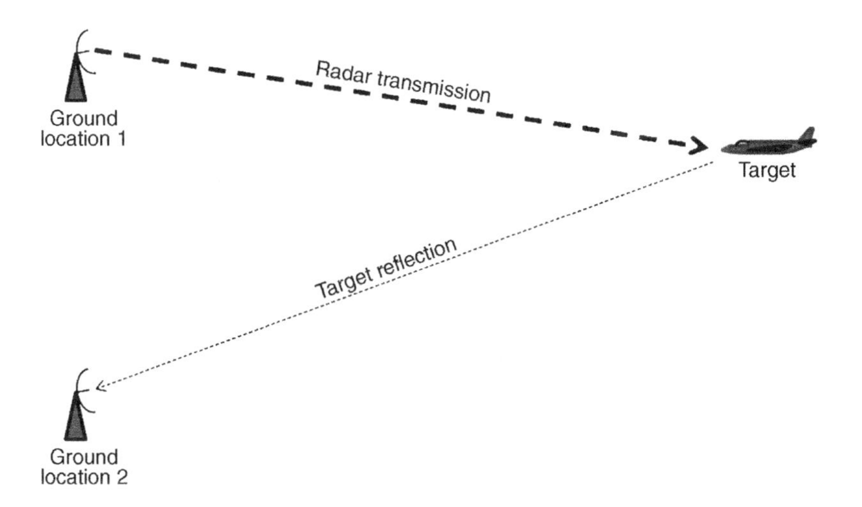

Figure 7.3 Mode of operation of a bi-static radar.

7.2 The Radar Equation

The radar equation most often quoted is one of the most well-known radio system equations in its basic form. However, it is very simplistic so we will start with a slightly different formulation after the form described by Skolnic. This formulation is more generic in that is does not assume that the transmitting and receiver antenna have the same characteristics and it includes terms to account for propagation conditions:

$$\frac{P_r}{P_t} = \frac{G_t G_r \sigma \lambda^2 F_t^2 F_r^2}{(4\pi)^3 R^4}$$

where

P_r = Received signal power at antenna terminals.
P_t = Transmitted signal power at antenna terminals.
G_t = Transmit antenna gain.
G_r = Receive antenna gain.
σ = Radar target cross section.
λ = Wavelength.
F_t = Propagation factor for transmit antenna to target path.
F_r = Propagation factor for receive antenna path.
R = Radar to target distance.

Figure 7.4 shows how this relates to the mono-static case. The $(4\pi)^3$ term is to account for spreading loss and the wavelength of transmission is not shown.

The primary differences between this representation of the radar equation and the more often quoted one is the inclusion of the terms F_t and F_r, which account for the difference between the actual loss and free space loss between the radar antenna(s) and the target. As we have seen in the communications sections, free space loss is seldom a realistic description of the real world situation.

We can also convert this formula for the bi-static case by separating the range term R into the transmitter-to-target (R_{tt}) and target-to-receiver paths (R_{tr}).

$$\frac{P_r}{P_t} = \frac{G_t G_r \sigma \lambda^2 F_t^2 F_r^2}{(4\pi)^3 R_{tt}^2 R_{tr}^2}$$

We can make the basic formula more useful for determining maximum range. Rearranging the formula to solve for range for the mono-static case gives:

$$R = \left(\frac{P_t G_t G_r \sigma \lambda^2 F_t^2 F_r^2}{(4\pi)^3 P_r} \right)^{\frac{1}{4}}$$

For maximum range, we need to solve for $R_{\max}$ and $P_{r,\min}$.

$$R_{\max} = \left(\frac{P_t G_t G_r \sigma \lambda^2 F_t^2 F_r^2}{(4\pi)^3 P_{r,\min}} \right)^{\frac{1}{4}}$$

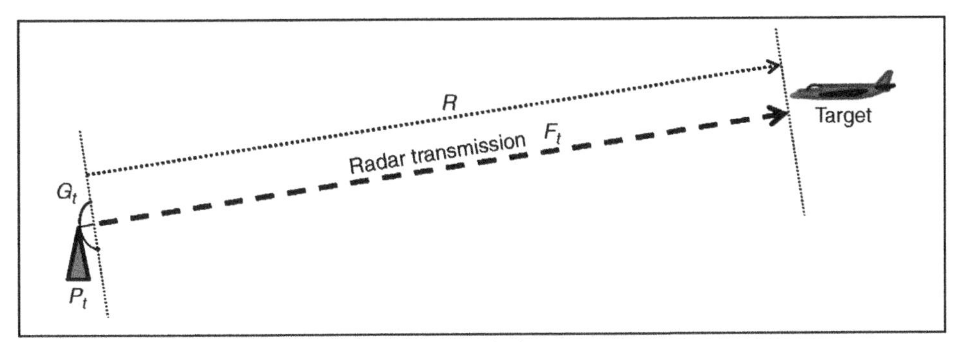

Radar transmit phase

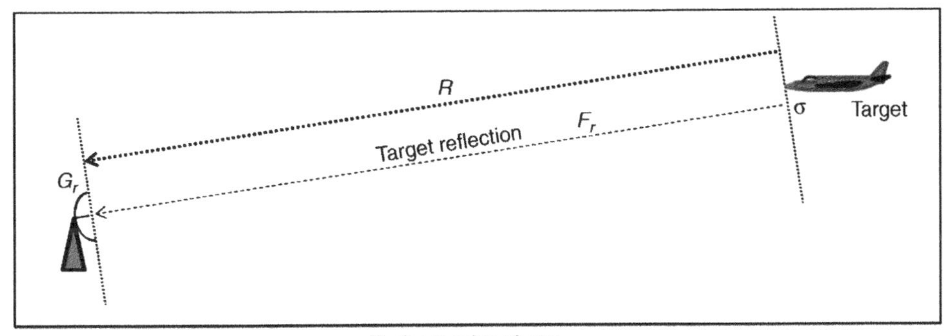

Radar receive phase

Figure 7.4 Radar equation terms displayed graphically. Wavelength and spreading loss terms have not been included in the diagram.

$P_{r,\min}$ is effectively the minimum signal that will work in the noise experienced by the receiver, which is

$$P_r = \left(\frac{S}{N}\right) P_n = \left(\frac{S}{N}\right) kT_s B_n$$

where
K = Boltzmann's constant = 1.38×10^{-23}W/K.
T_s = Receiver noise temperature in Kelvin (often = 290 K).
B_n = Noise bandwidth of the receiver.

The transmit power output by the radar generator may not be fully delivered to the antenna due to losses in the transmission lines. We can account for this by adding a term L to the equation. Also we can substitute for $P_{r,\min}$, giving:

$$R_{\max} = \left(\frac{P_t G_t G_r \sigma \lambda^2 F_t^2 F_r^2}{(4\pi)^3 kT_s B_n L}\right)^{\frac{1}{4}}$$

This equation is independent of the type of radar. To determine realistic values is it necessary to consider the type of radar and refine the equation as necessary. We can do this for pulse radars by considering the detectability of the signal and also by applying a correction factor for the case where the transmit power bandwidth is not the same as the received power bandwidth due to filter mismatch. The correction for mismatched filters depends on the type of pulse transmitted and the filter response to it. For matched filters, as would be expected in a modern system, this correction factor $C_B = 1.0$.

The detectability is expressed mathematically as:

$$D_0 = \frac{E_r}{N_0} = \frac{P_t \tau}{kT_s}$$

where

E_r = Received pulse energy.

N_0 = Noise power per unit bandwidth.

τ = Pulse length $= \frac{1}{B_n}$.

This gives the form of the formula for pulsed radars:

$$R_{\max} = \left(\frac{P_t \tau G_t G_r \sigma \lambda^2 F_t^2 F_r^2}{(4\pi)^3 k T_s D_o C_B L} \right)^{\frac{1}{4}}$$

Finally, we can manipulate the formula to express it in more useful terms and replacing wavelength by frequency, which is often more convenient:

$$R_{\max} = 239.3 \left(\frac{P_{t(kW)} \tau_{\mu S} G_t G_r \sigma F_t^2 F_r^2}{f_{MHz}^2 T_s D_o C_B L} \right)^{\frac{1}{4}}$$

The target cross-section is expressed in square metres (or when converted to logarithmic units, dBsm). However, it must be noted that for any complex structure such as an aircraft or ship, the cross section will vary substantially dependent on the angle of the incidence of the radar energy. Figure 7.5 shows an illustration of the type of variations of target cross-section we might expect from an aircraft or ship in the horizontal plane. The variations shown are not representative of any specific platform but rather are meant to be indicative. Aircraft may have a slightly higher cross-section when illuminated from the side, and this is normally more pronounced for ships. As the target is illuminated, changes in the target aspect will lead to large variations from illumination to illumination. For this reason, target cross-section is often determined and handled statistically.

Target strength not only depends on the physical size of the target but also on the materials of construction, the radar frequency, the target's shape and of course the

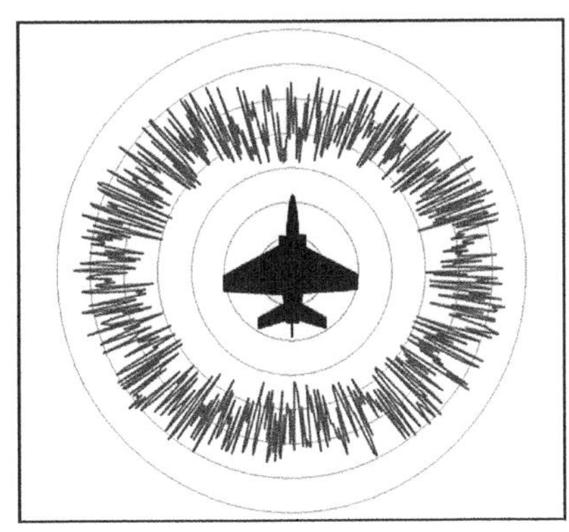

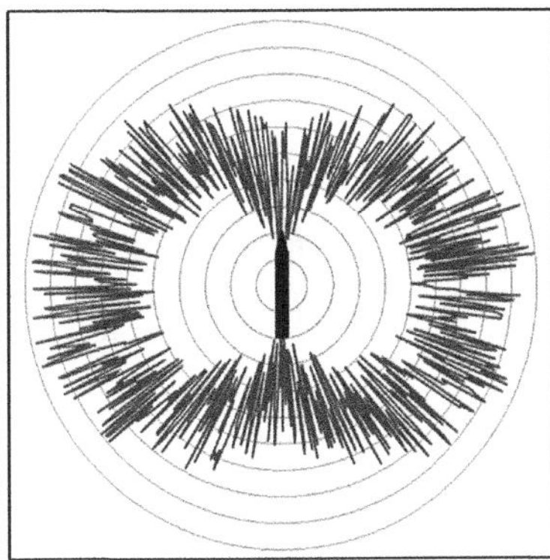

Figure 7.5 Indicative diagram showing variations of radar cross-section when the target is illuminated in different directions by a radar. Ships in particular show higher cross-section when illuminated from directly abeam. The variations shown will also vary with target tilt, so it can be seen that the expected response is very complicated and normally statistical methods are used to represent a target. Since target aspect may change during illumination, it is unlikely that the value of cross-section target strength will remain static.

aspect of the target illumination and the return signal, which will be very different in the case of bi-static radars.

The materials of construction, including the paint used to cover the load-bearing structure can be optimised to minimise radar reflectivity, as has been done for example for the US B-2 Spirit stealth bomber. Such materials are at present very expensive and difficult to manufacture and maintain.

The shape of the target is also a very important aspect of the radar cross-section. Most critical are features that are normal to the incidence of the radar signal. The F-117 and the B-2 both use computer-calculated surface shapes to reflect energy away from the transmitting antenna. Also, their air intakes and engine outlets are positioned above the aircraft so they are not visible to ground-based radar. Additionally, they use internal bomb bays so that the shape of the external bombs and their associated pylons do not add to the target strength.

Of particular note in reflectivity are salients known as 're-entrants'. These are best exemplified by aircraft air intakes and engine exhausts. However, another interesting area where this is important is the scenario where ships are attacked by a salvo of radar-guided missiles. In this case, once the first missile hits the side of the ship, the hole opened immediately acts as a major re-entrant. This is illustrated in Figure 7.6. The left hand figure shows the ship before impact and the right hand one shows it after the first missile has hit, opening up a large hole in the side. The hole causes a major re-entrant that massively affects the radar cross-section on that side. Missiles arriving on the side of the hole will tend to follow the first missile into the hole because the centroid of the radar reflection is so affected. This has been seen in many exercises when old ships have been used as targets for such missile salvos.

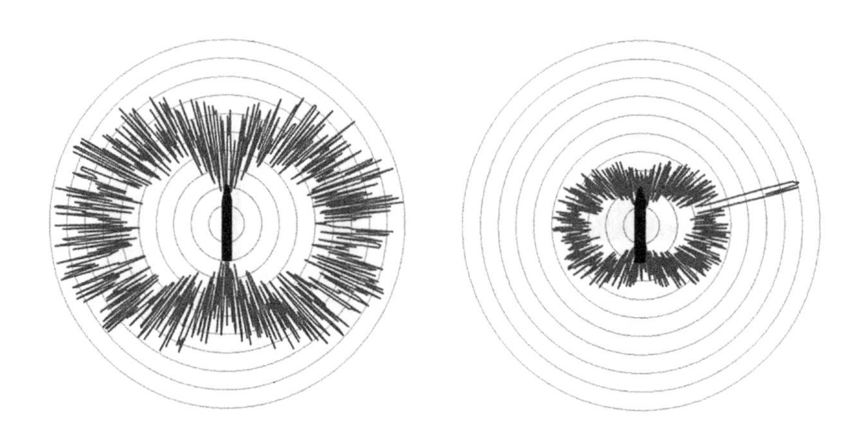

Figure 7.6 Ship radar cross-section illustration before and after a missile holes the ship on the starboard side. Radar-guided missiles arriving after this hole has been created will tend to follow the first missile into the hole.

At this point we have not considered the concept that a radar will not illuminate a target only once during a sweep. The number of successive pulses that the target will be illuminated by is dependent on the Pulse Repetition Frequency, the beamwidth in the direction of sweep and the speed at which the antenna rotates or repeats the sweep. For a mechanically rotating antenna, this can be expressed by:

$$M = \frac{\varphi(PRF)}{6(RPM)\cos\theta_e}$$

Or, for a mechanically rotated antenna with a vertical as well as horizontal sweep:

$$M = \frac{\varphi\theta(PRF)}{6\omega_v t_v(RPM)\cos\theta_e}$$

where

PRF = Pulse Repetition Frequency per second.
RPM = Rotation rate Per Minute.
φ = Azimuth beamwidth in degrees.
θ = Elevation beamwidth in degrees.
θ_e = Target elevation angle.
ω_V = Vertical scanning speed in degrees per second.
t_v = Vertical scan time in seconds (including dead time, if any).

This equation is in the direction of transmission, and is only valid when $\varphi/\cos(\theta_e)$ is less than 90 degrees. Both the horizontal (left) and vertical (right) movements are shown in Figure 7.7.

When the number of pulses arriving at the target is known, it is theoretically possible to determine the probability of receiving a target when looking over the received

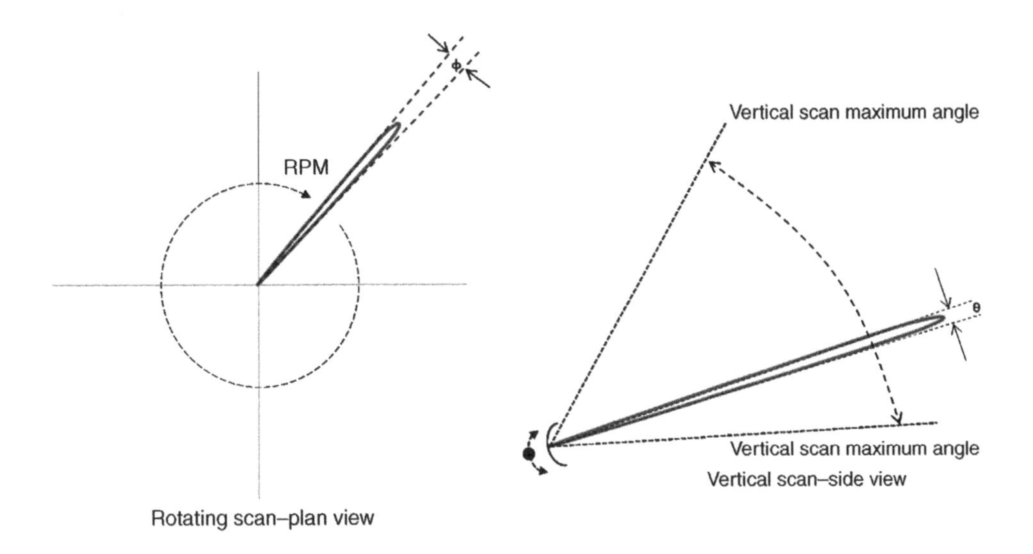

Figure 7.7 Rotating scan parameters (left hand side) and vertical scan parameters (right hand side). The angular velocity and the vertical scan time are based on the vertical scan distance shown.

responses of each by correlating them. In this case, the cumulative probability of getting detection is:

$$P_c(R) = 1 - \sum_{i=1}^{M} (1 - P_i)$$

This assumes that the target fluctuation is random between pulses. This may not be the true in practice, in which case more complex methods of correlation must be carried out, based on empirical data or complex calculation.

The equations in this chapter do not consider a number of other factors that will have an influence on the practical radar range. Chief among those is the returns generated by land, obstructions and atmospheric features. These returns will present possible false targets and may mask real returns. The treatment of managing false returns is dependent on the type of radar system, and we will now look at the main features of some of these.

7.3 Types of Radar

7.3.1 Basic Pulse Radar

Radars are typically described by their modulation scheme. Pulsed radar is the most common type and the simplest in concept. It works by sending out a short-duration, well-shaped pulse and then going into receive mode for the period of times between transmit pulses. The passive receiving period lasts until the time for a radar echo at the maximum range set will return in. This is illustrated in Figure 7.8. The transmission pulse, initiated by the electrical pulse, is transmitted and the signal returns to zero. On

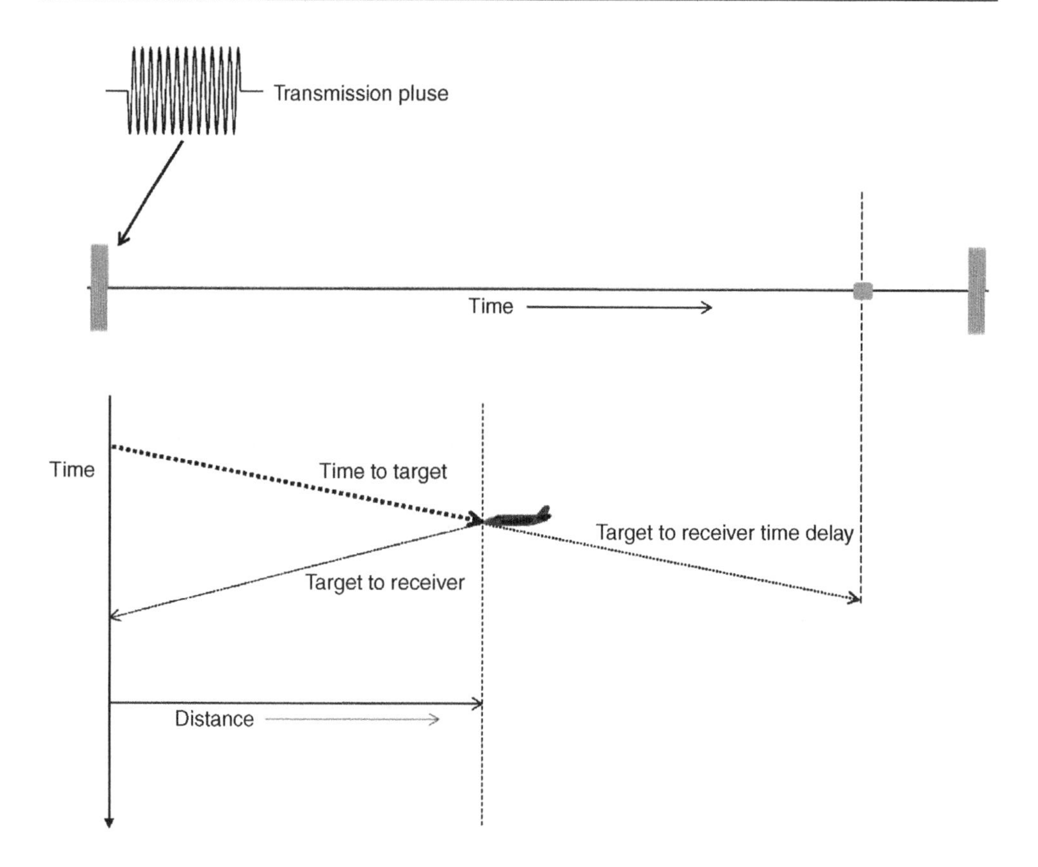

Figure 7.8 The minimum time between successive transmission pulses is dependent on the time delay for received targets from maximum range. This must include the time for the signal to illuminate the target and the time for the signal to reach the receiving antenna.

the time axis, the delay of the return signal is dependent both on the transmit time to the target and of course the delay from the target to the receiver. This is twice the distance to the target, and is the minimum time permissible between successive transmissions.

A block diagram of a simple pulse radar is shown in Figure 7.9. The same antenna used for transmission can be used for reception by using a duplexer that allows switching between the transmission circuit and the receiver circuit.

The range limits of Doppler radar are determined by the maximum interval between successive pulses. The ability to differentiate between two or more close targets is determined by the pulse duration. If the returns from two targets overlap, then they will form a single, longer pulse which the radar cannot differentiate from a single pulse.

7.3.2 Pulse Doppler Radar

Doppler radars are used to allow differentiation between ground returns and valid, moving targets. They are based on the principle that a target reflection will include an

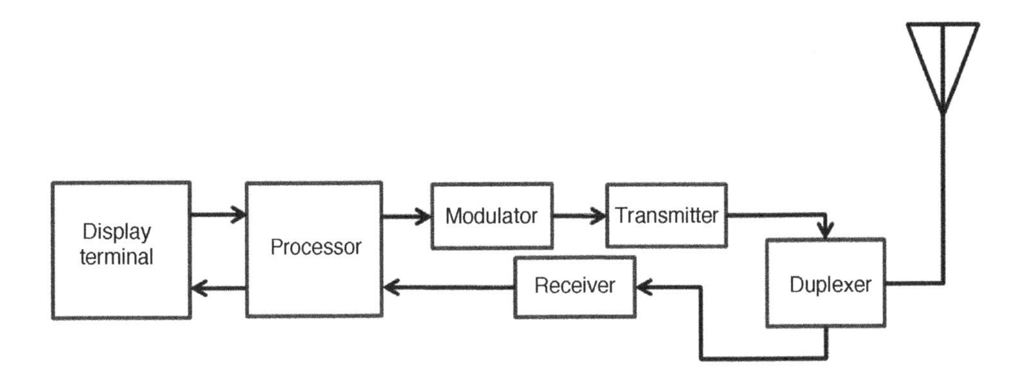

Figure 7.9 Block diagram of a simple pulsed radar.

offset due to its radial velocity with respect to a (fixed) receiver. This will affect the frequency returned by the following formula:

$$\Delta F = \left(\frac{2v}{c}\right) F_t$$

where
ΔF = Received frequency in Hz.
v = Instantaneous net velocity in m/s.
c = Speed of light in m/s.
F_t = Original transmission frequency in Hz.

Note that only the radial velocity is included; velocity orthogonal to the direction of movement will not cause a Doppler return. The type of response in the frequency domain is shown in Figure 7.10. Both returns might be arriving at the same instant. The large response on the transmission frequency may be caused by ground clutter or other non-moving reflectors. The smaller return at a higher frequency is due to a moving target, in this case moving towards the transmitter.

If the transmission frequency is 10 GHz, and the Doppler return is at 10.00000926 GHz, then the frequency difference is 9260 Hz. Evaluating from the Doppler formula gives us a radial target speed of approximately 139 m/s, or 500 kph.

A basic block diagram of a pulsed Doppler radar is shown in Figure 7.11.

Pulse Doppler provides the same range and bearing measurements as for normal pulsed radars, but in some cases there can be blind ranges and range ambiguities. These can be overcome by adjusting the PRF and other processing methods.

7.3.3 Pulse Compression Radar

One of the problems of pulse radar is that range resolution is limited by the length of the pulse and range itself is limited by the power that can be transmitted in the pulse. Both

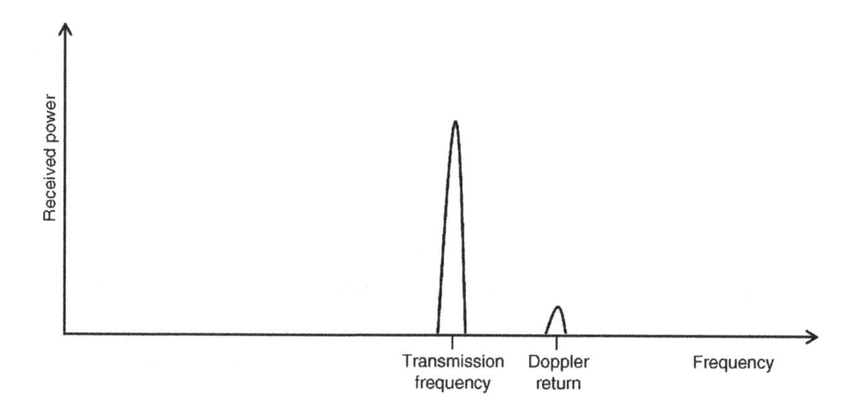

Figure 7.10 Return from a Doppler radar. The main response on the transmission frequency is caused by returns that are not moving with respect to the transmitter. The smaller response above the main frequency shows a target moving towards the transmitter. This return may be far smaller than that due to non-moving clutter, but since it is separated in frequency, it can be detected. The time taken for the return to get to the receiver provides the range information, as in the standard pulse radar case.

of these issues can be overcome by use of pulse compression. This works as shown in Figure 7.12. The original pulse of high power and short duration is expanded into a longer signal of lower power, as shown in Figure 7.12(a).

The transmit pulse is fed through an expansion filter, which extends the pulse duration and reduces the peak transmitted power. The received return, which has the same characteristics of the transmitted pulse, but obviously of lower power, can be fed through a filter that has the *complex conjugate response* of the time response of the one

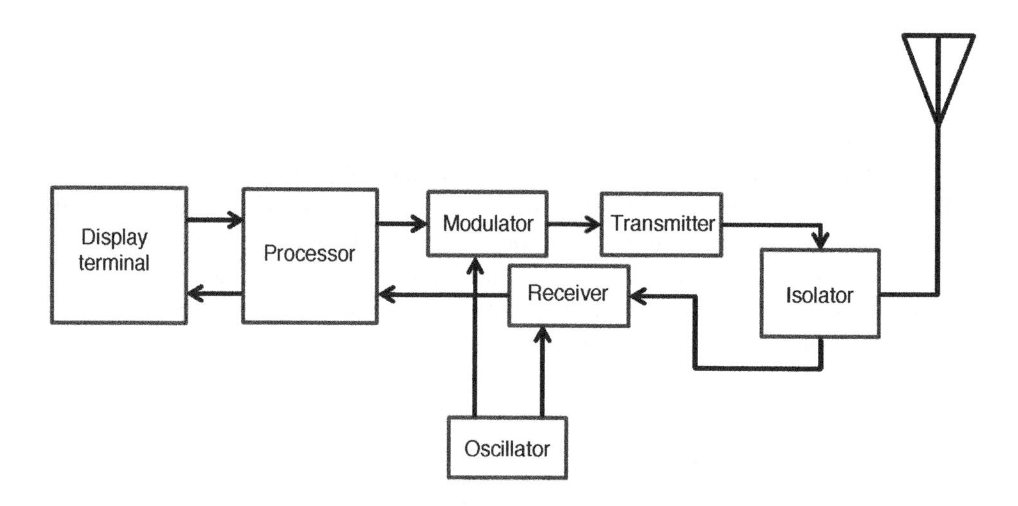

Figure 7.11 Simple block diagram of a pulsed Doppler radar. The oscillator is used to synchronise the transmitted and received pulses.

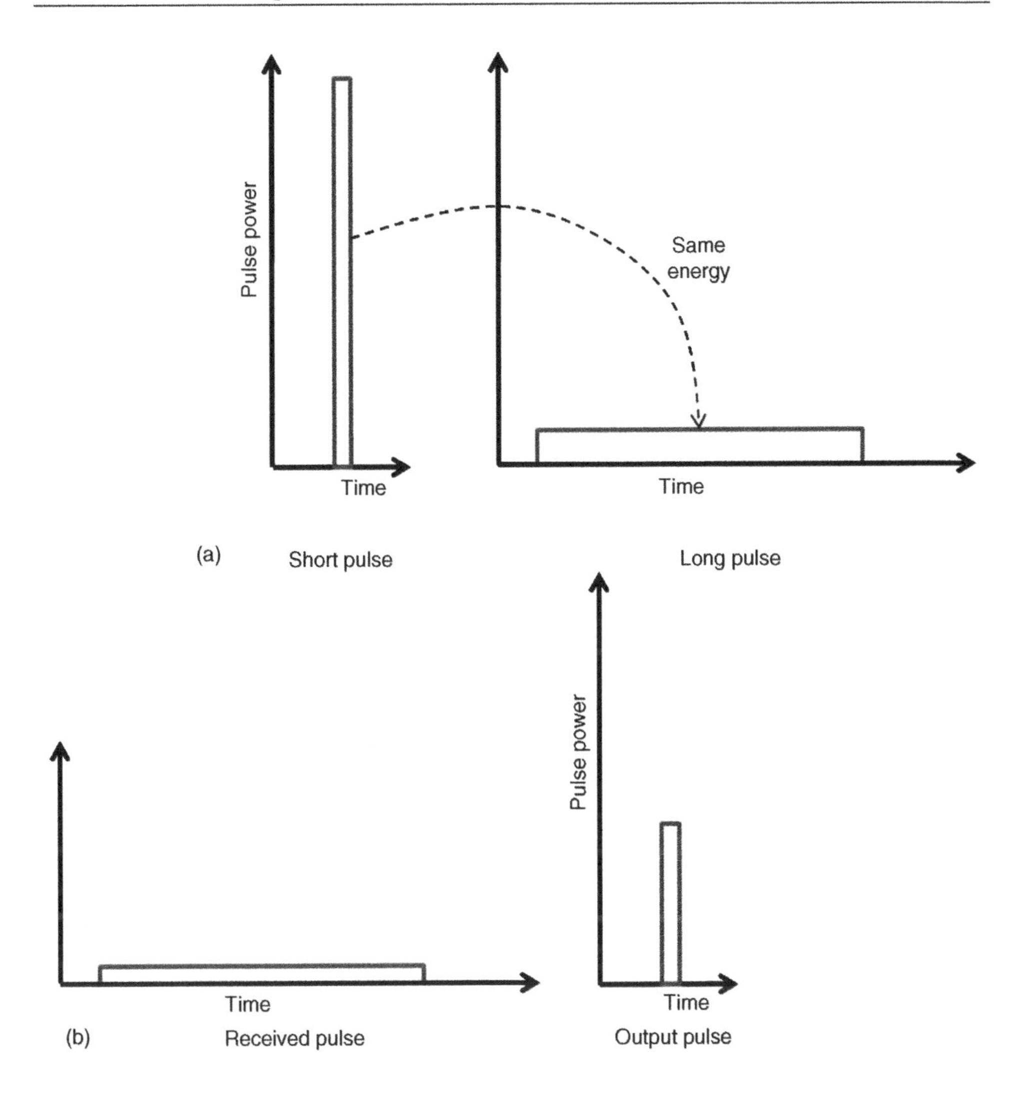

Figure 7.12 (a) The energy in a short pulse can be spread over time for transmission. (b) By compressing the received pulse back to the same duration, the original pulse can be re-generated.

used to generate the long pulse. This has the effect of providing an output of the same type as the original short pulse, of the form shown in Figure 7.12(b).

There will be other non-target related outputs that are normally reduced by weighting filters and by the choice of pulse compression technology.

7.3.4 Chirped Radar

Chirped radar works by modulating a pulse signal by a linear frequency varying signal. An example is illustrated in Figure 7.13. In this case, the frequency is linearly increased over the duration of the pulse.

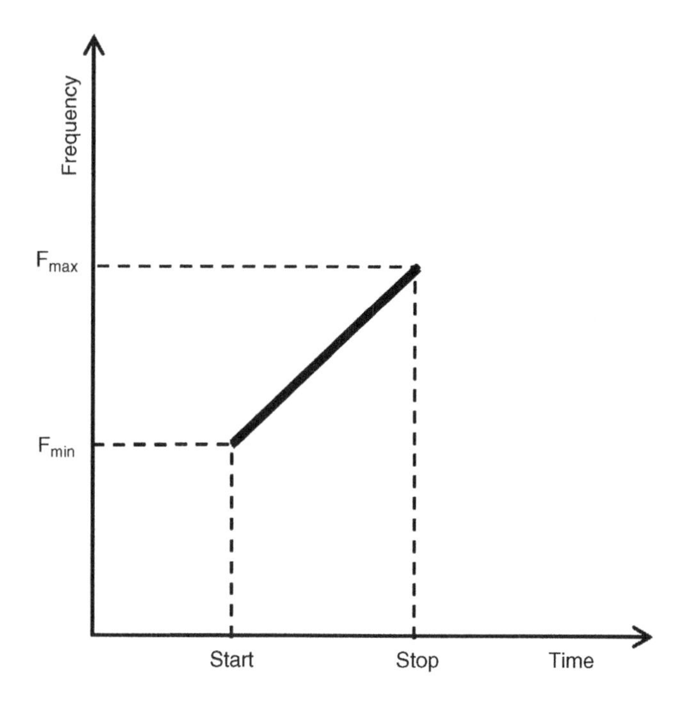

Figure 7.13 Chirped radar pulse, with linearly increasing frequency.

The system works in a very similar manner to that of compressive pulse systems, in that the receiver has a response that accounts for the change in frequency during the pulse. The receiver filter response correlates the output power over the receive time to produce a simple, short-duration pulse that includes most of the energy received from the target. Just as for the compressive pulse method, chirped radars have improved range ambiguity resolution.

7.3.5 Digitally Modulated Pulses

In compressive pulse and chirped radars, we used the concept of matched filters to retrieve a processed output better than the standard pulsed method. We can extend this idea from analog into digital methods by using digital modulation. Digital modulation implies adding a digitally-derived code sequence onto the transmitted pulses. On such method is to use phase modulation to alter the phase of the transmitted signal without reducing its power. Switching between a phase advance of 90 degrees and a phase with a delay of 90 degrees allows us to provide two distinct phase responses and these can be equated to the binary 0 and 1 symbols.

The process is illustrated in Figure 7.14, which shows a sequence of 101. Note that the phase for the bit carrying ‘0’ is inverted from the ‘1’ condition.

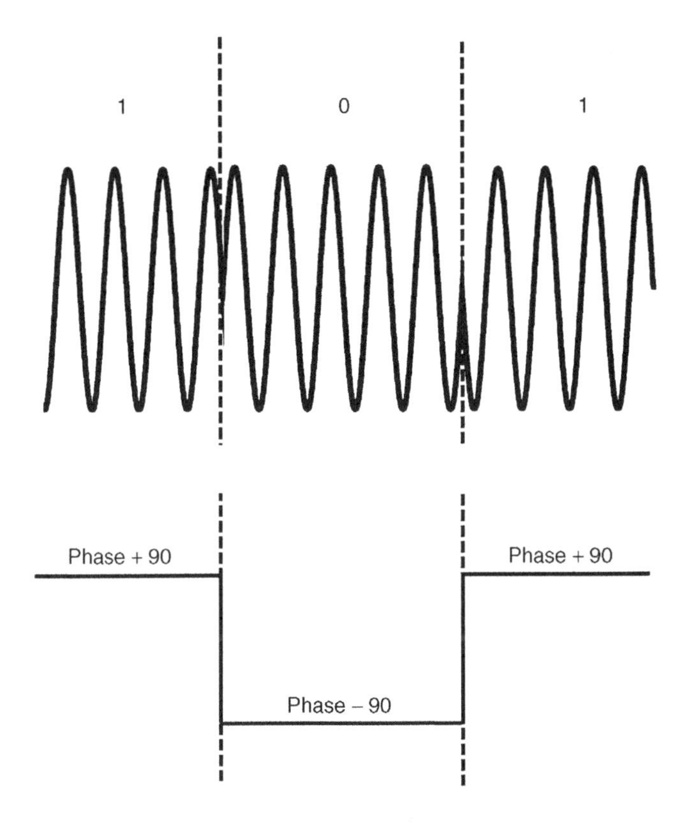

Figure 7.14 Digital modulation of a pulse carrier with phase difference characteristics 180 degrees apart. The two phase-states can be used to represent binary 0 and 1.

Figure 7.14 shows a very short sequence of only three bits. In practice, longer sequences provide a better advantage. For example, a series of 10 bits has a possible 1024 combinations. Being able to differentiate between the right code sequence and all of the incorrect ones allows us to better discriminate real targets against clutter return. This is effectively what we are trying to do with the original pulsed radar and digital modulation provides a further mechanism to do so. We can do this by using the digitally-modulated pulse stream and integrating over the duration of the entire modulated sequence. A 10-length randomly chosen code sequence can be used to modulate the transmission pulses, with phase changes to account for 1's and 0's. By integrating over the length of the sequence, it is possible to identify the transmitted code from all of the others by correlation.

Suppose we choose the following sequence at random:

1 0 1 1 0 1 0 1 0 1

Without, for the moment, considering losses through the channel, we can look at how this correlation process works. Consider Table 7.1, which shows the original

Table 7.1 Correlation of a digitally-modulated transmission with delayed versions of itself, wrapped to represent the initial code repeating itself. The correlation is carried out by summing the bits that are synchronised and subtracting bits that are not. Only when the full sequence repeats are all bits synchronised and the maximum correlation made

Bit Number	1	2	3	4	5	6	7	8	9	10	Sum
Original Transmission	1	0	1	1	0	1	0	1	0	1	−6
Delay (1)	1	1	0	1	1	0	1	0	1	0	
Correlation	*1*	*−1*	*−1*	*1*	*−1*	*−1*	*−1*	*−1*	*−1*	*−1*	
Delay (2)	0	1	1	0	1	1	0	1	0	1	2
Correlation	*−1*	*−1*	*1*	*−1*	*−1*	*1*	*1*	*1*	*1*	*1*	
Delay(3)	1	0	1	1	0	1	1	0	1	0	2
Correlation	*1*	*1*	*1*	*1*	*1*	*1*	*−1*	*−1*	*−1*	*−1*	
...											
Delay (10)	1	0	1	1	0	1	0	1	0	1	10
Correlation	*1*	*1*	*1*	*1*	*1*	*1*	*1*	*1*	*1*	*1*	

transmission and correlation with time delayed versions of itself. The correlation process is simply adding 1 for agreement between the transmitted bit and the delayed bit (i.e. if both = 1 or both = 0) and subtracting if they do not agree. In the table, the code repeats itself when it completes, so the missed first bits due to the delay are populated with the trailing bits of the sequence.

The sum values shown in Table 7.1 can be graphed as illustrated in Figure 7.15. This is a different sequence than that used earlier, but the principle of finding the match with the highest correlation is shown where the correlation is 10 – a perfect match. However, some of the other delays have reasonable correlation – up to 6 in three places.

The correlation shown in Figure 7.15 is not ideal, but this can be addressed by means of using a carefully selected sequence. One family of such sequences are known as 'linear maximals'. These have a number of useful properties. One of which is that they create the maximum possible length of sequence without repeating possible (2^n–1, where n is the number of bits). The other main one is that they have unique correlation processes, as shown in Figure 7.16.

Linear maximals share the property that when synchronised, they correlate completely, and when they do not, a constant value of −1 is produced. The one problem with linear maximals is that they belong to a relatively small group of potential solutions and it is possible for enemies to detect, identify and then 'spoof' the transmission if their desire and capabilities allow it.

So what are the benefits of digital modulation? It provides a far better way of countering false returns from complex reflectors, which will not provide a return replicating the original transmission. Returns that do not match the transmitted signal are treated as noise, and the way the de-correlator works is to suppress noise of whatever

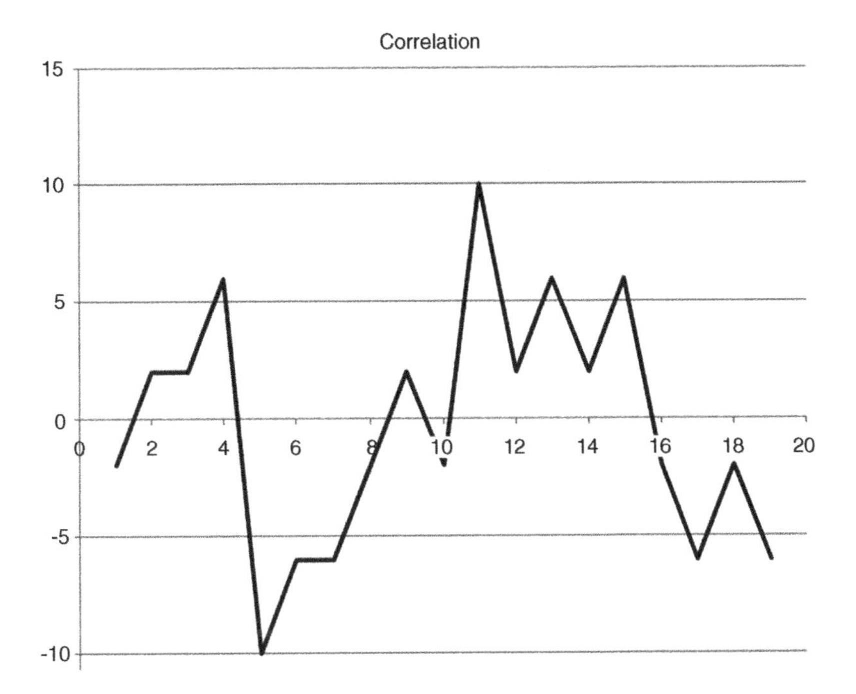

Figure 7.15 Correlation graph for a randomly-chosen series of 10 bits. The maximum correlation is shown as 10 on the graph, but other delays achieve a score of 6 in three places. This is not ideal for correlation, where it would be better if the wanted peak is far more obvious.

source. There is an effective processing gain due to this, which can be as much as the length of the maximal sequence. Thus, for a linear maximal of 10-bits, with a maximum length of 1023 bits, the possible gain in dB is $10log(1023) = 30.1$ dB approximately.

The length of the code sequence does however affect the radar's ability to discriminate between close targets since the pulse train is longer than a pulse of the duration of a single bit.

7.3.6 Continuous Wave Radar

Pure Continuous Wave (CW) radar emits a continuous carrier at a transmission frequency. Because the transmit antenna is used continuously, a second antenna must be used for return signals. A basic block diagram is shown in Figure 7.17. Pure CW cannot differentiate range, but is good for determining the Doppler response of targets.

CW radar is very simple in concept, but it does pose particular design issues related to spectral efficiency and designing the system to allow it to determine range as well as target radial velocity. As an example of an approach that can be taken, consider when the CW transmission is moderated by a linear FM sweep (FM-CW radar) as illustrated in Figure 7.18.

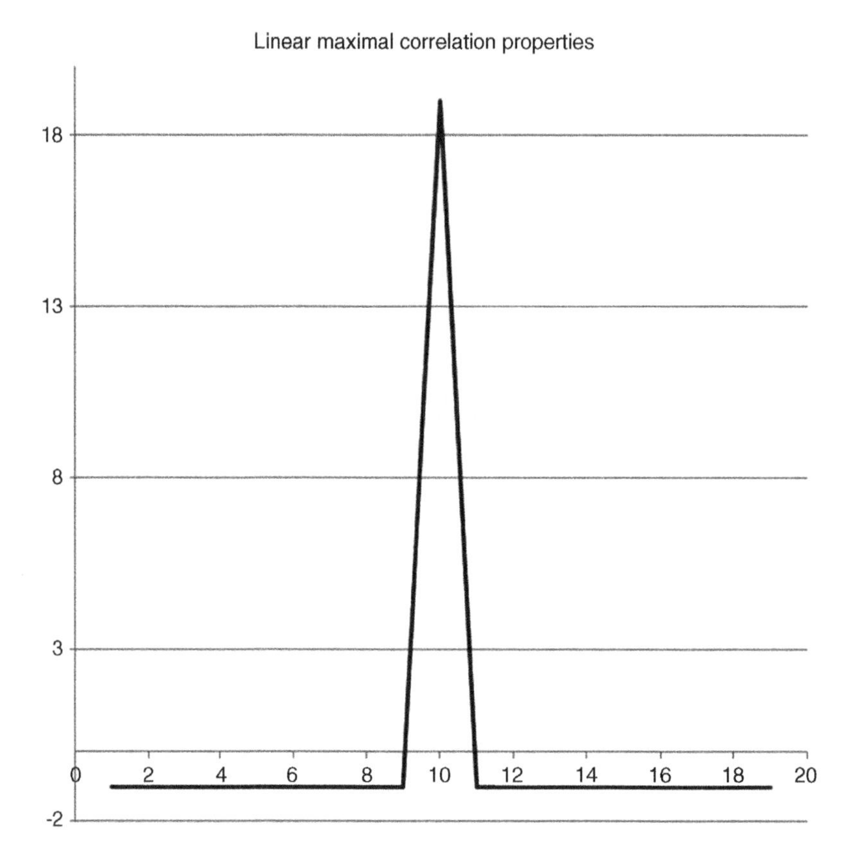

Figure 7.16 The correlation properties of a linear-maximal binary sequence. The correlation is at maximum when the signal matches the transmitted parameters, and at a constant −1 for all other delays.

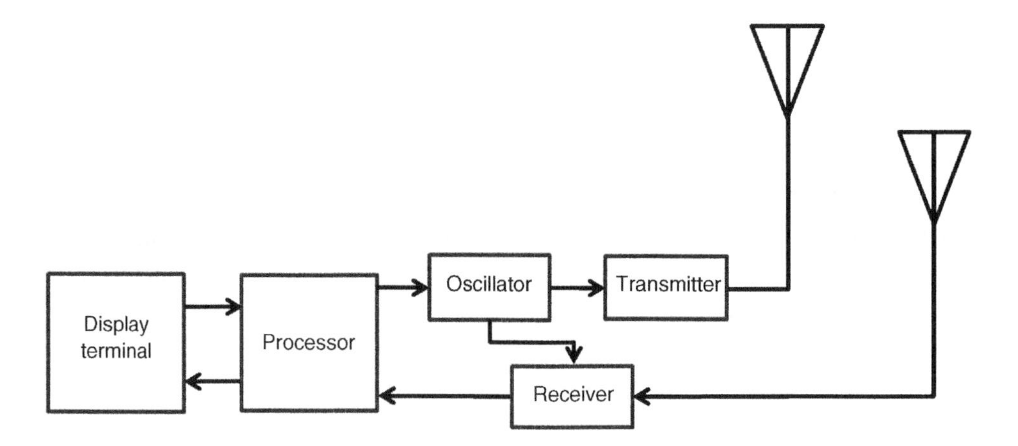

Figure 7.17 Basic block diagram of a CW radar system. The oscillator provides the CW wave for transmission and also reference for the receiver to allow Doppler analysis.

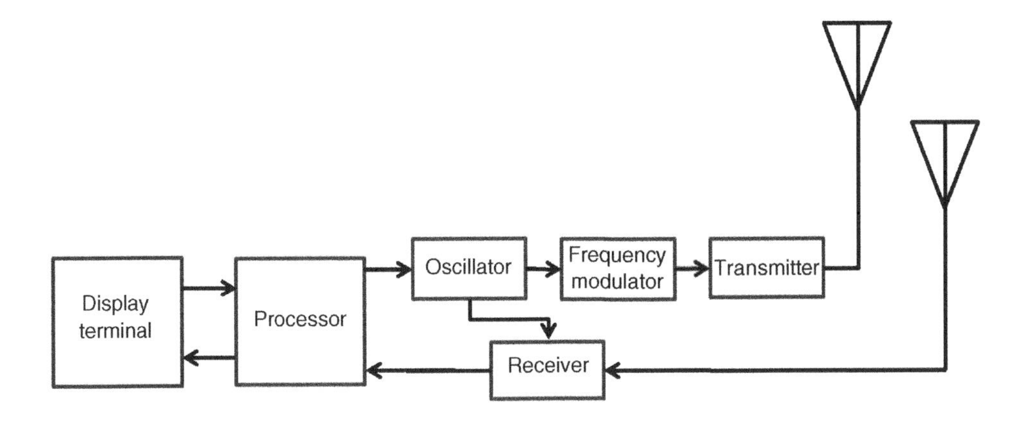

Figure 7.18 Simple block diagram for a FM-CW radar system. The frequency modulator ramps the frequency over a pre-selected range. Because the receiver input is from a time-delayed reflection from the transmission, the frequency is not the same as that being currently transmitted. The receiver can compare the receive time to the time that the specific frequencies were being transmitted, thus providing the time delay and hence range of the target.

The frequency modulator provides a linear sweep in frequencies, with a repeat cycle that prevents aliasing by being sufficiently long enough to prevent range ambiguities. For example, assume that an FM-CW radar has a repeating frequency sweep with a cycle time of 1 ms; this means that the maximum target range without ambiguity of $c \times t = 150\,000$ metres, or 150 km (remember we must include both the time to the target and back to the receiver). If a receiver obtains a portion of response at, say, 10.035 GHz and we know from the internal clock that this was sent 0.25 ms earlier, we know the time delay is 0.25 ms. The time for the return to come in gives a total travel time for the transmitted signal and reflection of 75 km and thus the target must be half of this, 37.5 km away in the direction of transmission.

Without proper processing and design, CW does not necessarily result in improvement in radar performance over other systems. However, with appropriate design and advanced processing, CW radar can be very effective.

7.3.7 Moving Target Indicator Radar

Moving Target Indicator (MTI) radar is designed to isolate moving targets from clutter. Figure 7.19 shows the way in which the radar coverage is split into cells for distance and angle. The angle resolution is determined by the radar horizontal beamwidth. The range resolution is again determined by pulse length.

MTI is a digital technology, and each portion of the coverage area is split into discrete cells that describe limits of angle and distance as shown in the figure. The system treats each cell as unique and as a single entity. It then looks for moving targets in each cell using Doppler techniques. It also uses memory to determine the return from the previous illumination of each cell. This is illustrated in Figure 7.20. The left image

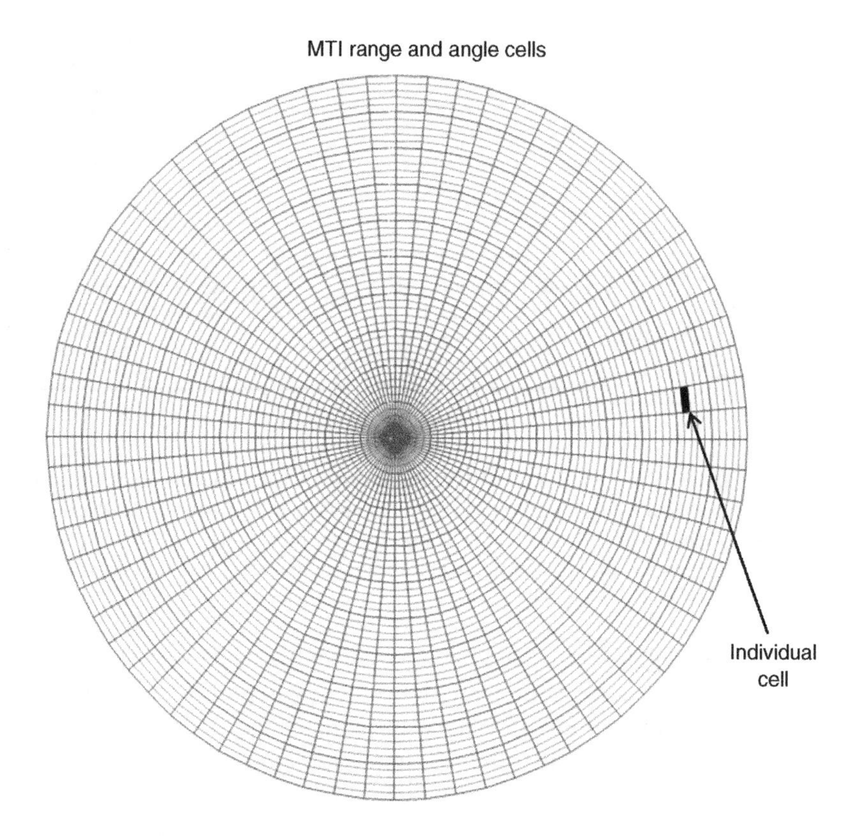

Figure 7.19 MTI radar cell structure.

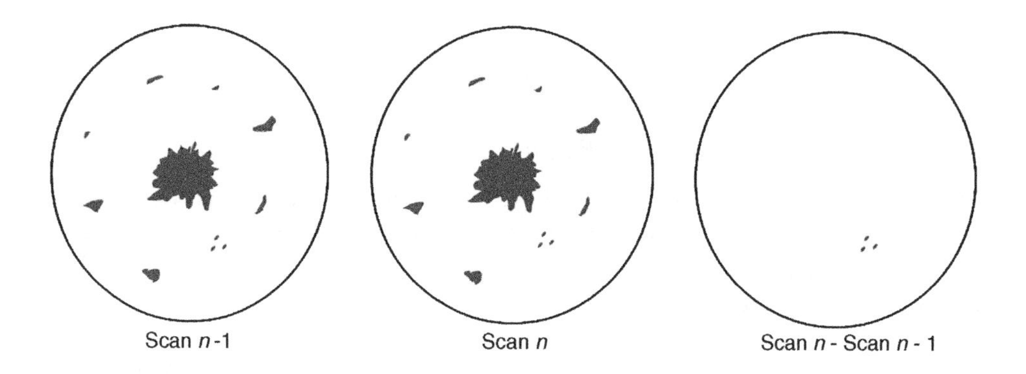

Figure 7.20 MTI radar example showing how clutter from successive sweeps can be removed by comparing the returns from successive sweeps and removing those (fixed) returns due to clutter. This allows the three moving targets to be distinguished more clearly.

shows the previous scan, including returns from targets and clutter. The middle image shows the successive sweep, which contains much the same data as the first one. By retaining the original scan in memory and then removing the duplicates, the three targets can be clearly differentiated from the clutter as illustrated in the right hand display.

MTI radar is also used from airborne platforms. This adds a degree of complexity because the radar platform is moving and therefore the Doppler shift is not with respect to a speed of zero, as it would be in a fixed installation. However, with suitable computer processing, the effect introduced by the platform's movement can be overcome.

7.3.8 Phase Array Radar

Phased array radars use sophisticated transmitter and receiver arrays composed of individual elements. The array face consists of a number of elements in both the horizontal and vertical directions. The instantaneous beam is formed by electrical steering using phase delays to determine direction; the number of elements in the horizontal plane to form the horizontal beam pattern and the number of elements in the vertical plane doing likewise for the vertical beam pattern. This is illustrated in Figure 7.21.

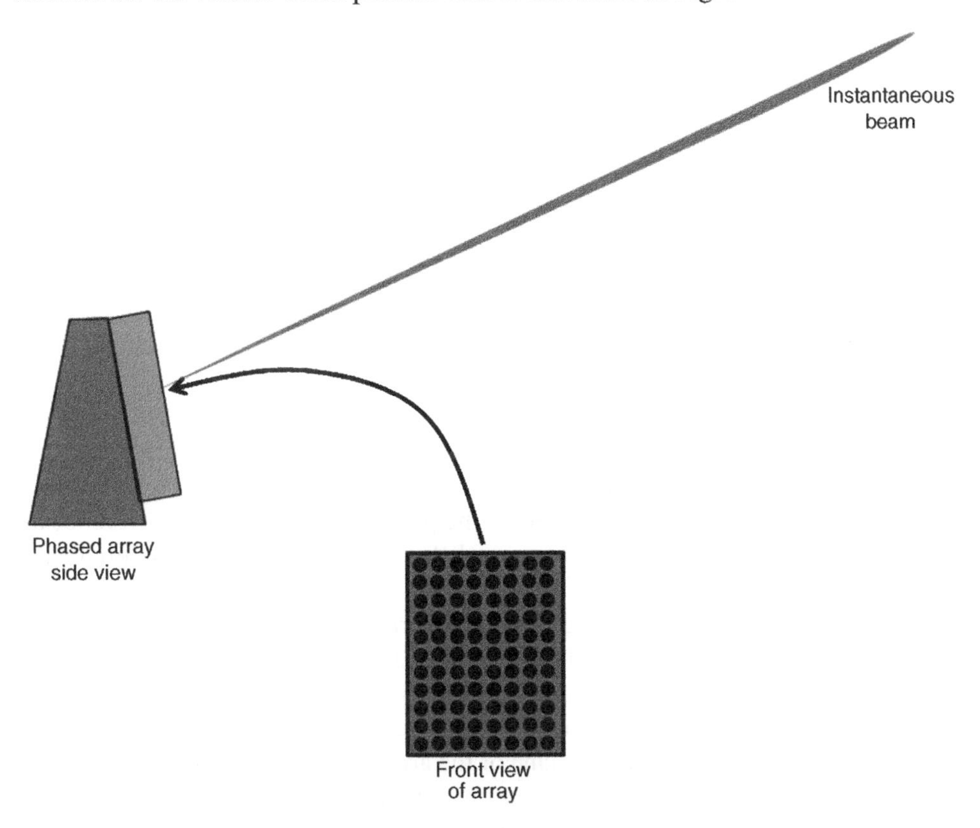

Figure 7.21 A phased array radar.

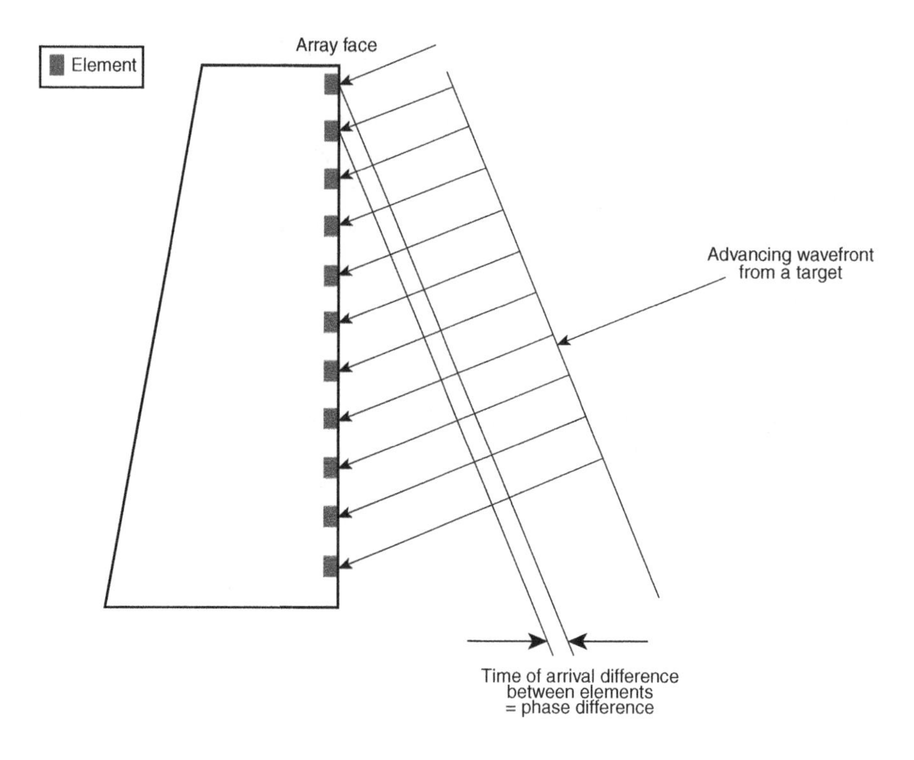

Figure 7.22 A waveform arriving at the face of a phased array. The advancing waveform hits the top element first since it is the closest. The wavefront then reaches each individual vertical element with a fixed time (or more importantly, phase) delay.

The total coverage of the radar is limited by the ability to steer the beams to the edge of the coverage area.

The beamforming of the phased array is illustrated in Figure 7.22. This shows a wavefront arriving from a specific target at an angle in the vertical direction. The wavefront first arrives at the top element first, followed by all the others, with an additional time delay per element based on the perpendicular distance difference as shown. The time difference is small enough for the delay to be considered as a phase delay.

Calculation of the delay time allows beamforming for both the transmit and receive signals. This is illustrated in Figure 7.23, which shows the system in receive state.

Phase delays are used in the correlation process to generate beams. Adding successive delays to inputs across a plane of the array directs the beam in the desired direction, and the line of sensors in the array acts to focus the beam to the desired beamwidth in that plane. Waveforms arriving from different directions are not

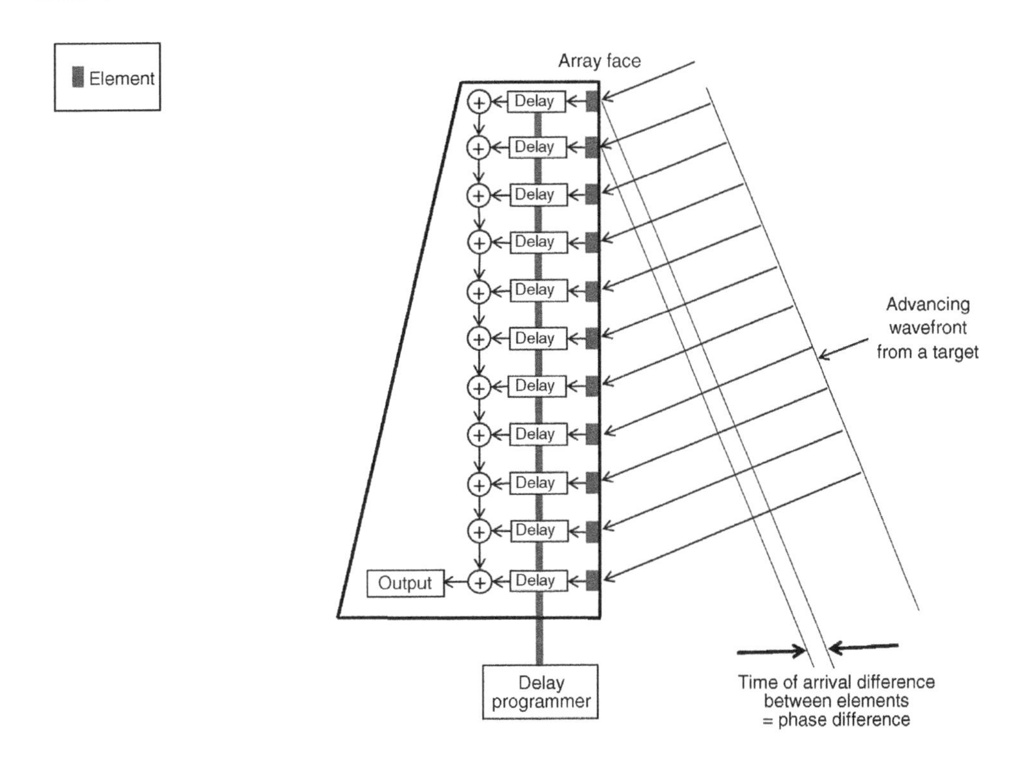

Figure 7.23 The use of programmable delays for beamforming in phased array radar. This sophisticated approach allows flexible beam production for both the transmit and receive cycles.

correlated and the phase differences between the components of signal arriving tend to cancel each other out. When the delay is known, the offset angle for both vertical and horizontal can be calculated. If the array is constructed correctly, the beams formed can be very narrow.

Another feature of electronic beamforming is that there is no need to follow a prescribed scan pattern. Computer programming can allow virtually any scan pattern to be created.

Phased array radars work within a relatively small frequency range because the ability to generate accurate beams depends on each element being spaced half a wavelength apart. Changing the frequency would of course upset this.

Phased array radars are relatively expensive, and the sophisticated array and associated electronics provide limitations on size and weight. However, they represent the state-of-the-art in long-range, wide coverage radar systems and are used by large fixed installations such as ballistic missile detection systems, onboard large surface ships, especially those designed for air defence and portable (although still large) battlefield systems.

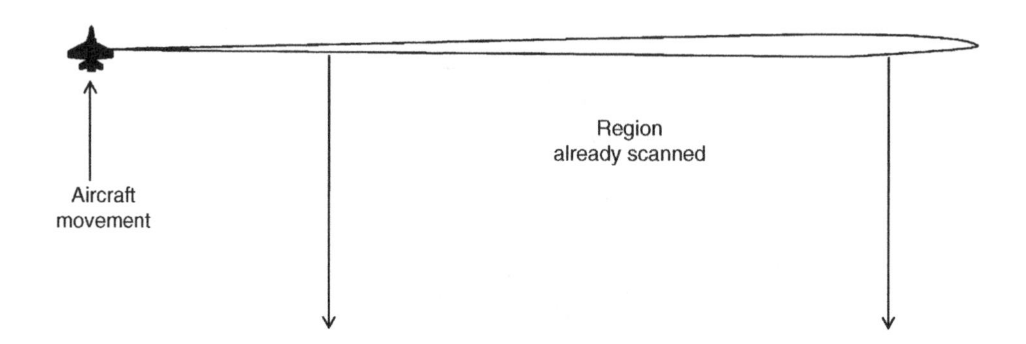

Figure 7.24 SAR radar in an aircraft, scanning to one side. The narrow beam passes over the ground to the side of the aircraft as it moves, building up a picture of targets.

7.3.9 Synthetic Aperture Radar

Synthetic Aperture Radar (SAR) is used in aircraft to provide the benefits of a phased array without the weight of a scanning system. This is shown in Figure 7.24.

Just as in phased arrays, the scanning range is split into range bins, which are populated with any targets that may be present. The beamwidth is dependent on the number of elements in the SAR array, and the rate at which new data is added depends on the speed of the aircraft.

7.3.10 Broadband (LPI) Radar

Military radar systems are subject to detection and counter if they are detected. There is therefore a desire to make such detection difficult for enemy forces. This can be achieved if the radar transmitter generates a signal that is difficult to detect against noise or other transmissions. There are a number of methods to achieve this. These include:

- frequency agility, so the enemy does not know what frequency the radar will be transmitting on;
- spoofing, in which the radar pretends to be another, non-threatening radar type;
- broadband radar, in which the transmitted energy is low but covers a wide bandwidth.

In this section, we will be concerned with broadband radar, using methods we have already introduced. The primary method of generating broadband transmissions is to use a higher code rate (called a 'chip' rate) to spread the original system. Since the higher-rate modulation has a higher switching rate it necessarily has a wider bandwidth. However, the same energy is being transmitted the level of the

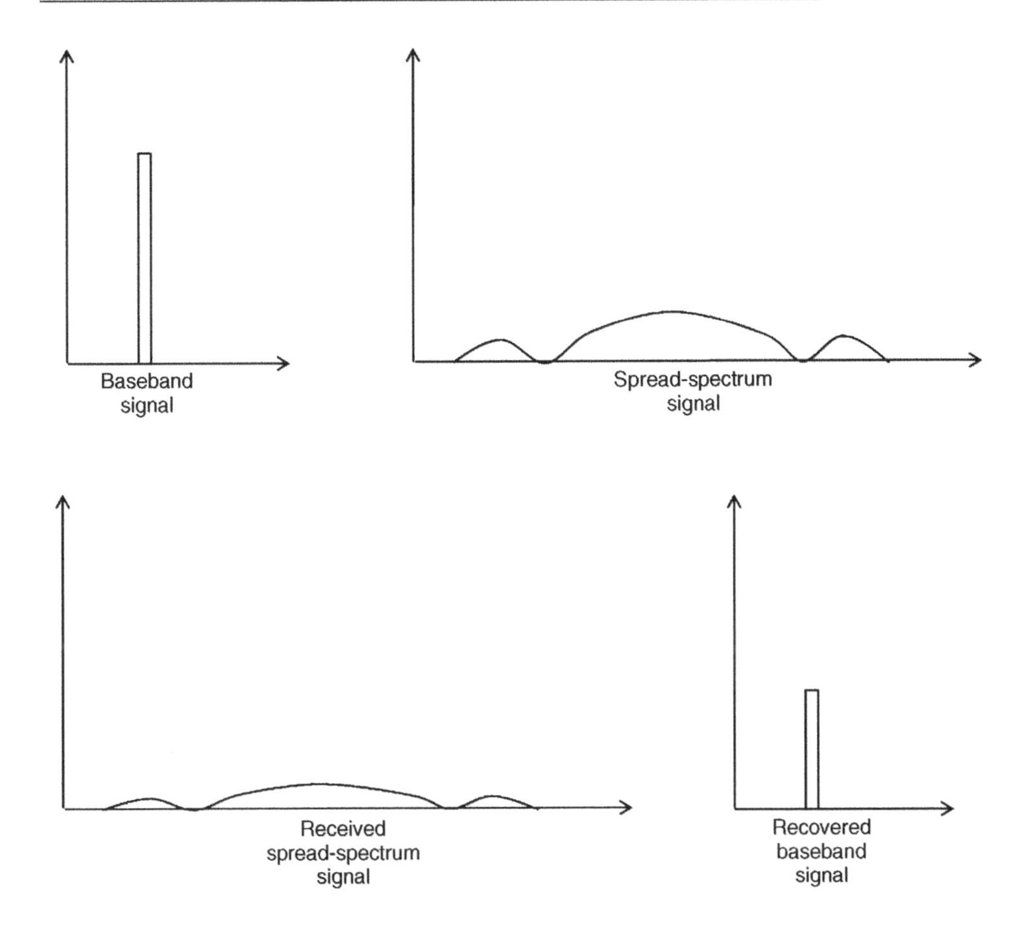

Figure 7.25 The top half of the diagram shows the original baseband signal. This is modulated by a spreading code, which produces the spread-spectrum signal shown on the top right-hand side. This is of lower power than the original signal within the same bandwidth, but transmits the same RF power over a wider bandwidth. The bottom half of the diagram shows the received signal at the radar receiver. This has the same characteristics as the transmitted band, but when modulated by the original spreading code, the baseband information is recovered. Other receivers that do not know the original spreading code will be unable to reproduce this process.

energy over the transmitted bandwidth is substantially lower. This is illustrated in Figure 7.25.

The energy produced has two main advantages; the first is that the transmitted energy is lower in power and spread over a wider band that is difficult to detect. The second is that the processing of the received signal benefits from a processing gain that counteracts against noise and interference.

The spread-spectrum response can be generated by using pseudo-random coding sequences as described in Section 7.3.5. However, in this case, the chip rate of the spreading code is significantly higher than bit rate of the baseband signal.

7.3.11 Secondary Radar

Before completing the section on radar systems, it is necessary to consider a slightly different form of radar that is in use throughout the world. This is known as secondary radar and it works on quite different principles to that of the other radars we have so far considered. Secondary radar is used with primary radar for air traffic control. However, instead of receiving a reflection of the airborne target, the purpose of the radar transmission is to trigger a pre-coded response (known as a 'squawk'). In the case of civil air traffic control, the secondary radar pulse leads to a return message consisting of the aircraft identity and some flight parameters. In military terms, the system can be used in an IFF (Identify, Friend or Foe) system. In the military case, the aircraft transponder responds with a code identifying it as friendly. If no response, or a false response, is made, then the illuminated aircraft is considered an enemy.

The basic process is shown in Figure 7.26. The ground-based radar (in this case) sends out the interrogative pulse. When this is received by the transponder on the

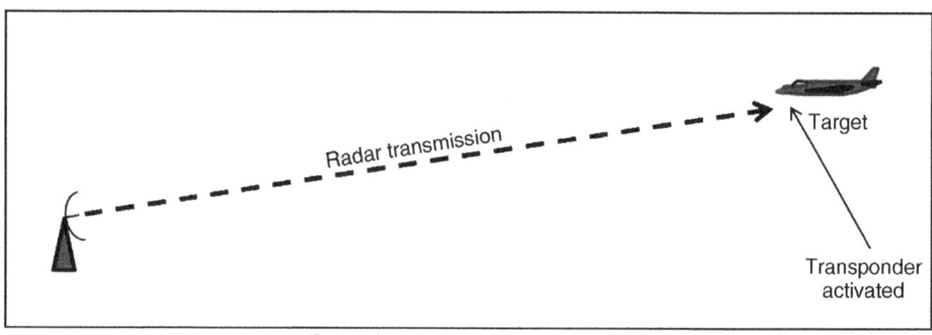

Secondary radar transmit phase

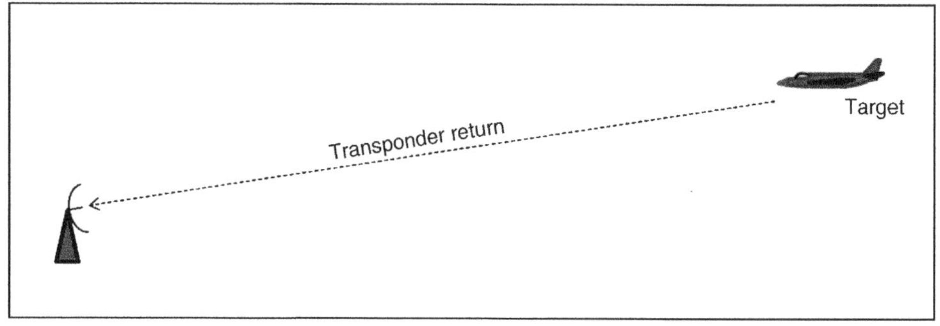

Secondary radar receive phase

Figure 7.26 Illustration of secondary radar operation. The radar transmits on a given frequency. The aircraft receives this transmission and it triggers the onboard transponder to transmit. This transmission is received and processed by the receiving radar antenna.

aircraft, it sends back pre-formatted information to the radar. The transponder return is received and processed by the ground radar.

The transmit power of the transponder, unlike that of a normal radar return, is not determined by the strength of the power transmitted by the radar system. This means that secondary radar is dependent on two separate paths, not unlike those of a radio communications system. It also means that secondary radar can be treated as a special case of radio communications, rather than as a classic radar.

7.4 Radar Jamming Techniques

Radar jamming is a very different proposition compared to jamming communications systems. Often, jammers emitting noise over the victim bandwidth (referred to in radar circles as cover jamming) is a highly effective method for communications jamming; however, this is less effective in general against modern radar systems. Instead, the jammer normally attempts to replicate the return the victim radar is expecting but with false characteristics. This is known as deceptive jamming. In all cases, the Jamming-to Signal ratio (J/S) is as important for radar jamming radars as it is in communications systems, but this will be described in Part Two. This section focuses on the basic jamming methods used for radar systems.

The purposes of radar jamming can be wide and varied. They include the following objectives:

- Platform self-protection:
  - defeating surveillance radars from identifying attacking platforms;
  - defeating tracking radars from being able to track targets and launch or accurately position hard-kill options such as missiles and intercept aircraft;
  - defeating missile and automatic gun guidance systems.
- Stand-off jamming:
  - denying the enemy the opportunity to detect other attacking platforms;
  - denying the enemy sufficient information to organise defence against attacking platforms.
- Tactical picture denial:
  - preventing the enemy from understanding the nature of the attacking force;
  - introducing uncertainty as to where and what the attacking force is focussing on;
  - decoying the defending force to the jammer platforms.
- Strategic picture denial:
  - jamming strategic defence systems, such as satellite launch detection systems, in order to produce confusion;
  - using jamming decoys to change the enemy perception of the actual threat.

Methods for radar jamming are categorised as follows:

- Range-Gate Pull-Off (RGPO). This uses pulses generated in the jammer to mimic the transmission characteristics of the radar system and then to mask the real return by higher power signals. Once this has been achieved, the jammer can change the timing of returns to return a false range.
- Velocity Gate Pull-Off (VGPO). This is used to confuse Doppler radar systems by altering the frequency or phase of the signal received at the radar to alter the apparent velocity of the target.
- Inverse-gain jamming. This is used to deny the radar directional data by returning a signal that appears to originate from all directions at once, or a wide arc.
- Angle stealing. This is a method using high-power jammers to generate signals that appear to show the target on a different bearing than its true one.
- Cover jamming. This uses noise to decrease the range that a radar can discriminate targets against by increasing noise at the receiver location and hence de-sensitising it.

The tactical methods of deployment are also split into different categories:

- Self-protection jamming. This is where a platform has its own organic jammer. This also includes the use of towed decoys where the jammer is towed behind the platform.
- Stand-off jamming. This is where a dedicated jammer is positioned on a separate platform to defend attacking aircraft or ships. The jammer is normally well separated from the protected platforms and from the victim radar.
- Stand-forward jamming. This is where expendable jamming systems such as UAVs or artillery shells are used close to the victim radar system to protect forces that are further away. The need for expendable systems is important since the jamming platform will be within range of enemy threat systems.
- Escort jamming. Escort jamming occurs when dedicated jamming platforms are mixed with an attacking force.
- Mutual support (also known as cooperative) jamming. This is the coordinated use of platforms to achieve the jamming objective. The use of multiple platforms increases the available power, but also it allows various tricks to be performed such as 'blinking'. This involves jammers from different platforms within the radar's beamwidth switching on and off. This produces an artificial 'glint' in the radar tracking system which can result in it breaking angle track.

We will start with the simple case of a single jammer against a single radar system.

The normal basic radar scenario was shown in Figure 7.1. Figure 7.27 shows an effective jammer scenario where a jammer on an aircraft sends back a jamming signal that is carefully generated in terms of temporal and spectral form, time and power. In this case, the jammer has managed to fool the radar both in terms of direction and range. This is complex to achieve in practice but is the ideal scenario.

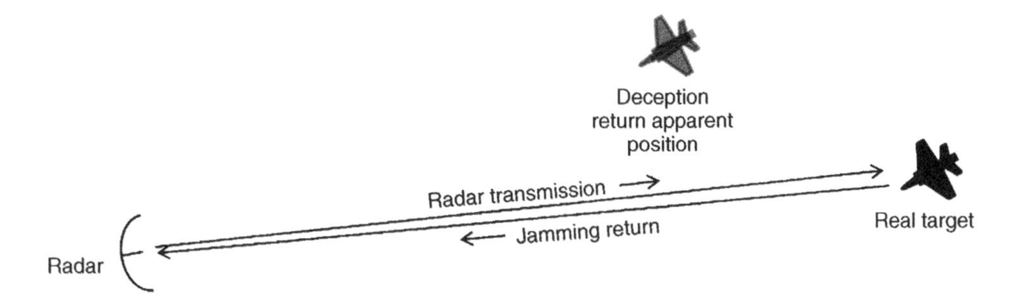

Figure 7.27 The purpose of jamming is to make the victim radar believe that the target is not in its real location.

If we assume for the moment that we are dealing with a single pulse radar system, we can use it to illustrate some common methods of jamming clearly. Consider Figure 7.28. Diagram (a) shows a representation of an original transmit pulse and a return pulse that indicates the presence of a real target some time later. As we know, the time of the return gives the range of the target. However, in diagram (b) the illuminated target has produced a false return covering, and at a higher power level than, the real return. At this point, the jammer return is made at the exact time that the next pulse arrives. This may seem pointless, but in fact the stronger jammer signal causes the gain control in the radar receiver to adjust itself to the strength of the returning signal. This is

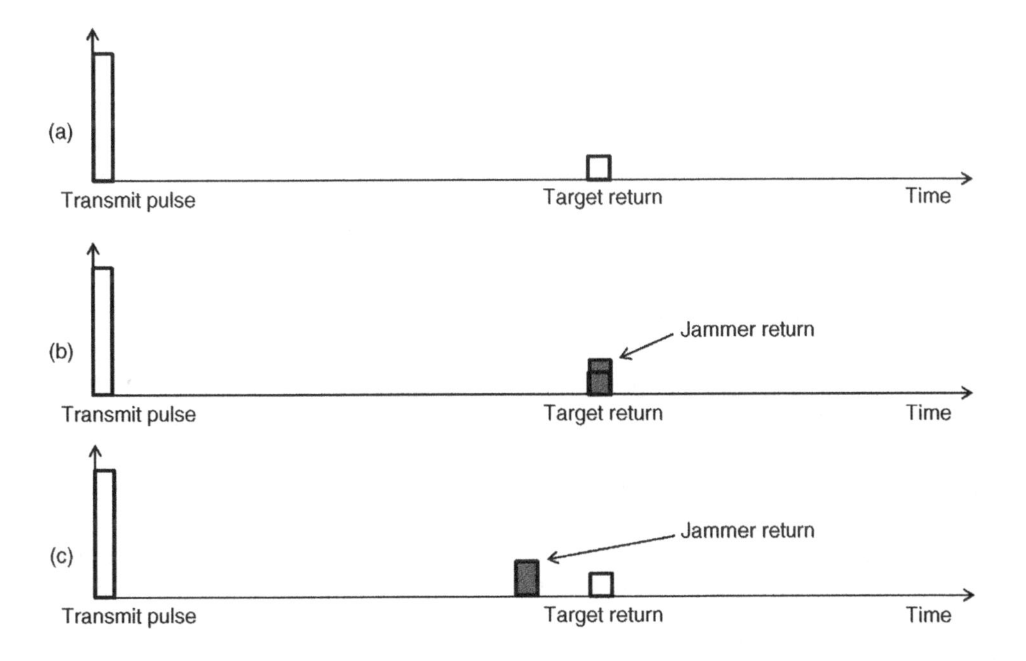

Figure 7.28 A simple deceptive jamming scenario (see text).

necessary because the signal variation between close and far and different targets, may be very large and the dynamic range of the receiver processors are inherently limited. In diagram (c) the jammer has transmitted a return before the actual return is created. Because the radar is de-sensitised to the real target return, it instead responds to the jammer return. Thus, the jammer has deceived the radar into believing the target is closer than it actually is. The same process can be followed but with additional delays to make the target appear further from the radar than it is. This is the range gate pull-off method.

This is all very well, but the radar still sees the target on its true bearing. Thus, defensive measures such as missile or aircraft engagement are still possible. However, the practical limitations of such systems need to be taken into account. For example, many missiles systems use active seeker heads switched on in the optimum position to engage a target at a specific range. Before the time that the missile is expected to be in seeker range, it is not active. If the target passes the missile (or more accurately, its seeker envelope passes the target) before this occurs, then the missile will never detect the target. If the seeker head is activated before the target comes into range, the missile is also likely to miss. For beam-riding systems that require both front-lock (the missiles seeker to the target) and rear-lock (the missile to the guiding radar) to be active, loss of front-lock will normally cause the missile to self-destruct. If the target is detected too late, even if the target is tracked, the slew rate of the missile might cause rear-lock to fail due to the rear-facing receiver losing its link with the tracker. In this case also, the missile will be assumed to be out of control and will also self-destruct. Aircraft attempting to intercept a target with incorrect location may not be able to close with the target effectively, especially if they are vectored onto a location astern of the attacking target. If they are vectored ahead of the target, the enemy aircraft may be able to engage the interceptor before it has localised the target.

It would be better to confuse the radar as to range but also height and direction. Of the two, direction provides the better counter to enemy systems. There are two methods of achieving this. So-called 'inverse gain jamming' denies the enemy with directional data. Angular stealing (to be described more in Part Two) provides a mechanism for actively forcing the enemy radar system to believe the target is in a different direction completely.

Inverse gain jamming is a method to deny a radar directional data. It is most often used against conical-spiral tracking radars. This is based on an understanding of the radars directional gain and then supplying a jamming signal that varies in power according to the gain in the direction of the jamming platform to counter the directionality of the victim radar. This is illustrated in Figure 7.29. Diagram (a) shows the relative strength envelope of the radar from the target's perspective. Pulses arriving will vary in strength according to which part of the antenna's gain response they are emerging from. Diagram (b) shows the envelope of the return generated by the jammer on the target platform. Diagram (c) shows the victim radar receiver screen without jamming. The target is seen in both range and direction as the black dot. With

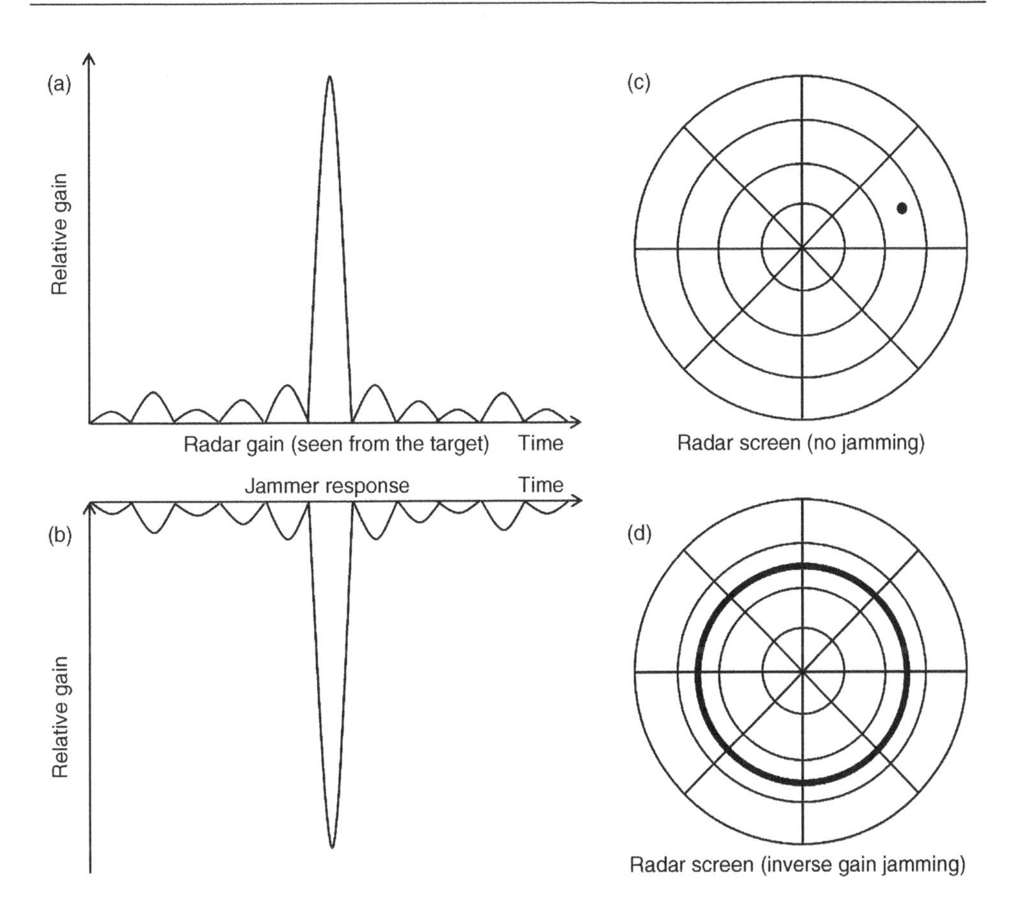

Figure 7.29 Inverse gain jamming denies the victim radar from determining the direction of the target. Diagram (a) shows the radar pulse strength as seen from the target. Diagram (b) shows the response the jammer sends in return. Diagram (c) shows the victim radar screen without jamming, and Diagram (d) shows the jammed radar screen. The target shown in (c) is now spread over all directions.

the addition of inverse gain jamming, the direction information is lost because of the jammer's response. The radar screen now shows a ring rather than the resolved target bearing, as shown in Diagram (d).

Figure 7.29 shows the effect of inverse gain jamming against rotating radar with a human interpreting the results in order to show how the system works. However, in practice, it is most often directed against systems using electronic detection methods, such as missile trackers. When used in this way, the effect is to push the radar seeker away from the target rather than towards it.

Inverse gain jamming is difficult to achieve because of the power required. The jamming is not only ineffective against other radar systems that may be present in the area, but the radiated power may make the jammer platform more detectable and hence more vulnerable to other assets in the area.

Cover jamming uses transmission of noise to raise the noise floor of the victim receiver. Because the noise has no distinguishing features from noise, it is difficult for the victim radar system to determine that it is being jammed at all. This is similar to the case of communications jamming.

As well as the active jamming techniques described, decoys are also used. This ranges from the Penetration Aids (Penaids) used by ballistic missile re-entry systems, in which multiple decoys indistinguishable from the real warheads, to the use of chaff to produce false targets. These are described in Part Two.

Table 7.2 Common ECCM techniques and the part of the radar system used to counter jamming methods

		ECCM method countered				
		Noise	False target	RGPO	VGPO	Angle stealing
Radar sub-system	ECCM method					
Antenna	Reduce sidelobes	x	x			
	Monopulse angle tracking					x
	Minimise cross-polar response					x
	Sidelobe blanking	x	x			
	Sidelobe cancellation	x				
	Electronic scan		x	x		x
	Adaptive receive polarisation					x
	Cross-polarisation cancellation					x
Transmitter	Increase power	x				
	Pulse compression	x				
	Frequency diversity	x				
	Frequency agility	x	x			
	PRF jitter		x	x		
	RPGO memory nulling			x		
Receiver	Increase bandwidth		x		x	
	Beat frequency detector	x		x		
	Cover pulse channel processing		x			
	Home-on-jam	x				
	Leading/trailing edge track			x		
	Narrowband doppler noise detector	x	x			
	Velocity guard rates	x	x			
	VGPO reset		x		x	
Signal processing	Coherence processing		x	x	x	
	CFAR		x	x	x	
	Doppler/range rate comparison			x	x	
	Total energy test	x				

7.5 Radar Jamming Mitigation Techniques

Since the introduction of military radar systems, efforts to prevent them achieving their objectives have been developed by jamming or decoying them. However, another race to prevent denial of radar performance due to enemy actions has also been occurring. These have resulted in methods that are termed Electronic Counter-Counter Measures (ECCM).

Some typical techniques derived from Skolnik (2008) are shown in Table 7.2. These should only be seen as a subset of the available techniques. Skolnik notes that 150 ECCM techniques are quoted in the available literature.

Another important concept in ECCM is that of burn-through. This is the distance where the radar receives adequate signal quality to overcome the jamming effect. It occurs closer to the radar when the jamming effect is negated by the robustness of the radar two-way link. This will be discussed further in Part Two.

References and Further Reading

Ademy, D. (2001), *EW101: A First Course in Electronic Warfare*, Artech House, MA, USA, ISBN 1-58053-169-5.
Peebles, P.Z. (1998), *Radar Principles*, Wiley-Blackwell, ISBN 978-0471252054.
Schleher, D.C. (1999), *Electronic Warfare in the Information Age*, Artech House, MA, USA, ISBN 0 890 0652 68.
Skolnik, M. (2008), *Radar Handbook*, McGraw-Hill, USA, ISBN 978-0-07-148547-0.

8

Radio-Controlled Improvised Explosive Devices

8.1 The Poor Man's Weapon of Choice: IEDs

Improvised explosive devices have been used as long as explosive weapons have existed. The principle of using explosives in a manner other than they were originally developed for has been used for sabotage, terrorism, insurgency and ad hoc military uses is both well established and simple to understand. In recent history, they have been used in conflicts in Vietnam, Northern Ireland, Chechnya, the Middle East, Afghanistan and Iraq as well as for terrorism carried out across the world. IEDs form part of a group of unconventional weapons including:

- Improvised Nuclear Device (IND);
- Improvised Radiological Device (IRD);
- Improvised Biological Device (IBD);
- Improvised Chemical Device (ICD);
- Improvised Incendiary Device (IID).

Fortunately, some of these types of device remain unrealised.

IEDs are used as an effective form of asymmetric warfare. Asymmetric warfare occurs where both sides have different strategies and tactics, and it is used by the weaker side to provide a method of attack that does not meet a superior force head-on. Conceptually, it is similar to the ideas of guerrilla or Special Forces operations; it is a method of achieving effect on opposing forces without taking them on in areas where they are strongest and where, therefore, the risks are too high.

Communications, Radar and Electronic Warfare Adrian Graham

It is possible to categorise devices into a number of different types, such as:

- Homemade devices, made from readily available components. These include fertiliser bombs and fire bombs (so-called Home Made Explosives [HMEs]). This type of weapon was used extensively by the Provisional Irish Republican Army (PIRA). Also, more recently, Al Qaeda has shown initiative in developing new and more powerful HMEs, as seen in their attempted attack on aircraft in 2006. The attack was foiled but it led to the introduction of severe limits for liquids carried on commercial aircraft to prevent further attempted attacks.
- Demolition charges made from materials to hand, such as adapting explosives stores to blow up bridges etc. This may occur when a side is retreating quickly and needs to demolish bridges and other structures. In this case, any unexpended ammunition – particularly artillery shells and grenades – are detonated by an initiation charge. These can also be considered as Improvised Munitions
- IEDs based on unexpended ammunition, intended as booby traps (so-called 'victim operated devices'). In regions experiencing conflict or having recently experienced such a conflict, unexploded and unused ammunition is plentiful and easily converted into an improvised device. One simple such device could be made by balancing an object of interest above a primed grenade. When the object is picked up, the grenade is initiated. Often these are triggered by unintended victims, particularly children, and they continue to do so.
- Custom-built devices. This is a newer trend seen in Iraq and Afghanistan, where specialist devices such as armour-penetrating explosively-formed penetrators are designed and built by off-site specialists and delivered by combatants into the area of conflict. It is questionable as to whether these are indeed improvised or a new form of intentionally designed device for a very specific application. However, from the viewpoint of the victim, such distinctions are moot.

In this analysis, we will restrict our focus to IEDs, since we are principally in the triggering of such devices and their inhibition rather than the device effects. Triggering of IEDs can be initiated in a number of ways, each with their own advantages and drawbacks.

- Victim Operated Improvised Explosive Devices (VOIED)In this case, no externally activated trigger is necessary; the victim performs an action that triggers the device. This could be breaking a trip-wire, opening a door, picking up an object, driving over a pressure pad or passing a magnetic sensor, such as a proximity fuse. The advantage of this approach is that once set, the system can be left unattended without risk to the person placing the device. The disadvantages are that once a few such attacks have occurred, the opposing force will learn to avoid triggering such devices. Also, the device is indiscriminate and cannot differentiate between legitimate target and innocent civilian.

- Delayed operation, by a timing circuitThese are relatively easy to produce but are hit-or-miss affairs. The device will be unattended until it is time to explode. Within this timescale, the opposing force may find and disable the device or circumstances may change making the device counter-productive. It is also indiscriminate.
- Command wire systems. This is the use of a command wire from a command or observation post directly to the device to be detonated. Its advantages are that the victims can be selected by the commander based on visual or other queues and that the device is passive in terms of emitting detectable transmissions. The drawbacks are the time to prepare the placing of the device and command wire, the relative shortness of any command wire which leaves the operator vulnerable to counter-attack, and the fact that the command wire and associated equipment is likely to be found and to provide forensic information to the opposing side.
- Radio Controlled IED (RCIED). Radio controlled systems involve using an RF device to initiate a device. These will be discussed at length in the following sections. RCIEDs have the advantage of flexibility; there is no physical link between commander and device. The range between the device and commander can be longer and the commander may select an Observation Point (OP) advantageous to their detection and identification of potential targets. However, radio signals are vulnerable to detection, localisation and inhibition using radio jammers.

The following sections will look in more detail at RCIEDs, how they are used and how they can be detected and defeated.

8.2 Radio Control for IEDs

The basic problem of creating an RCIED mission is illustrated in Figure 8.1. In this case, the intended target is a patrol. The RCIED commander has selected a location where the road goes through a narrow gap between hills. The hills will prevent the patrol from going off-road to avoid the device. However, the location is so obvious that it is likely that the patrol will be alerted to the possibility of attack and should be on guard.

The RCIED commander has selected a potential location to place an RCIED device near the road. The RCIED device contains a radio receiver, with the commander having a corresponding transmitter to trigger the device. The commander has selected an observation point where he can see the approach to the device and the device location, and where he also believes he will be able to successfully send a triggering message to the planted device.

At this point in the device commander's planning, the device team are vulnerable by having to identify and reconnoitre the proposed attack location. The team may be exposed via a number of means, such as:

- detection by regular patrols;

- detection by stand-off systems such as UAVs, reconnaissance and intelligence collection aircraft and systems, or specially mounted patrols;
- detection through failure to maintain operational security, for example by communicating via insecure means;
- exposure by local people, who may be for or against the bombers. Also, a change in local mood may be detected by alert patrols;
- HUMINT methods, for example penetration of terrorist cells by opposing spies and informants.

This is therefore a fraught time for the device team. The device may also be detected after placement due to such factors as obvious excavation marks, unexplained changes to the local environment or new items such as abandoned cars, or exposed device components. Clearly, it is in the bomber's interest to minimise such disturbance and to leave as small a physical, social and technical footprint as possible. However, placement of the device is crucial for a successful attack and therefore it is impossible to omit this crucial step without running the risk of an ineffective attack.

Once the device location has been selected and the device with associated radio receiver has been planted, a suitable OP must be selected so that the commander can observe potential targets before or as they approach the device location. The

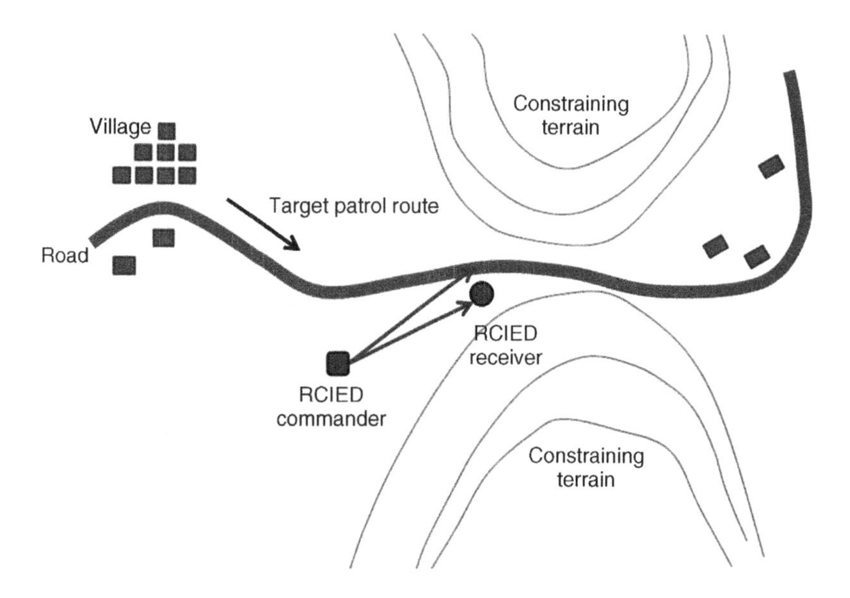

Figure 8.1 Typical environment for an IED-based ambush, in this case using a radio controlled device. The RCIED commander has visual oversight over the attack point and a radio to trigger the device. The commander would have planned this ambush based on observing previous patrols and identifying the constraints imposed on the patrol by the terrain.

transmitter used to trigger the RCIED must also be within range of the receiver. This will depend on a number of factors, as explained below.

The key radio parameters are:

- frequency of operation;
- transmitted power;
- transmit and receive antenna heights;
- antenna gain/loss;
- noise floor at the receiver;
- sensitivity of receiver device.

Because of the topology of the scenario, it is likely that the transmitter and device receiver will be within a few hundred metres of each other. The exception to this can be where existing networks are used, such as GSM mobile phone systems. In this case, the base station used to relay the trigger command may be many kilometres away.

Since the range is typically so small, it is possible to identify the critical factors that will affect performance of the radio link.

- Terrain effects will not normally vary significantly within the operational range, although of course there are exceptions to this.
- Atmospheric variations will in general be negligible.
- Local clutter such as buildings or dense vegetation may be the dominant attenuation factor.
- Ground conductivity may be an important factor within this regime, since the antennas will typically be so low above local ground (less than one metre high typically).
- Local fading will be an important factor within the nominal coverage range.

In terms of radio prediction and simulation, it is usually not possible to determine where precisely an attack may occur and therefore from a device inhibition (i.e. jamming) point of view, a site-general model approach is necessary. US forces adopted a simple model with a correction for ground conductivity. This is used to represent both target and jammer links. The model is shown below. It assumes that the jammer is for VHF communications links only, that the effective Jammer to Signal Ratio (JSR) is approximately 8:1 (linear terms) and that the links have no or minimal intervening terrain.

The model provides two formulae, the first one to determine the necessary jamming power:

$$P_j = P_t \cdot K \cdot \left(\frac{H_t}{H_j}\right)^z \cdot \left(\frac{D_j}{D_t}\right)^n$$

And the second, which determines the maximum effective range of a jammer:

$$D_j = D_t^n \sqrt{\frac{P_j}{P_t \cdot K \cdot \left(\frac{H_t}{H_j}\right)^2}}$$

where
P_j = Minimum jamming power required.
P_t = Effective power output by the enemy transmitter.
H_j = Elevation of jammer above sea level.
H_t = Elevation of enemy transmitter above sea level.
D_j = Jammer-to-receiver link distance in km.
D_t = Enemy transmitter-to-receiver link distance in km.
K = Jammer tuning accuracy:
 2 for FM receivers in VHF range.
 3 for CW or AM in the VHF range.
n = Terrain and ground conductivity factor:
 5 = Very rough terrain; poor conductivity.
 4 = Moderately rough terrain; fair to good conductivity.
 3 = Rolling hills; good conductivity.
 2 = Level terrain; good conductivity.

The performance of the power prediction model is illustrated in Figure 8.2 for different categories of ground conductivity. The figure shows the power required to jam enemy transmission links of given distances. The enemy communications link distance is shown on the bottom axis. The jamming link distance is a constant 10 km.

Notice that the formulae are independent of frequency and also of minimum receive sensitivity of the enemy receiver. Also notice that for short ranges, the jammer power is very high and probably unrealisable. This is not unreasonable because if the enemy communications are closer together, the link is far more difficult to jam.

It must be noted that there are several caveats to the use of this model:

- Firstly, it assumes no significant obstructions between transmitter and receiver.
- Clutter is not accounted for in this model.
- The value produced is a nominal figure to which fading characteristics must be applied.
- The model is fairly coarse.

Why use a coarse model when terrain is available for most of the world? The answer is that a prevention system must be workable in any situation. Describing and analysing a unique set of circumstances using actual terrain and clutter values will result in a site-specific model that cannot be generalised to different scenarios. This is acceptable for

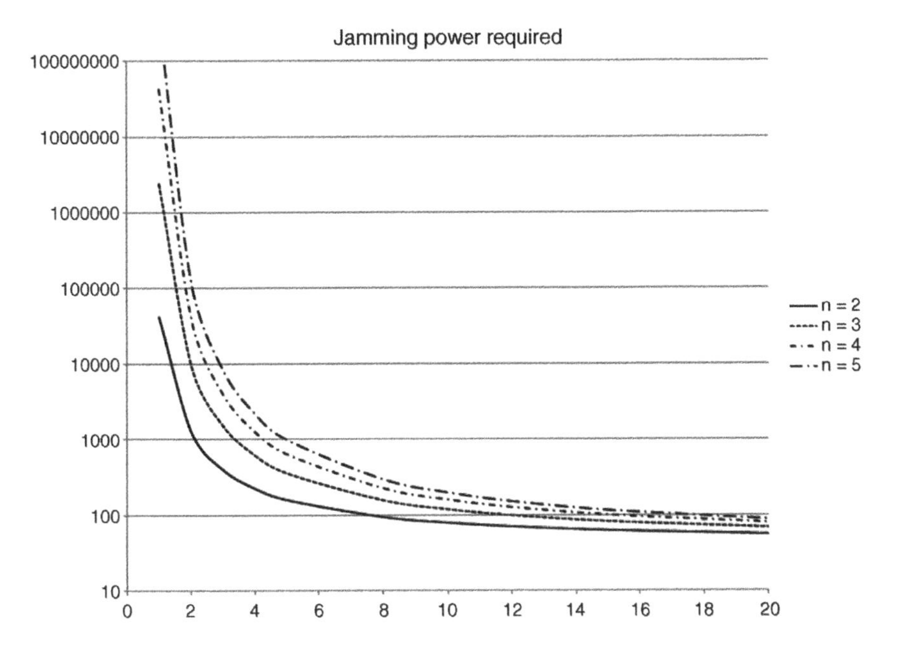

Figure 8.2 US jamming model for different ground conditions (see text).

simulations but not for mission planning tools, which must be suitable for any scenario likely to be encountered.

In their paper *RF Propagation in Short Range Sensor Communications*, Dapper et al. (2007) demonstrated an important aspect of modelling for Unattended Ground Sensors (UGS) is the consideration of propagation in circumstances where the antennas are much closer to the ground than is generally the case. This is also true for RCIEDs and therefore it is not too great a leap to utilise the same model for modelling the transmitter to device receiver and for the jammer to the receiver.

The typical fading characteristics of a signal over a few wavelengths are shown in Figure 8.3.

As previously described, the variation of signal level for a heavily cluttered environment will generally by Rayleigh in nature, and may be Ricean in more open areas (with a variable *k* factor).

Note that there are potentially deep fades, particularly since the antennas are likely to be close to the ground and on the same level as many obstructions. If the receiver antenna happens to be in a deep fade location with respect to the transmitter, then the triggering signal may well not be received. The commander can in theory check the device by testing the receiver beforehand – so long as the explosives are disconnected as has not always been the case in previous cases, where 'own goals' have been the height of that particular bomber's achievement.

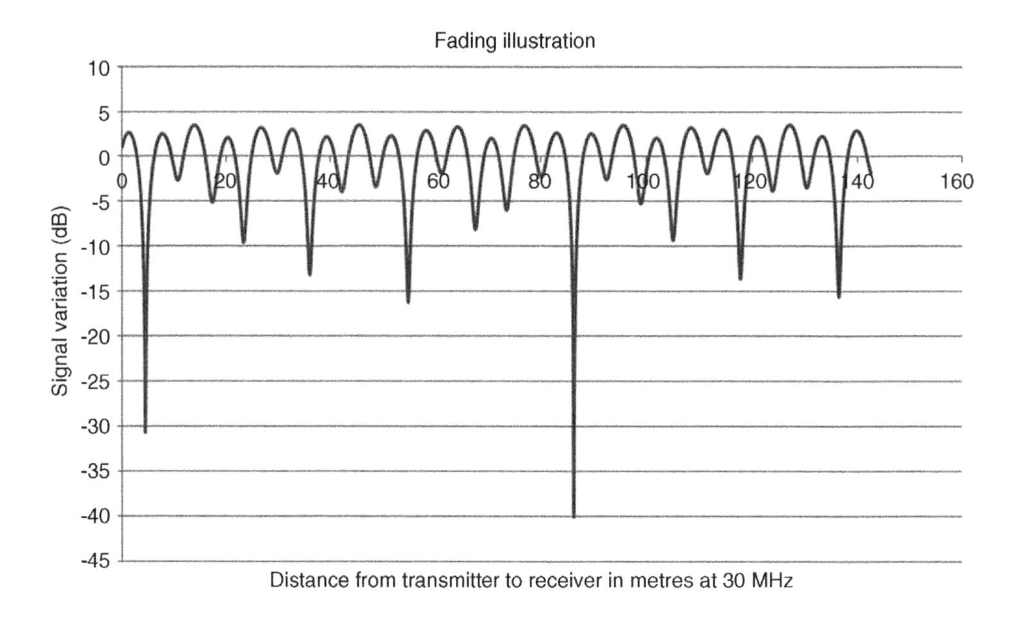

Figure 8.3 Illustration of fading of a radio signal with multiple reflectors.

Recent years have also seen scenarios where a primary explosive device is backed up by one or more secondary devices designed to maximise the impact of an attack, thus what is in some ways the worst case scenario for RCIEDs is illustrated in Figure 8.4.

In this figure, there is one primary and two secondary devices placed around the initial attack location. The intention is to use these devices to add harm whichever route the patrol takes to extract itself from the situation.

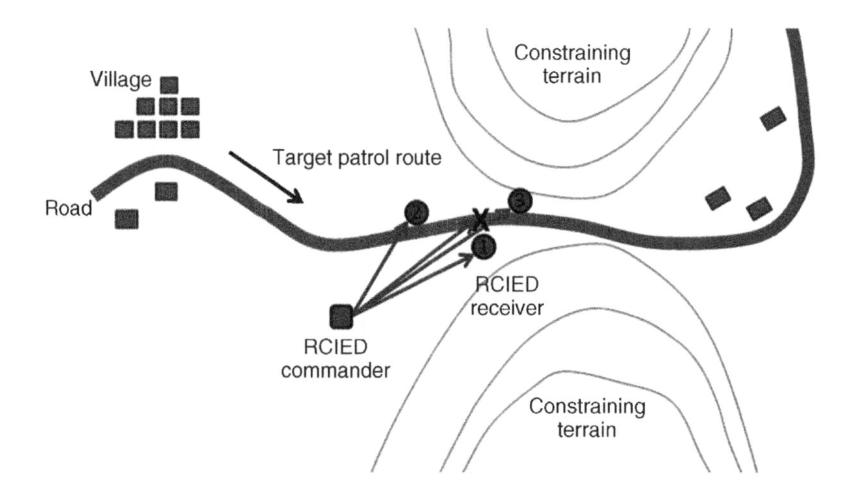

Figure 8.4 RCIED scenario.

Table 8.1 Methods of countering IEDs

Method	Objective
Intelligence-led raids	Disrupt enemy operations and safe houses based on specific intelligence
Aggressive missions	Disrupt enemy operations, capture enemy combatants, inflict combatant casualties, destroy enemy equipment based on operational objectives
Patrolling	Disrupt enemy operations, show presence to locals, identify changes in mood and local routines
Engaging with local leaders	Isolating extremists, building trust
Provision of medical services to locals	Isolating extremists, winning hearts and minds
Assistance to local civil power	Provide local security, assist in projects to improve locals' lives, reduce hold of extremists on locals
Disruption of supply chain	Where devices are built elsewhere and moved into their operational area
Strategic intelligence collection	Build picture of opposition strategies and future activities
Theatre-wide intelligence collection	Generate theatre picture, build enemy ORBAT, determine force disposition and activities (OPTEMPO)
Local intelligence collection	Determine opposition in locality, identify movements, safe houses, planned and actual operations

We will be using this scenario to look at detection and countering this type of attack.

8.3 Detection of IED Radio Control Systems

While the RCIED and his team are planning their attacks, the opposing forces will be looking at ways of countering them and their operations. This is not merely a technical exercise, although we will be focussing on radio jamming techniques. However, it is important to recognise that effective countermeasures must be carried out over a wide front. Techniques that are typically used in parallel are illustrated in Table 8.1.

The emphasis is on the use of intelligence to focus military operations and looking after the people who require protection from the extremists who will use RCIEDs.

In technical terms, it is necessary to determine an approach to counter RCIEDS using their inherent vulnerabilities. However, given that the radio emissions from the system itself may only occur during initiation, how can this be achieved? A number of possibilities are listed below:

- Detection of unusual movements or absence of normal movements in a suspicious location.
- Detection of associated operational communications.
- Mechanical disruption of device (e.g. bulldozer, mine clearance vehicle).
- Pre-emptive initiation using high-power jammer before its intended use.

- Force protection jamming around patrol.
- Explosive Ordnance Demolition (EOD) of located devices.
- Technical exploitation of captured devices and associated equipment.
- Questioning of captured RCIED devices.

In Part Two, we will look at the RF methods of detection, first by examining the characteristics of the links that are vulnerable to detection, interception and localisation.

References and Further Reading

All internet references were correct at time of writing.

Dapper, M.; Wells, J.S.; Huon, L. (2007), *RF Propagation in Short Range Sensor Communications*, http://www.nova-eng.com/downloads/wp_rfprop.pdf.

Part Two

Practical

9

Predicting HF Radio

9.1 Propagation at HF

9.1.1 Skywave

HF has been used by military forces since its introduction. For some time after the introduction of satellite systems, HF fell into abeyance for some time, but the costs associated with satellite systems and advances in HF have led to resurgence in interest for HF as a cheap, worldwide radio technology. HF is not however without its difficulties, which principally revolve around working the HF spectrum in order to achieve workable links over second, minutes, hours, days or months.

For reasons of application and operational considerations, HF is normally split into skywave and groundwave. Groundwave is used for relatively short links over a few tens of kilometres whereas skywave can be used worldwide, with care and in the right conditions. Groundwave is affected by the ground environment conductivity. Skywave HF is dependent on the structure of the upper atmosphere and the effects of the sun, which vary them.

A diagram of the relevant parts of the atmosphere is shown in Figure 9.1.

Radiowaves transmitted at the right angle and frequency will not escape the Earth but will instead reflect back into the atmosphere and can bounce from terrain to allow multiple skips. These skips are what allows worldwide transmission to be received.

The upper layers of the atmosphere vary by time of day, latitude and the output from the sun. Of particular note are sunspots, which are caused by variations in storm cycles on the sun. The variation of sun spots heavily influences HF propagation. Although the actual number of sunspots is variable, it does vary according to an underlying cycle. This means that although future sunspot numbers cannot be predicted, the likely values can be and are predicted. Figure 9.2 shows the predicted sunspot numbers for the period 2010 to 2015.

Communications, Radar and Electronic Warfare Adrian Graham

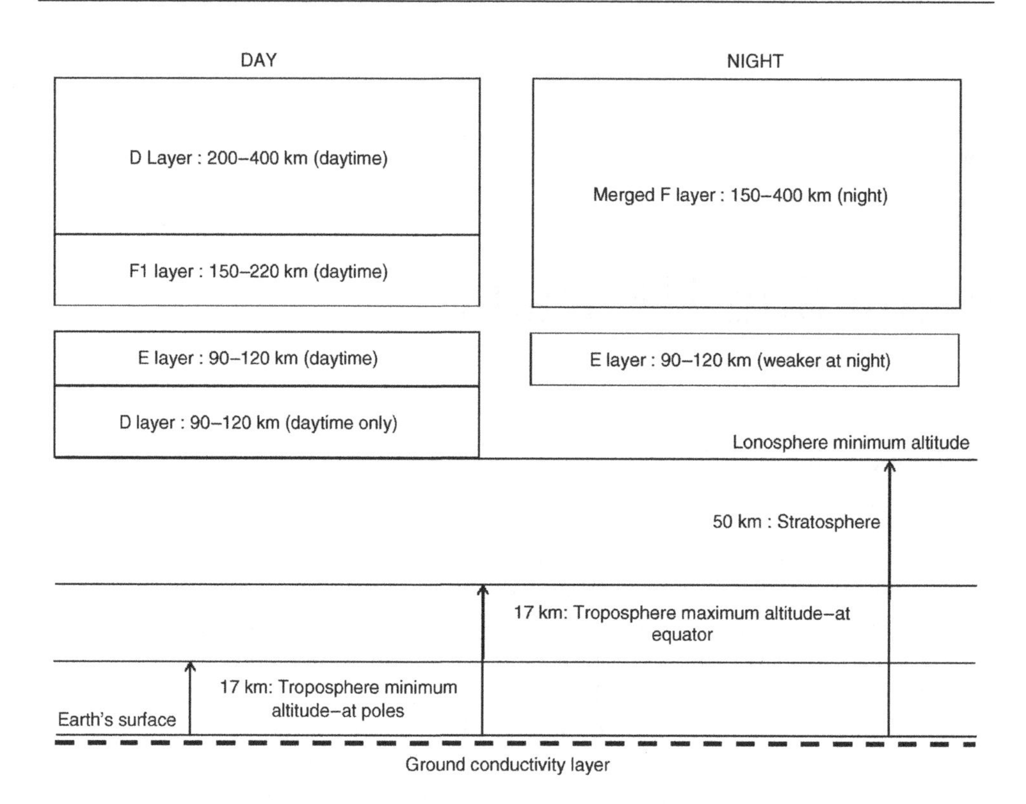

Figure 9.1 The ionosphere by day and night.

In order to take advantage of HF high atmosphere skips, radio energy must be inserted into the correct angle and within a specific frequency band so that it does not pass straight through the reflecting regimes, refract into the wrong locations or be absorbed by the D-layer. This is illustrated in Figure 9.3. The first transmission (1) is inserted at a high angle and above the highest refracting frequency. The highest usable frequency is more normally known as the Maximum Usable Frequency (MUF). Transmission (2) is inserted at a lower angle, but still does not refract sufficiently to return to the surface of the Earth. Transmission (3) is below the MUF and is reflected back towards the Earth's surface. The range achieved depends on the geometry of the link, so it can be seen that the angle of transmission must be determined in order to reach the intended target range.

HF energy refracted from the ionosphere back to the surface of the Earth will reflect back at the same angle. The energy is then refracted back down and so forth, producing multiple 'hops'. In between these areas are regions where little or no energy is present. These are known as 'skip zones'. This is illustrated in Figure 9.4.

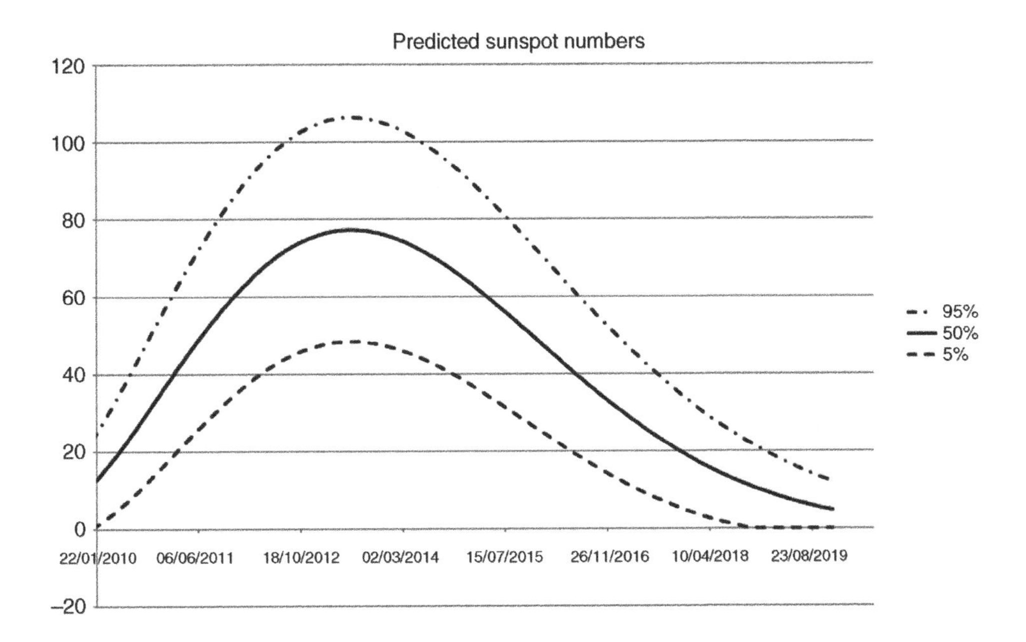

Figure 9.2 Predicted sunspot numbers, obtained from http://www.swpc.noaa.gov/ftpdir/weekly/Predict.txt.

We are not just concerned with the MUF; we also need to consider the Lowest Usable Frequency (LUF). This is dependent on the absorption caused by the D-layer, as illustrated in Figure 9.5.

Thus, we have to consider the MUF and the LUF, and ideally find the Frequency of Optimum Transmission (FOT). We have identified that the MUF and LUF depend on the strengths of the layers of the ionosphere. During the night, the D-layer dissipates. This means that lower frequency transmissions are possible during the night. Also, the E- and F- layers are also less pronounced during the night time. This means that the MUF and LUF change between the day and night, and thus the FOT will change during this time. Figure 9.6 shows the typical form of a diagram for LUF, MUF and FOT for a 24-hour hour period. This is sometimes known as an 'eye diagram', because it often looks somewhat like a human eye. The units along the bottom are hours in local or universal time code.

In most cases, it will not be possible to use a single transmission frequency for the whole 24-hour period. Often in practice, transmissions are broken into four-hour slots (known as a 'schedule'), during which different frequencies are used to provide a link over an entire 24-hour period. A typical example is shown in Figure 9.7.

The mathematics of HF skywave operation is very complex and it would take a concerted effort for an HF planner to calculate the right frequencies, the antenna

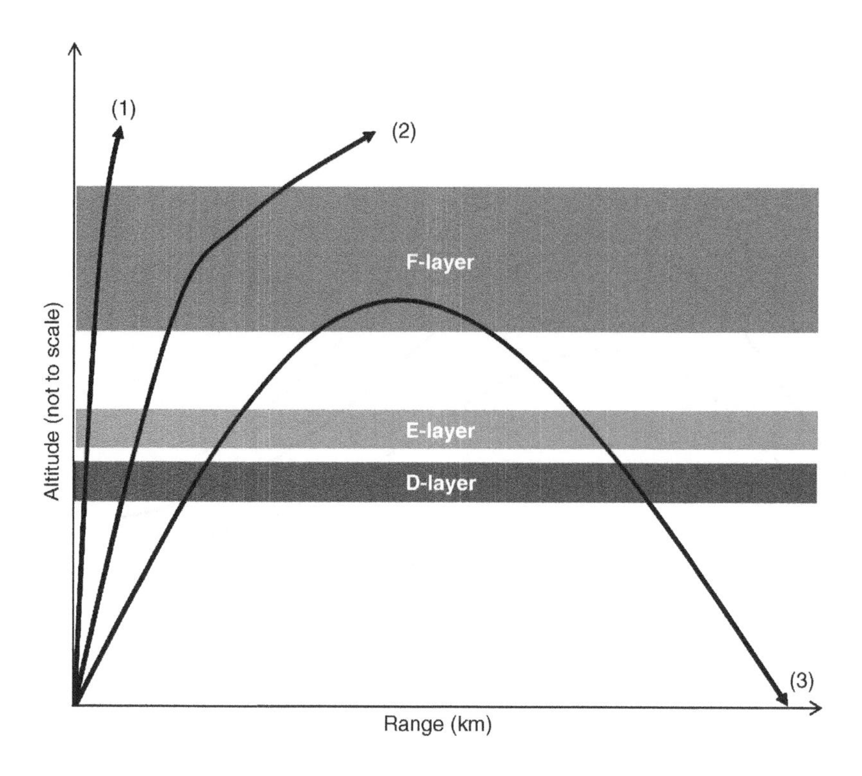

Figure 9.3 Insertion of HF transmissions into the ionosphere. Transmission (1) is inserted at a high angle and is above the Maximum Usable Frequency (MUF). Transmission (2) refracts, but not enough to return to the Earth. Transmission (3) is below the MUF and refracts back to the Earth, with the geometry of the refracted energy determining where it lands. Note that although these transmissions are shown as lines, in practice, there will be spreading of the energy so the range covers an area, not a point as shown.

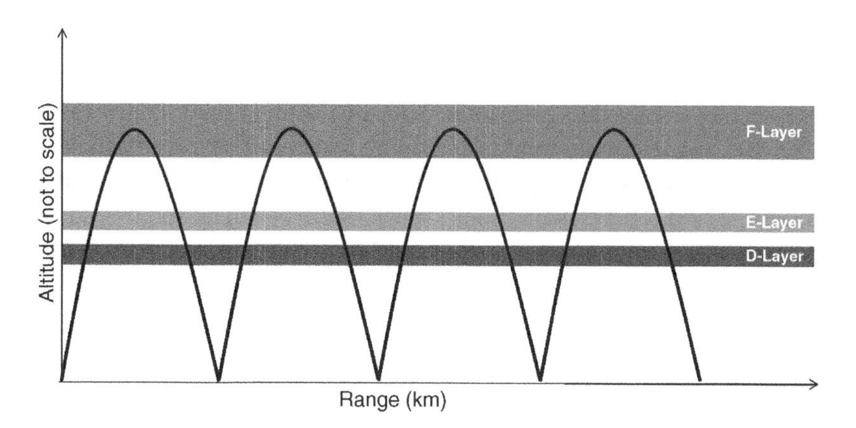

Figure 9.4 Multiple hops for an HF transmission. Where the energy bounces off the surface of the Earth, the signal level is at its highest. The regions in between are known as skips zones, where little or no signal is present.

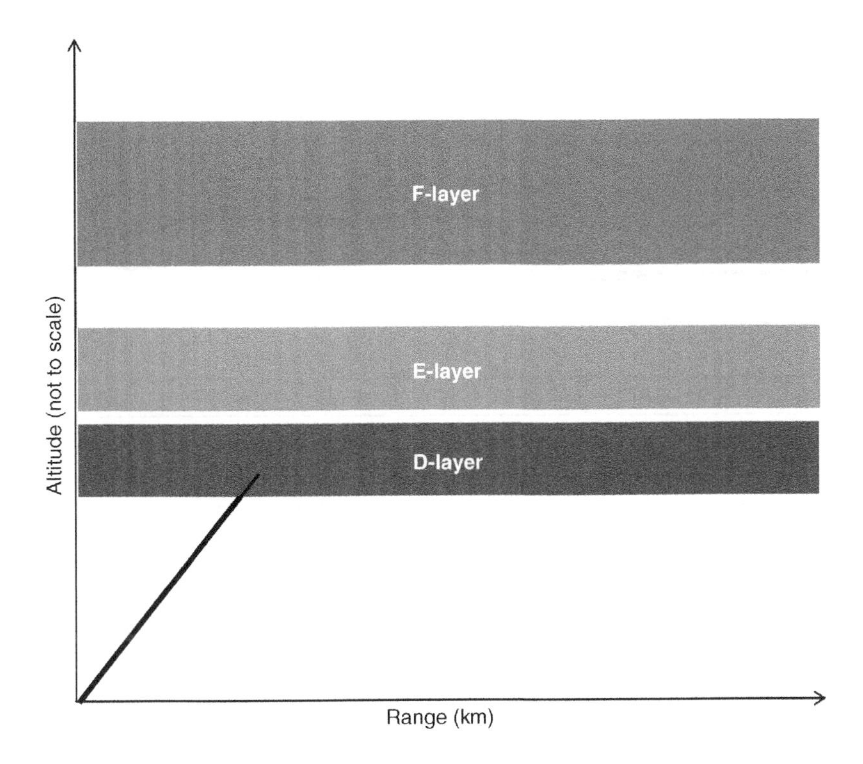

Figure 9.5 Absorption by the D-layer where the frequency used is below the Lowest Usable Frequency (LUF).

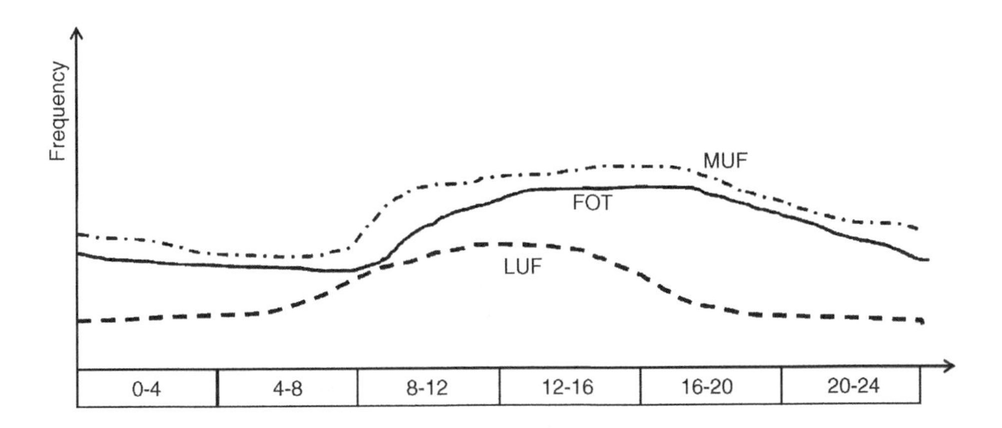

Figure 9.6 Variations in LUF, MUF and FOT over a 24-hour period. This diagram is indicative only; in practice, it will vary from location to location and between successive days, weeks and months.

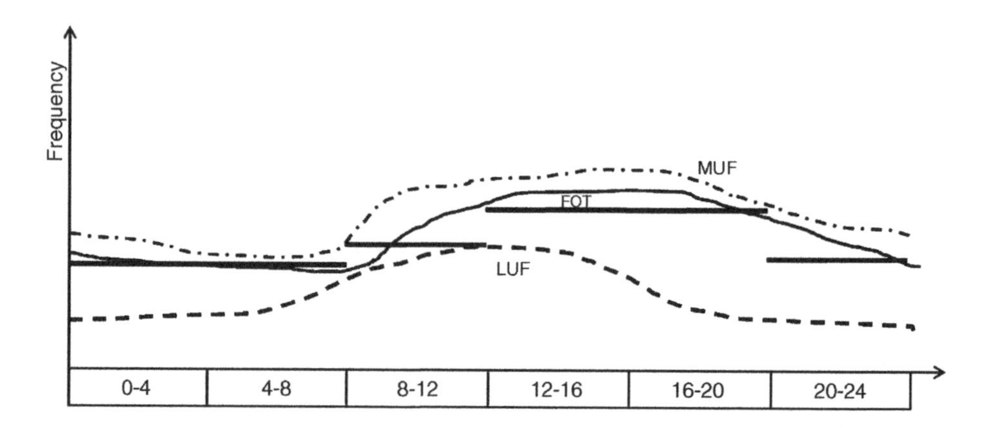

Figure 9.7 An HF transmission schedule. The transmit frequencies must be promulgated to the receiver in advance. The schedule shown uses the minimum number of channels to complete the 24-hour circuit. It could instead use different frequencies per 4-hour to provide some protection against eavesdroppers.

configuration and the link performance. It is far better to use one of the several freely-available HF skywave programs instead.

Some of the most well-known models are VOACAP, ICEPAC and, with its new HF component, AREPS.

VOACAP was developed on behalf of the US Department of Commerce by the Institute for Telecommunications Services (ITS), at the time of writing, is available from http://elbert.its.bldrdoc.gov/hf.html. IONPAC is also available from the same location. The Advanced Refractive Effects Prediction System (AREPS) was produced by the US Space Warfare and Naval Warfare Systems Centre, Atmospheric Propagation Branch. Users must register to obtain a copy, but the software is free. The downloaded software also includes a user manual and other supporting documentation.

The AREPS HF sub-system is shown in Figure 9.8. The original system is in colour, not black and white as shown. This makes it easier to interpret.

The user can select from pre-configured communications systems and enter the other appropriate data relating to antenna locations and the environment. The system can read current conditions from online resources where an internet connection is available. The top left display shows the currently selected results in a map display. Below that is the eye diagram showing usable frequencies as described above.

Radio equipment parameters can be configured by a separate equipment editor. This can be used to enter system characteristics including the antenna pattern. For ease of use, a number of default antennas are supplied as standard. The communications editor screen is shown in Figure 9.9.

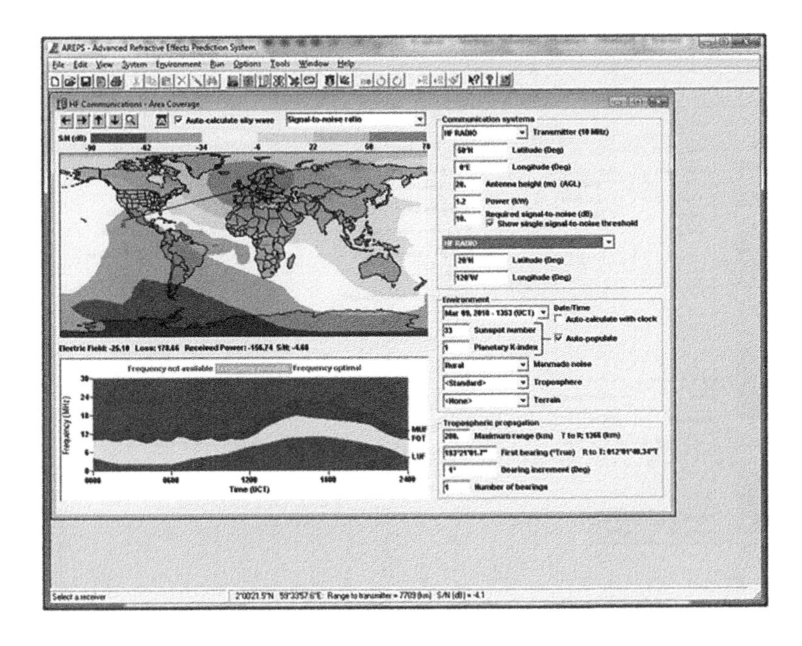

Figure 9.8 The AREPS HF interface. Reproduced by permission of SCC Pacific and the Atmospheric Propagation Branch.

9.2 HF Skywave Link Budgets

In order to perform accurate HF link performance assessments, we need to consider the following factors:

- transmitter and receiver antenna characteristics;
- transmitter and receiver locations;
- transmit frequencies (each one considered separately);
- transmit power;
- receiver location noise;
- ionospheric conditions.

All of these elements are important in order to get the right answers out of any of the available HF prediction systems, because they are all considered in the predictions. Using unrealistic values for any will affect the results provided by the system. To illustrate this, we shall look at some potential pitfalls. We will do this rather than examining how the system is used correctly because proper use can be determined by reading the manual for the system concerned (although it is well-known that engineers rarely read the manual!).

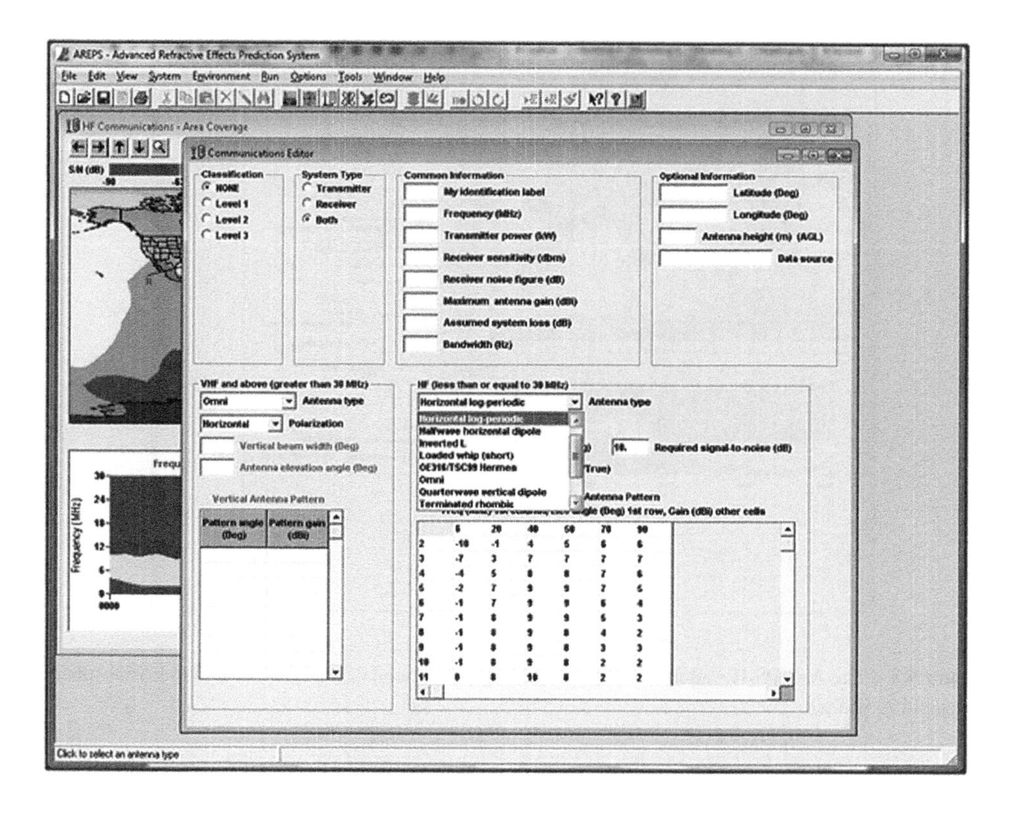

Figure 9.9 AREPS radio system data entry screen. There are two options; one for VHF and above and the other for HF systems. The system includes some default antenna types as shown in the drop list. Reproduced by permission of SCC Pacific and the Atmospheric Propagation Branch.

In the first example, shown in Figure 9.10, an inappropriate frequency has been selected. The system is set to show signal-to-noise ratio. The transmitter is based in the UK and the frequency selected is 4 MHz. For the time shown, the eye diagram shows that 4 MHz is below the LUF. The coverage area where the required signal-to-noise ratio is met or exceeded occupies a very small area around the UK and none elsewhere. In this case, the results display shows the effects of the problem clearly by the paucity of coverage required.

The solution to the problem shown in Figure 9.10 is to use the eye diagram to determine the optimal frequencies to use.

Figure 9.11 shows a more insidious error. Suppose we have a receiver, which is in an urban area, with all of the associated noise that implies. However, the default value is 'rural'. If we forget to change the entry, the result is as shown in the figure. This is however misrepresentative of the situation.

Figure 9.12 shows the coverage when the receiver location is correctly set to 'Business'. The coverage area is far smaller, but it gives us a better estimation of the expected level of performance for the link.

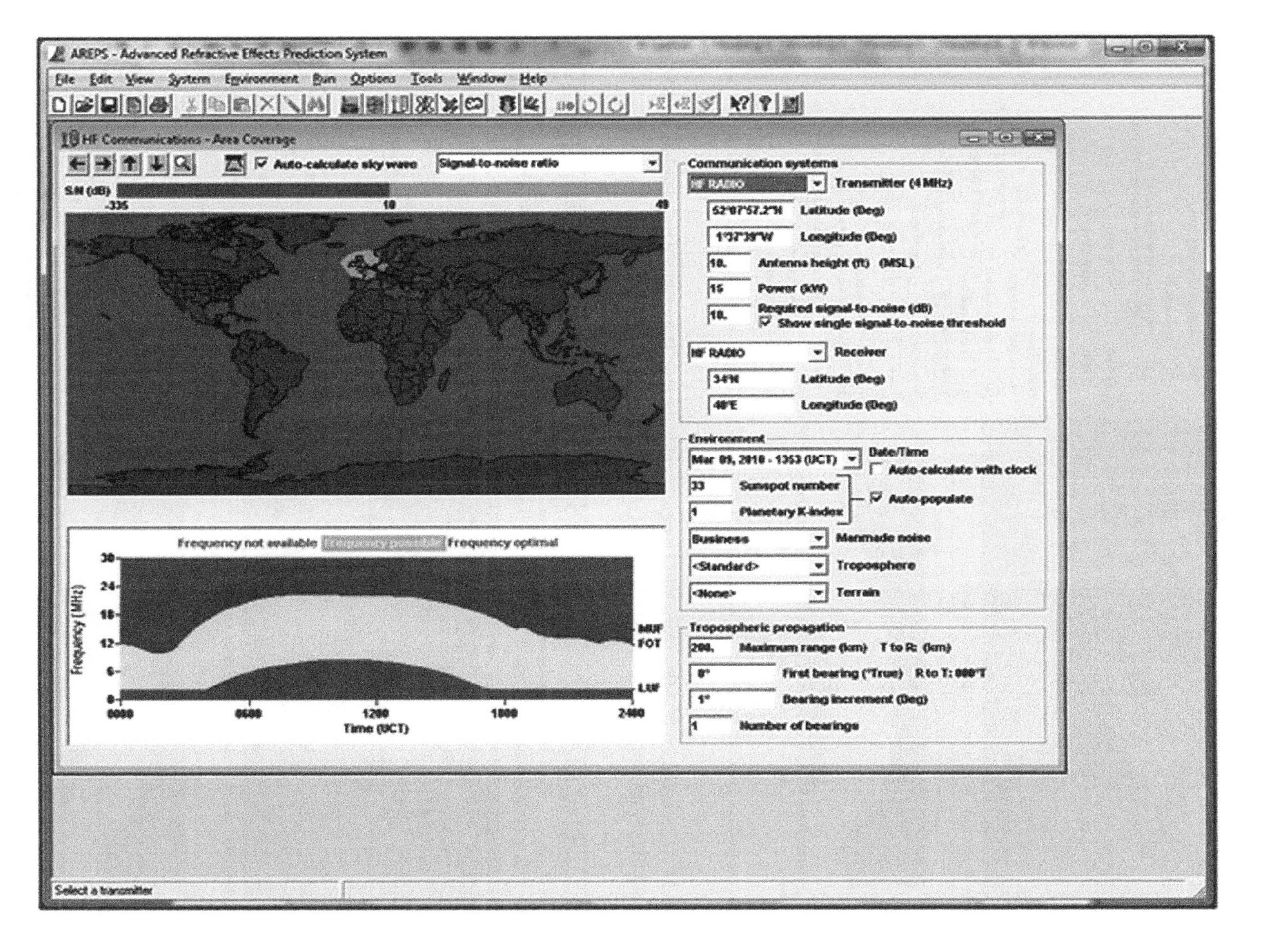

Figure 9.10 AREPS display when an inappropriate frequency has been entered. The resulting coverage area is very small, providing a strong visual clue that all is not well. Reproduced by permission of SCC Pacific and the Atmospheric Propagation Branch.

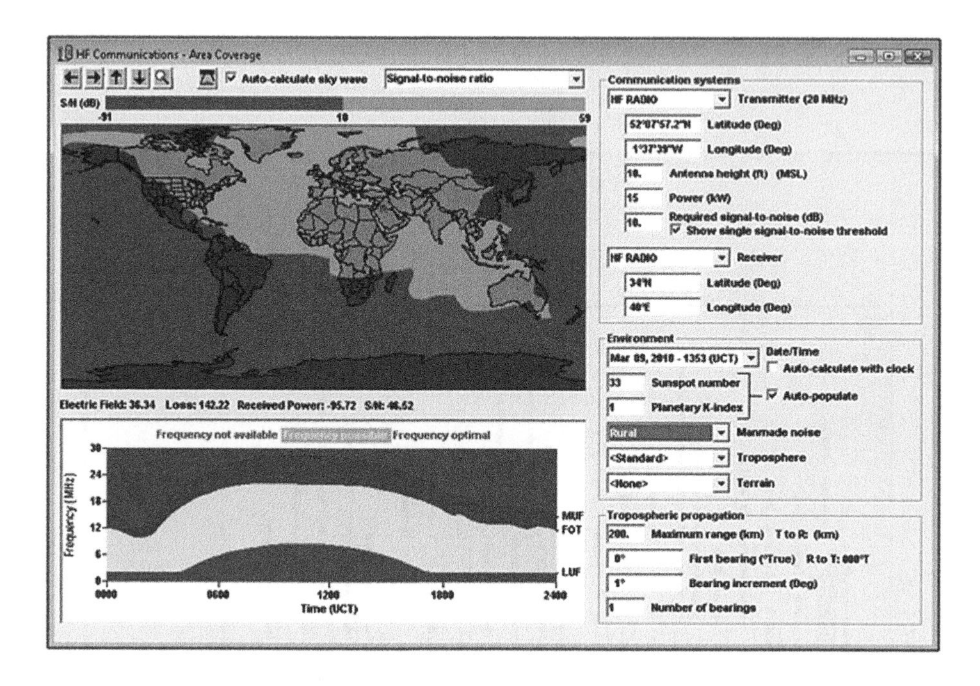

Figure 9.11 Coverage for 20 MHz using rural receiver environment. Reproduced by permission of SCC Pacific and the Atmospheric Propagation Branch.

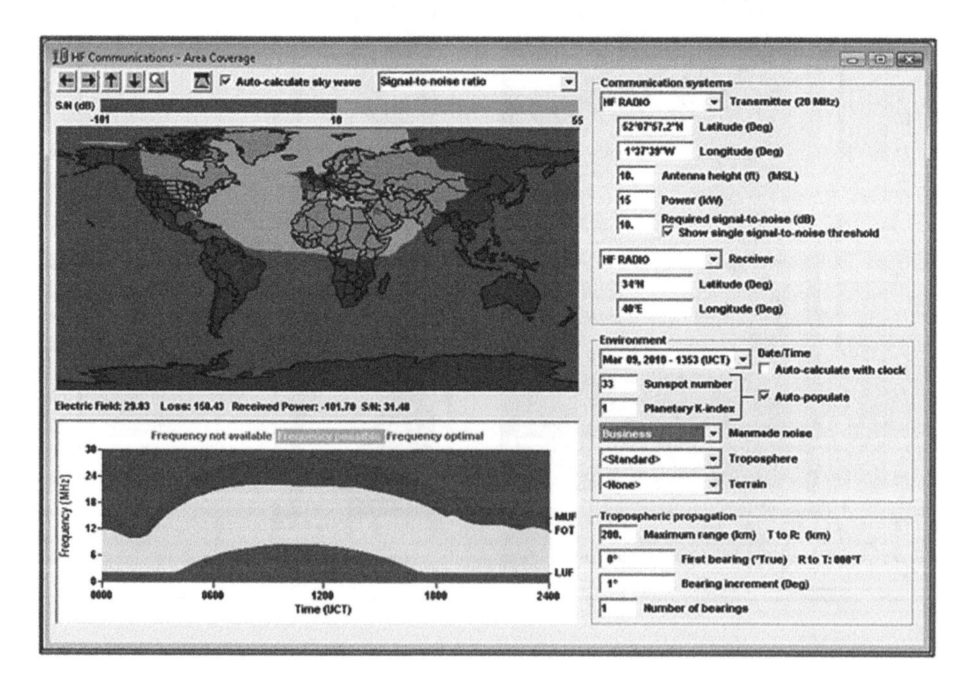

Figure 9.12 Coverage prediction for a receiver in a business environment. Reproduced by permission of SCC Pacific and the Atmospheric Propagation Branch.

Remember that if we need to use several frequencies to allow communications over a 24-hour period or longer, we need to perform a link assessment for each different frequency to be used, for the time of day each is to be used. In HF terms, we can consider each individual frequency as a channel and the totality of the link including all frequencies as the 'circuit'. It is the circuit that we need to design, so consideration of each frequency used is simply one part of the overall circuit design.

9.3 Groundwave

HF groundwave does not, as the name suggests, depend on the Ionospheric conditions but rather the conditions through which the ground and surface waves travel. They principal factors are the ground conductivity and permittivity. Over good ground, the signal will travel further but poor ground will lead to shortened communications links. For tactical applications, HF groundwave is often considered as being restricted to line-of-sight, but in practice it often travels further. The ITU has a model called ITU-R P.368, which can be used for terrain-independent estimation of groundwave range. The ITU also issues coarse worldwide maps of ground conductivity (with some holes) that can be used for planning purposes. Figure 9.13 shows a typical output of AREPS for an HF groundwave prediction.

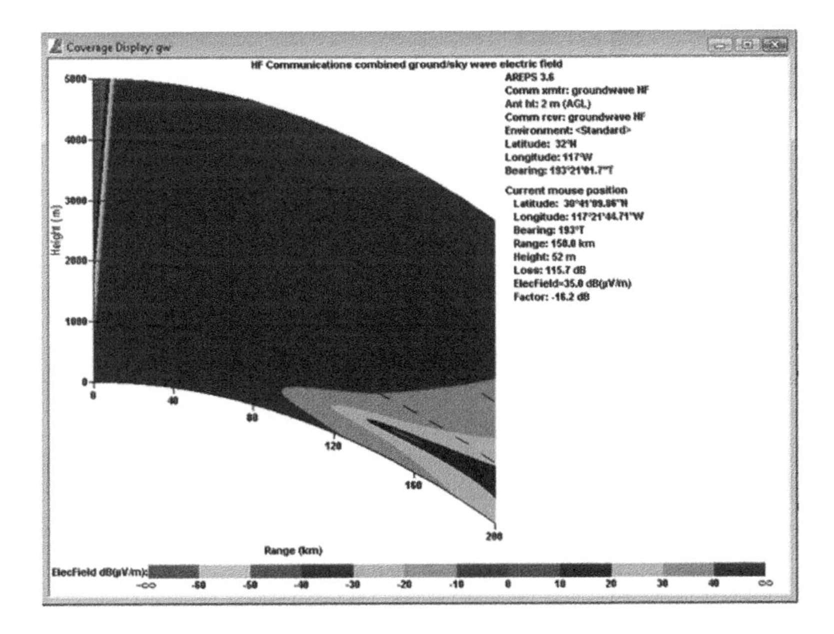

Figure 9.13 Groundwave prediction from the AREPS system. Reproduced by permission of SCC Pacific and the Atmospheric Propagation Branch.

References and Further Reading

All internet references were correct at time of writing.

The AREPS software was developed by SSC Pacific and the Atmospheric Propagation Branch. They can be contacted at the following website/email:

http://areps.spawar.navy.mil/.

e-mail: areps@spawar.navy.mil.

Recommendation ITU-R P.368: Ground-wave Propagation Curves for Frequencies Between 10 kHz and 30 MHz.

Recommendation ITU-R P.527: Electrical Characteristics of the Surface Of The Earth.

Recommendation ITU-R P.533: HF Propagation Prediction Method.

Recommendation ITU-R P.832: World Atlas of Ground Conductivities.

Recommendation ITU-R P.842: Computation of Reliability and Compatibility of HF Radio Systems.

Recommendation ITU-R P.1060: Propagation Factors Affecting Frequency Sharing in HF Terrestrial Systems.

Recommendation ITU-R P.1239: Reference Ionospheric Characteristics.

Recommendation ITU-R P.1240: Methods of Basic MUF, Operational MUF and Ray Path Prediction.

Recommendation ITU-R SM.1266 Adaptive MF & HF Systems.

10

VHF to SHF Radio Prediction

10.1 Propagation above HF

10.1.1 Introduction

Propagation at VHF and above depends more on tropospheric conditions rather than on the ionosphere, although of course satellite to ground and vice versa system will need to be able to pass through the ionosphere. The troposphere is the lowest part of the atmosphere and is defined as the part of the atmosphere that supports life. It does not suffer as much from space effects such as sunspots, although severe disruptions in the upper atmosphere can exceptionally have an effect. Far more common is effects due to weather and, for above about 6 GHz, atmospheric attenuation and losses due to precipitation. For many terrestrial links, the dominant factors are due to terrain and radio clutter.

In this chapter, we will be looking at prediction of radio propagation at VHF and above. We will introduce the main mechanisms, their effects, and will also examine some common propagation models. We will split links by frequency and also by their length, since these require different modelling techniques.

10.1.2 Short-Range VHF and UHF Links

Short range links such as those commonly encountered at VHF and UHF are more affected by terrain and clutter effects than they are by the atmosphere. In this context, this means links in the region of tens of kilometres. We can analyse the most important mechanisms and identify their specific contribution to link loss. These are:

- distance (link length);
- reflections;
- refraction;

Communications, Radar and Electronic Warfare Adrian Graham

- scattering;
- diffraction;
- absorption.

In this section, we will look at the basic mechanisms, and in later sections of this chapter, we will look at how they are accounted for in common models and modelling methods.

10.1.2.1 Distance

The greatest single factor is distance. While free space loss is generally a poor predictor of link performance except in very specific conditions, it is important because many of the propagation models we will discuss describe the excess loss above free space loss. Readers may recall that free space loss for an isotropic antenna is defined in the most useful units as:

$$L = 32.44 + 20 \log f + 20 \log d$$

where
L is loss in dB.
f is frequency in MHz.
d is distance in km.

This formula shows that loss increases with frequency as well as distance.

10.1.2.2 Reflections

Reflections occur when radiowaves encounter objects that are large compared to their wavelength. At 30 MHz, the wavelength is 10 metres and at 300 MHz, it is 1.0 metres. Surfaces do not need to be flat, but the flatter they are, the greater the coherent reflection effect. Perfectly flat ground or a body of calm water offer the greatest chance of a good reflecting surface. When the surface of water such as the sea is disturbed by wind, the reflections become less coherent and their influence on the received signal strength is therefore less. The classic reflection scenario is shown in Figure 10.1.

The reflection will change the phase of the incident RF energy by 180 degrees. The energy arriving at the receiver is the vector sum of all incident energy and thus, if the path length difference between the direct and reflected wave are exactly the same strength and 180 degrees out of phase, the resulting energy would be zero. However, normally the phase cancellation will not be perfect in practice, but it can still result in reception of a very small signal. As shown in Figure 10.1, some of the energy may be trapped in the surface of the Earth and will travel along radiating upwards so there may

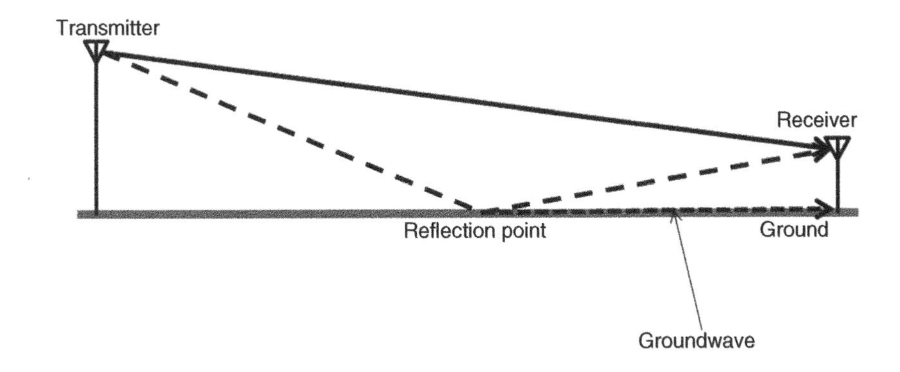

Figure 10.1 Reflection of radio energy from a flat surface (compared to the wavelength of transmission).

be a component of this energy at the receiver as well. Additionally, some of the RF energy may be absorbed at the reflection point.

10.1.2.3 Scattering

Scattering occurs when the RF energy bounces from small objects, typically in many different directions with different strength. At optical frequencies, scattering is the mechanism we use to see; the sun is a (large) point source. We can see the reflection of the sun on a flat sea clearly. However, we can see the sea itself and all other parts of the environment due to scattering from all of the visible features. RF scattering works in much the same way, except that radio antennas are not able to distinguish between energy arriving in the same way as our eyes; they are far less complex in this manner. Scattering causes multiple receptions at the receiver that are usually of random strength and phase, although their strength is likely to be lower than that caused by reflections. It is the mechanism that leads to fast fading, and which is often accounted for in propagation models by statistical means. In some cases, ray tracing methods can be used to explicitly model scattering, although this is often difficult because of the amount of data required to accurately model the environment and all the changes that may occur due to moving objects including people, vehicles, wind effects (by causing object movement) and so on. We will focus on statistical methods of accounting for scattering later in this section.

10.1.2.4 Refraction

Refraction is a very important concept in radio prediction for VHF and above. In normal conditions, atmospheric pressure falls off with altitude. The reduction in pressure also causes a change in the refractive index for radiowaves. The effect of this is that compared to optical waves, radiowaves transmitted near the surface of the Earth

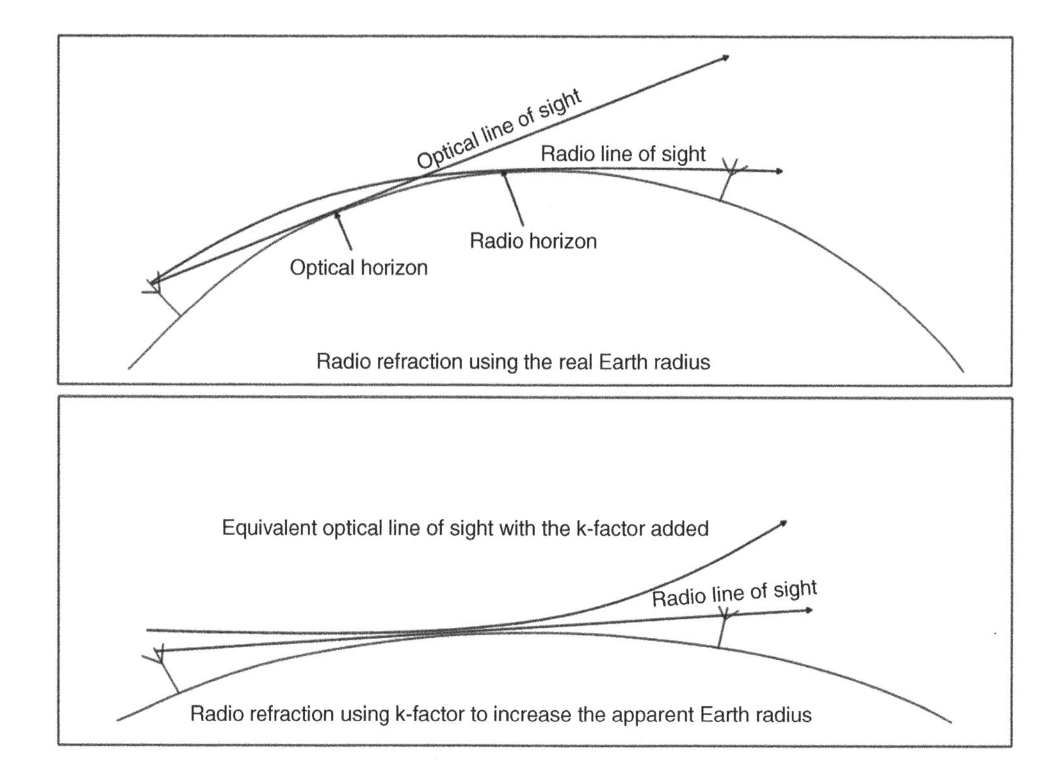

Figure 10.2 Accounting for the variation in radio refractive index by applying the concept of 'effective Earth radius'. This is a correction to the actual radius of the Earth to effectively 'straighten out' the most often encountered conditions due to pressure reduction with increasing altitude.

will tend in most conditions to bend downwards back towards the surface. We can apply a correction to account for this phenomenon. This is illustrated in Figure 10.2.

The correction term is known as the 'k-factor'. The most often quoted value for k is $4/3$, which gives an effective Earth radius of $4/3 \times 6{,}370$ km, which is approximately 8,500 km. However, it is important to recognise that this is a nominal figure that can vary according to the conditions, and in different parts of the world, it can be very different from this. The ITU produces maps for the median refractive index, from which the k-factor can be derived.

Variations in the k-factor can result in very different propagation conditions, including sub-refraction, and super-refraction as illustrated in Figure 10.3.

In Figure 10.3, the typical refraction path (4/3 Earth) is shown, along with two other conditions. The first is super-refraction, where the RF energy bends far more than normal back towards the surface of the Earth. This causes the signal to reflect at a point before the radio horizon. If the energy is absorbed when it reaches the ground then no communications at longer ranges are possible. Often, the energy will be reflected back into the atmosphere and then super-refracted down again to form hops and skips as are

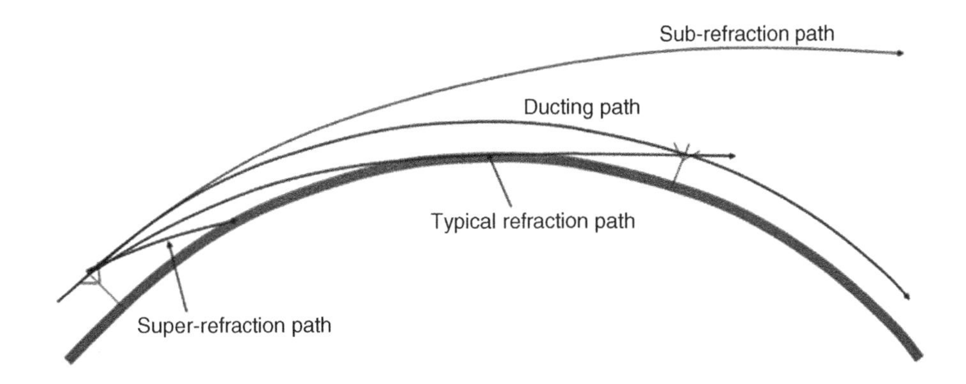

Figure 10.3 Different refraction paths caused by different atmospheric conditions. Super-refraction occurs when the signal is bent down so severely, it does not even reach the horizon, making long range communications difficult. The energy in this case can also be reflected back into the atmosphere and back down again, causing similar skip zones to those encountered in HF communications. A sub-refraction path leads to a reduction in downward bending, which may mean that terrestrial receivers may not be able to receive the signal properly. Of particular interest is the condition where the refractive index causes energy to radiate parallel to the surface of the Earth. This is known as 'ducting' since such ducts are not normally very wide and the act like pipes (or ducts) for the energy travelling along them.

familiar for HF skywave but at shorter ranges. Sub-refraction means that the energy is less refracted than normal. For a microwave link, this can cause the highly-directional radio signal to pass above the receiver and thus disrupt communications. Also shown is the condition where the refractive index results in a path parallel to the surface of the Earth. Such ducts are not normally present at all altitudes, but rather within a vertical part of the atmosphere. Ducts at ground level, common over sea paths due to evaporation of water from the surface of the sea, are known as ground ducts. Ducts that form higher up are known as elevated ducts. Figure 10.4 shows two sample atmospheric columns, the left hand one showing a surface duct and the second showing an elevated duct. In both diagrams, the important factor is the negative gradients that form the ducts.

The standard atmosphere for radio propagation prediction does not include ducts but rather a refractive index that decreases with height. The absolute refractive index is a function of the dry air component and the wet air component, and is described by:

$$N = N_{dry} + N_{wet} = 77.6\frac{P}{T} + 3.732 \times 10^5 \frac{e}{T^2}$$

where
P is the atmospheric pressure (hPa).
e is the water vapour pressure (hPa).
T is absolute temperature (K).

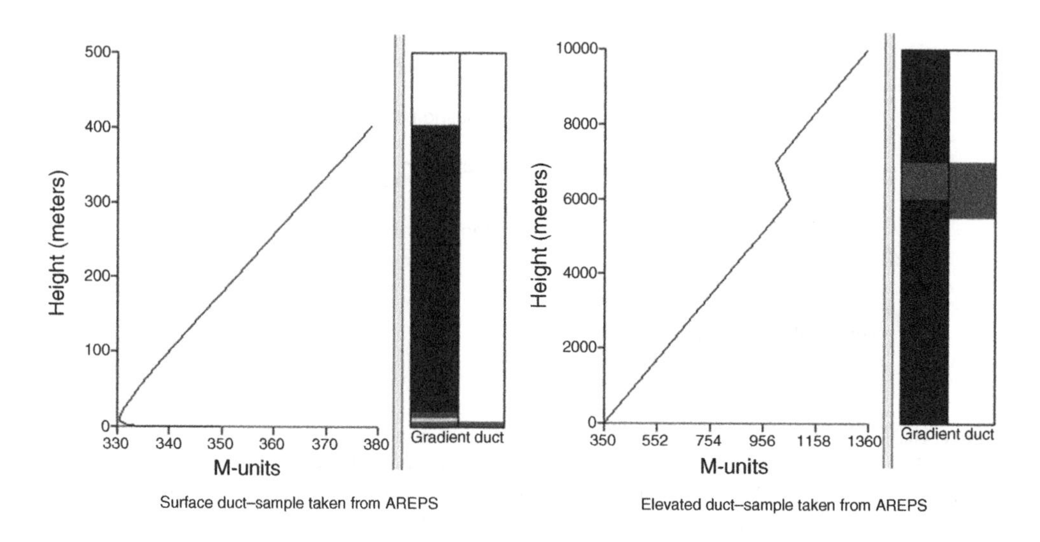

Figure 10.4 Example ducts taken from the AREPS program. The left hand picture shows a surface evaporation duct. The right hand picture shows an elevated duct. In both cases, the ducts are formed by negative gradients in the M-units (see text). Reproduced by permission of SCC Pacific and the Atmospheric Propagation Branch.

The long-term (i.e. standard) height dependence on the refractive index is given by:

$$N_s = N_0 \exp\left(\frac{-h_s}{h_0}\right)$$

where
$N_O = 315$.
$H_O = 7.35$ km.
$H_s =$ height in km.

In terms of understanding the refraction of radiowaves, the variation of N is important and is termed ΔN. A value of $\Delta N = -40$ for the kilometre above ground is equivalent to a k-factor of $4/3$. The relationship between ΔN and k is:

$$k = \frac{157}{157 + \Delta N}$$

Note that a value of $\Delta N = -157$ results in an infinite k-factor. This is where the refractive index is parallel to the surface of the Earth – as though the Earth's radius were infinite.

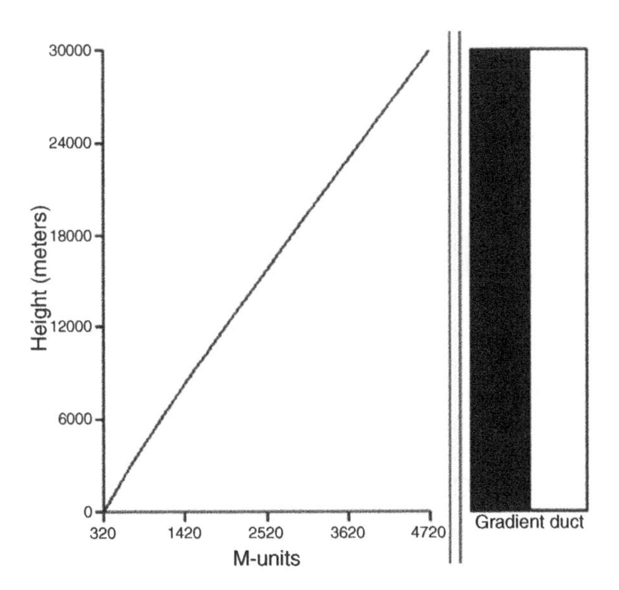

Figure 10.5 Standard atmosphere as shown in the AREPS software. Reproduced by permission of SCC Pacific and the Atmospheric Propagation Branch.

Many propagation models use these parameters to determine the paths of radiowaves and in this case, there is normally a method of entering or editing either the k-factor or ΔN.

The standard atmosphere as shown in AREPS is shown in Figure 10.5. In this case there is no negative gradient and hence no ducting at any level.

Note that AREPS uses M-units rather than N-units. The conversion is straightforward:

$$M = N + 157h$$

where h is altitude in kilometres. If h is in metres, then the h coefficient is 0.157 to reflect the change in units.

10.1.2.5 Diffraction

For many terrestrial radio and radar systems, terrain is one of the most important features. A typical terrain profile, generated in AREPS is shown in Figure 10.6. If the terrain is between the transmitter and the receiver, then the received signal will be significantly lower than the value that would be obtained from a free space loss prediction.

When used for communications prediction in AREPS, this terrain is displayed as shown in Figure 10.7. The darker shade (originally red in the program) shows where

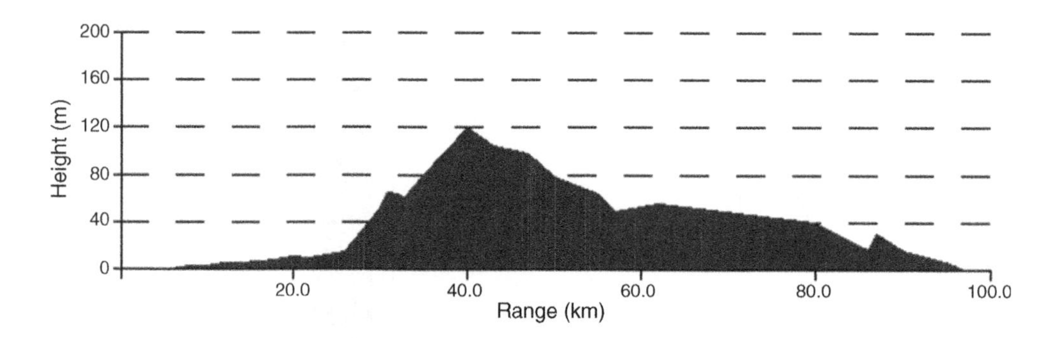

Figure 10.6 A terrain profile as generated and displayed using the AREPS software. Reproduced by permission of SCC Pacific and the Atmospheric Propagation Branch.

communications is possible. Close to the ground and beyond the peak of the hills, the shade is lighter, showing that communications are impossible. This is because the terrain is blocking much of the signal. The energy that does arrive at a receiver close to the ground in this region is diffracted over the terrain.

The scenario is shown in Figure 10.8. A transmitter on the left of the picture is radiating at VHF omni-directionally. The receiver is on the right hand side. There is no

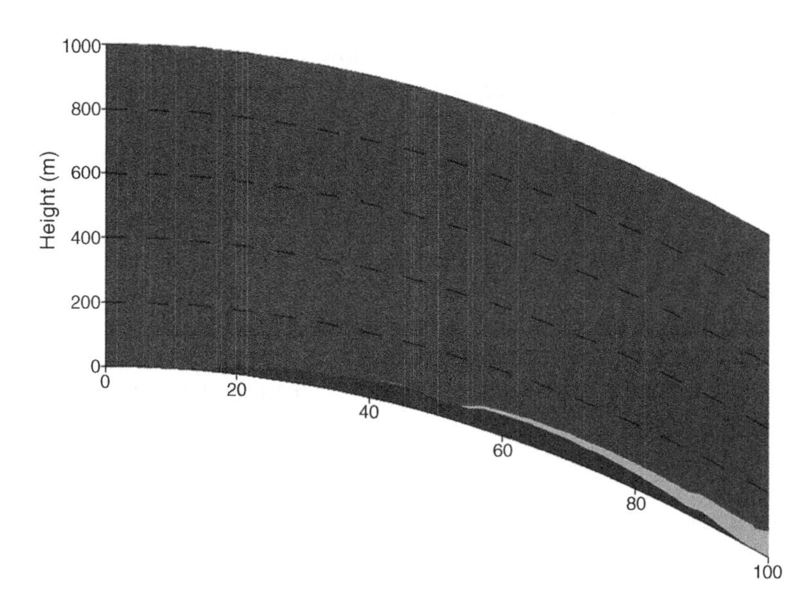

Figure 10.7 This shows a communications plot from AREPS using the terrain shown in Figure 10.6. It looks different because of the slant range display. The lighter colour shows that communications are not possible for a receiver close to the ground. This is not due to range; otherwise the darker colour that shows effective coverage would not be present for higher altitude targets. Instead, this is due to the terrain blocking the signal. The only RF energy available in this regime is due to diffraction over the terrain. Reproduced by permission of SCC Pacific and the Atmospheric Propagation Branch.

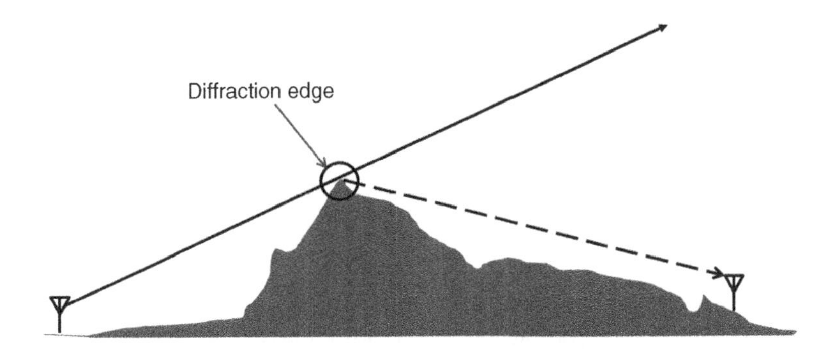

Figure 10.8 Diffraction over a single main diffraction obstacle.

direct line of sight because of the terrain. The edge of good coverage is shown as the solid black line. The long dashed line shows the energy in the direction of the receiver. This is energy diffracted over the main diffraction obstacle (circled).

Notice in Figure 10.8 that although the terrain is complex, the diffraction edge is only a single point along it. The received signal strength can be estimated by adding the loss due to distance to the loss due to diffraction over the obstacle. This can be simplified to the form shown in Figure 10.9. Note that the Earth bulge (due to the Earth's radius) must also be included in the calculation for the height of the intrusion above the link terminals (it must be subtracted from the actual height for the equation that follows to be accurate).

The distances d_1 and d_2 are the distances from the transmitter and receiver respectively. All of the units should be self-consistent, thus if the distances d_1 and d_2 are in kilometres, the height value must also be expressed in kilometres. From

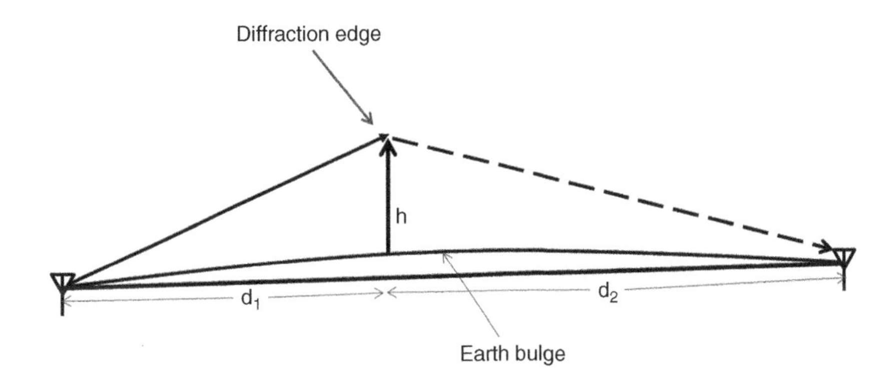

Figure 10.9 Simplified path profile, showing only the diffraction edge and the terminals and omitting the other terrain points.

Figure 10.9, we can determine a unit-less parameter v which we can use to assess the loss due to diffraction.

The value v can be expressed in a number of ways, including:

$$v = h\sqrt{\frac{2}{\lambda}\left(\frac{1}{d_1} + \frac{1}{d_2}\right)}$$

This expression can be used for both positive and negative values of h. If h is negative, then the intrusion is below the direct radio line of sight. The value v is used in a complex mathematical formula known as the Fresnel integral:

$$F_c(v) = \int_0^v exp\left(j\frac{\pi s^2}{2}\right)ds = C(v) + jS(v)$$

where

$$C(v) = \int_0^v cos\left(\frac{\pi s^2}{2}\right)ds$$

$$S(v) = \int_0^v sin\left(\frac{\pi s^2}{2}\right)ds$$

The value of diffraction loss in dB can be determined by using the real and imaginary parts of the Fresnel integral using the following formula:

$$J(v) = -20\,log\left(\frac{\sqrt{[1 - C(v) - S(v)]^2 + [C(v) - S(v)]^2}}{2}\right)$$

This is graphed in Figure 10.10 for values of $v = -3$ to 3. The loss values are in dB.

This is known as knife-edge diffraction. For values of v $>= -0.7$, the following simpler formula, which is easier to solve:

$$J(v) = 6.9 + 20\,log\left(\sqrt{(v - 0.1)^2 + 1} + v - 0.1\right)$$

where again the loss value $J(v)$ is in dB.

Terrain obstructions are seldom similar in form to the knife-edge object shown in Figure 10.9. However, this simplification has been used in many propagation prediction methods, and with additional corrections, it can be made to account for the thickness and shape of obstructions as well, which we will see when we come on to practical propagation models.

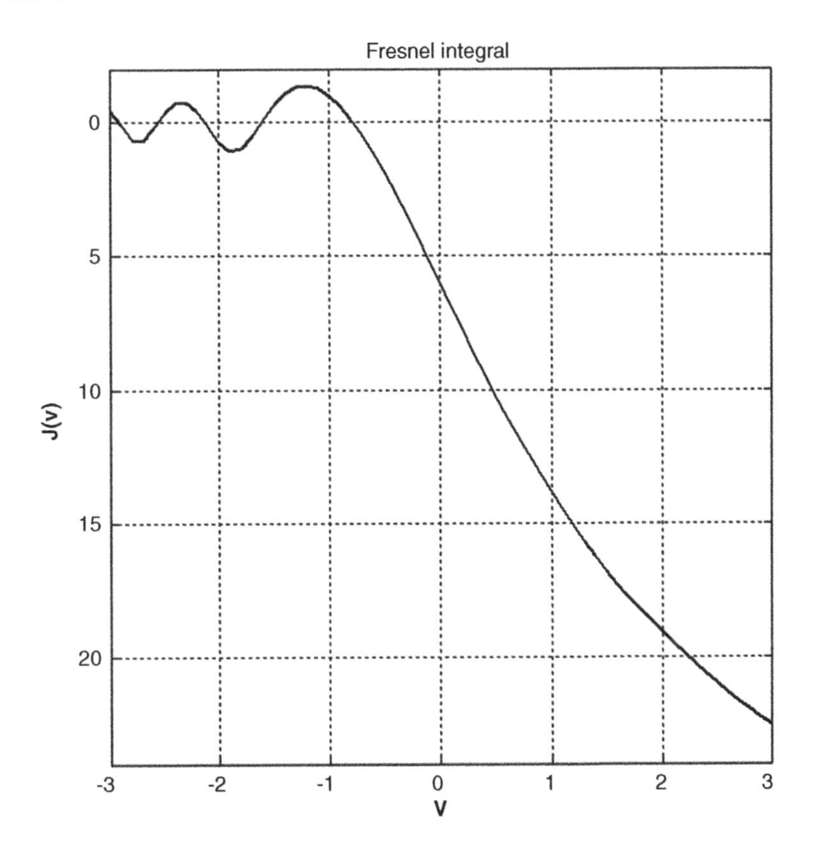

Figure 10.10 Graph of diffraction loss against the unit-less intrusion parameter ν.

10.1.2.6 Absorption

Radiowaves can be absorbed by a wide variety of structures found in normal life. This includes building materials, dense vegetation and even the human body. The extent of absorption depends on the frequency of transmission, the type of material and its thickness. Many papers have been published giving measured values at specific frequencies for loss due to various structures, although most of these are too specific to be generally applicable. ITU propagation recommendation ITU-R P.1812-1 gives the interim figures shown in Table 10.1, which provide a basic guide, but must be used with care since the values are fairly general and further work is required to provide figures that are more reliable. These figures are additional losses to those applied to account for clutter when compared to radio receivers positioned in a built-up area but out in the street.

Clearly, loss will increase in large buildings where there may be many walls between an outside transmitter and a receiver inside a building.

Table 10.1 Interim loss figures for building penetration, based on ITU-R P.1812-1

Frequency	Median loss (dB)	Standard deviation (dB)
200 MHz	9	3
600 MHz	11	6
1.5 GHz	11	6

Absorption is also an issue for body-mounted antennas. Figure 10.11 shows figures derived from ITU-R P.1406 for antennas worn on the body at head height and waist height. Losses will depend on the relative orientation of the body-worn antenna and the far antenna but this is often difficult to predict in advance and thus these figures provide a reasonable guide.

Modern antennas for mobile phones and other applications where it is to be expected that the antenna will be close to the body often include the effects of body loss in their specifications and indeed in their design.

10.1.3 Long Range VHF and UHF Links

Long range VHF and UHF links such as those used for ground-air and air-air links will also be affected by atmospheric effects even where terrain and clutter are not an issue.

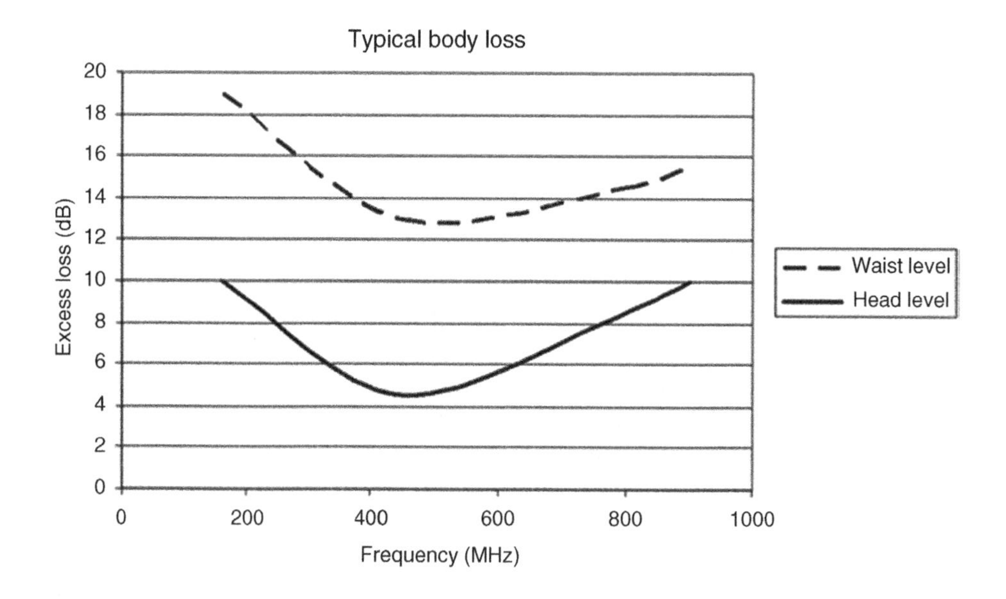

Figure 10.11 Body loss figures derived from ITU-R P.1406.

There can be as many as three different dominant propagation regimes for such long links. These are:

- line of sight propagation mechanisms (free space loss for highly directional antennas, two-ray for most links);
- diffraction;
- tropospheric scatter.

There will also be variability due to atmospheric variations such as:

- variations to the k-factor;
- ducting and other anomalous conditions;
- variations in troposcatter response.

To account for these, propagation models designed for long range prediction at VHF and UHF will express the expected loss for specific percentages of time, with 5%, 50% and 95% being the most often quoted. Models like ITU-R P.528 (aeronautical propagation) are based on long-term measurements that by their nature include all of the atmospheric variations likely to be accounted for these percentages of time.

10.1.4 Variability at VHF and UHF

Figure 10.12 shows a graph derived from ITU-R P.1406, which shows the standard deviation for variability for VHF and UHF links over land and sea paths. The values

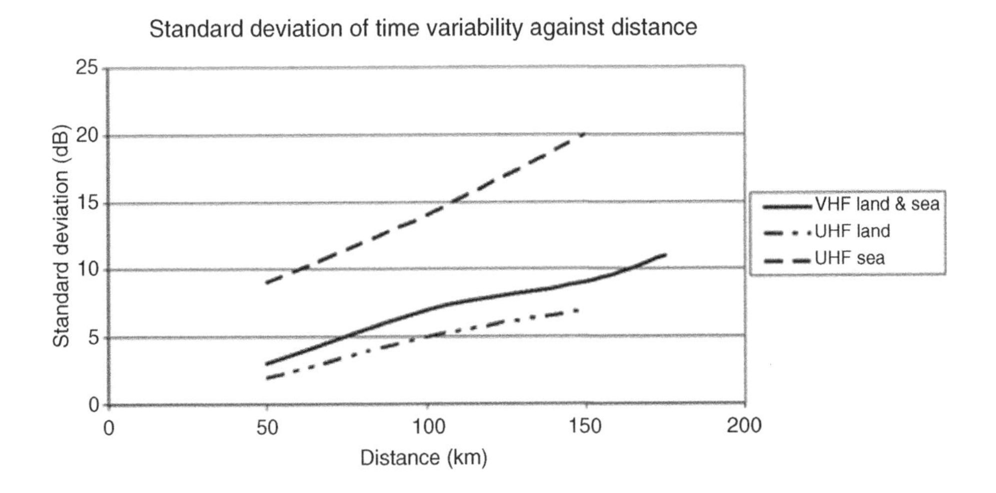

Figure 10.12 Standard deviation for links as VHF and UHF over land and sea paths, derived from ITU-R P.1406. The variation increases with distance, which is an intuitive result.

increase with distance as would be expected; the longer the path through the atmosphere, the more effect that atmospheric variations make.

At VHF and UHF, rainfall and other precipitation effects do not affect radio propagation (although it may affect link performance if the terminal equipment is not waterproof!).

10.1.5 SHF and Above

One of the key differences between V/UHF propagation and propagation at SHF and above is that the effects of attenuation due to the contents of the atmosphere along the link need to be taken into account. Figure 10.13 shows an approximation for gaseous

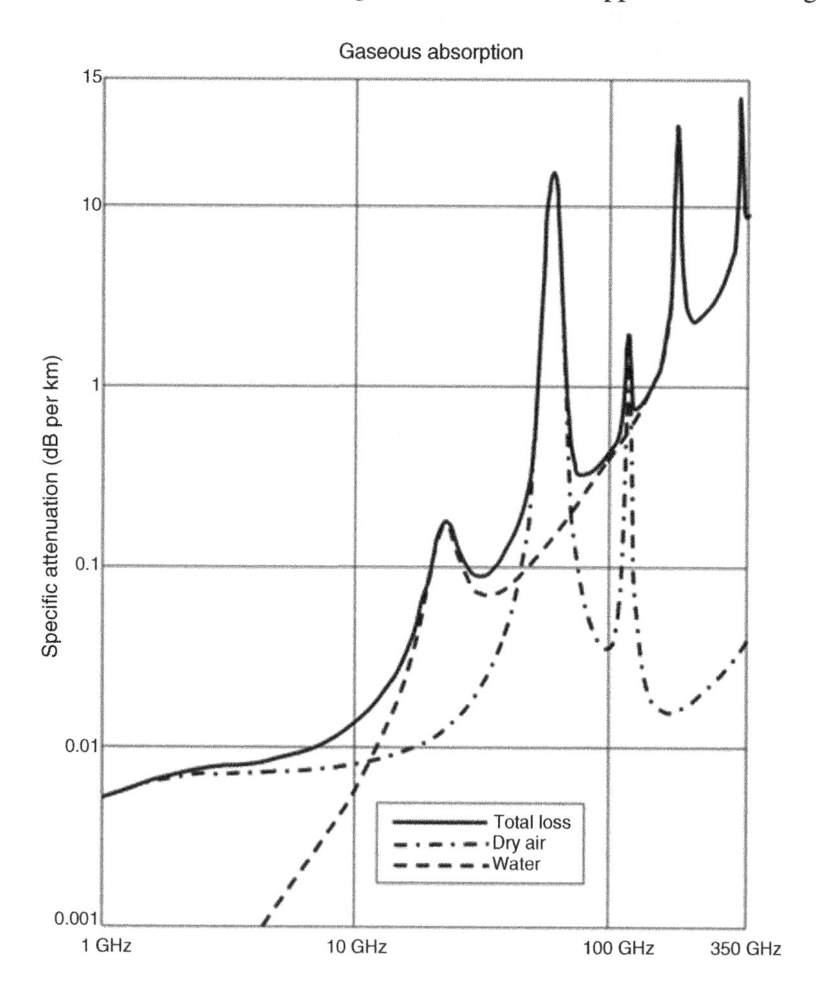

Figure 10.13 Approximation to gaseous absorption for the frequency range 1 GHz to 350 GHz in dB per km, based on a temperature of 15C, pressure of 1013 hPa and 7.5 g/m^3 of water vapour content.

absorption per kilometre over the range 1–350 GHz, for sea level paths (see ITU-R P.676 for more accurate methods of determining the actual values). The thick line shows total attenuation and the two dashed lines show the contribution due to dry air and water particles suspended within the air.

For example, the attenuation at 20 GHz, the attenuation per km is approximately 0.3 dB. Thus for a 10 km communications link, the loss along the link would be 3 dB, or for a radar system would be twice this to account for the send and receive paths, i.e. 6 dB.

Atmospheric attenuation is always present, but other effects are intermittently present such as rainfall, fog, ice and snow. Figure 10.14 shows some attenuation figures for different frequencies, polarisation and rainfall rate.

Rain, particularly heavy rain, is most often not present along the entire length of a link. Heavy rain regions are referred to as 'cells'. If, for example, there is a 5-km long cell of rain falling at the rate of 45 mm/h in the direct path of a link using 20 GHz with vertical polarisation then the attenuation per kilometre can be read from the graph to be approximately 4 dB per km. Thus the total attenuation for the cell is $5 \times 4 = 20$ dB. For a radar link in which the cell is in the path of both the transmit and receive paths, the total additional attenuation would be twice this; 40 dB. Similar calculations can be carried out for different types of precipitation. Rainfall and other precipitation may also affect the temperature and pressure at the altitudes between the clouds and the ground. This can lead to localised variations in the vertical refractive index, which will also influence propagation.

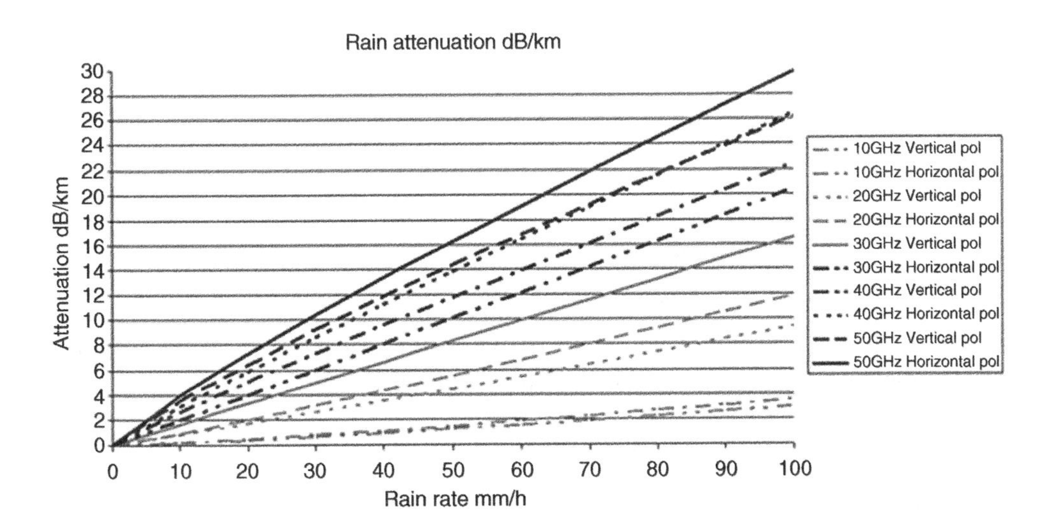

Figure 10.14 Graph of rain rate against attenuation for different frequencies and polarization.

10.2 Modelling Methods

Propagation modelling is an essential part of modern radio engineering, as well as being very useful for mission planning and design of radio networks. These models are used in computer-based predictions to assess expected system performance. The history of radio propagation prediction dates from as far back as the 1940s if not before, but it is only since the advent of advanced computers that their use has become indispensable.

Propagation models can be said to fall into one of two categories, or a combination of the two. The first is termed 'deterministic' and the second 'empirical'.

Deterministic models are based on an understanding of the underlying physics of radiowave prediction. We have seen the components of deterministic models when we have looked at free space loss and diffraction modelling using the Fresnel integral. Deterministic models are very useful for representing situations where all of the propagation factors are well understood. However, if the model omits some critical features, then the results may be inaccurate.

Empirical models are based on measurements taken at specific frequencies in particular environments. They have advantages in that if the measurements taken are highly representative of the scenario in which the real network will work, then the model will take into account all of the relevant parameters, even if they are not identified during the analysis. The danger is that the model may be inappropriately applied to a different scenario that does not replicate the original conditions for which the model was derived.

It can be seen that selection of appropriate models for given scenarios is important. We will now look at some models and their features. We will also look at some models that are combinations of both deterministic and empirical approaches.

We will also consider another way of describing models, which are 'site-specific' (aka point-to-point) models and 'site-general' (aka point-to-area) models. Both of these have application at different points in the design and deployment of radio systems.

10.3 Deterministic Models

10.3.1 Free Space Loss Model

The loss model for free space loss is easily derived from basic physics. The logarithmic form of the model using the most commonly used units are:

$$L = 32.44 + 20\log(d) + 20\log(f)$$

Where distance is in kilometres and frequency is in MHz.

In previous parts of the book, we have noted that free space loss is not often a good guide to the loss incurred in real links. However, there is one set of circumstances

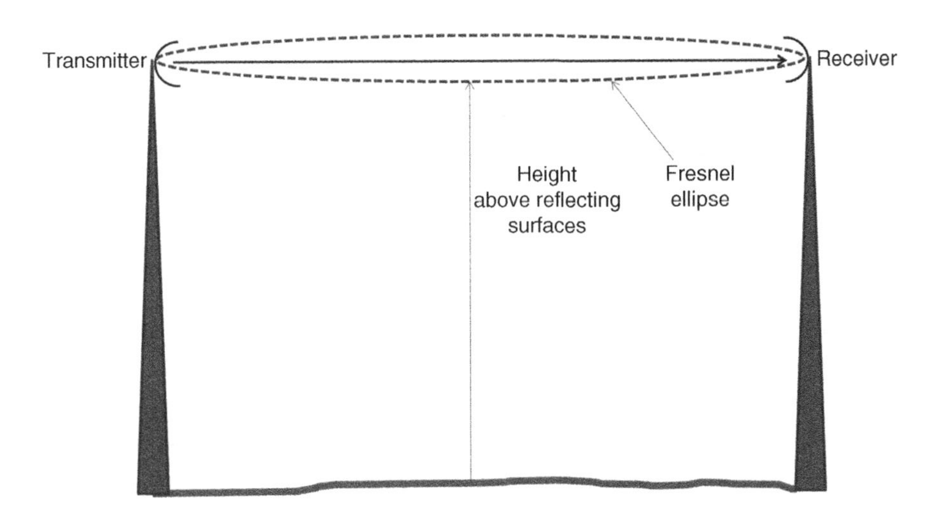

Figure 10.15 Free space loss can be used where both the transmitting and receiving antennas are highly directional, pointed directly at each other and well clear of reflecting surfaces or refraction regions.

where free space loss can be used. This is illustrated in Figure 10.15. This is where there are two highly-directional antennas well clear of any reflective surfaces or refractive regions where energy may be bent towards the receiver from off the main beam.

In the figure, the antennas are mounted on high towers, but in practice, they can both be on aircraft. Note that a ground to air link may not fall into this category since the ground terminal may be subject to ground reflections close to the antenna. To determine how well clear of the ground the antennas are, we can use the concept of the Fresnel ellipse. This is distinct from the Fresnel integral previously discussed. The Fresnel ellipse is a mathematical construction that describes loci of points around a central line, in this case the direct (radio) line of sight path.

The radius of Fresnel ellipses can be determined from the following formula:

$$R_n = 500\left[\frac{nd_1d_2}{f(d_1+d_2)}\right]^{\frac{1}{2}}$$

where

R_n = Fresnel radius at a point along the link.

n = Number of Fresnel ellipse.

d_1 = Distance from one terminal to the point where the radius is calculated.

d_2 = Distance from calculation point to the other terminal.

f = Frequency in MHz.

The most important Fresnel ellipse number is 1. The reason for this is that most energy transmitted from a transmitter to a receiver is contained within this region. If there are no reflecting, scattering, refracting, diffracting or absorbing elements within this region, then free space loss is a good approximation of the link loss.

10.3.2 Two-Ray Models

In most links, energy will be arriving at a receiver from a number of directions, but the main reflection is from the ground. This is illustrated in Figure 10.16.

The energy reflected from the ground will have a phase difference compared to the direct wave. For a perfect reflector, the phase of the incident wave will be inverted during reflection. At the receiver the energy from the direct and reflected waves will be vectorially added. This is illustrated in Figure 10.17.

Actual reflections will not be perfect, but instead are affected by ground conductivity and relative permittivity. In this case, the reflection coefficient can be determined from:

$$\rho = \frac{(\varepsilon_r - jx)sin\,\psi - \sqrt{(\varepsilon_r - jx)cos^2\psi}}{(\varepsilon_r - jx)sin\,\psi + \sqrt{(\varepsilon_r - jx)cos^2\psi}}$$

where

$$x = \frac{18 \cdot 10^3 \cdot \sigma}{f}$$

$$\psi = tan^{-1}\left(\frac{h_{tx}}{d_{tx}}\right)$$

h_{tx} = Height of transmitting antenna (m).
d_{tx} = Distance to reflection point (m).
σ = Conductivity (Siemens).
f = Frequency (MHz).

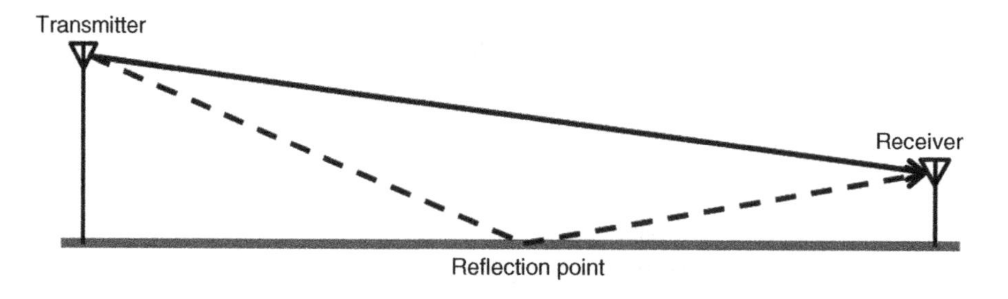

Figure 10.16 The principle of the two ray model.

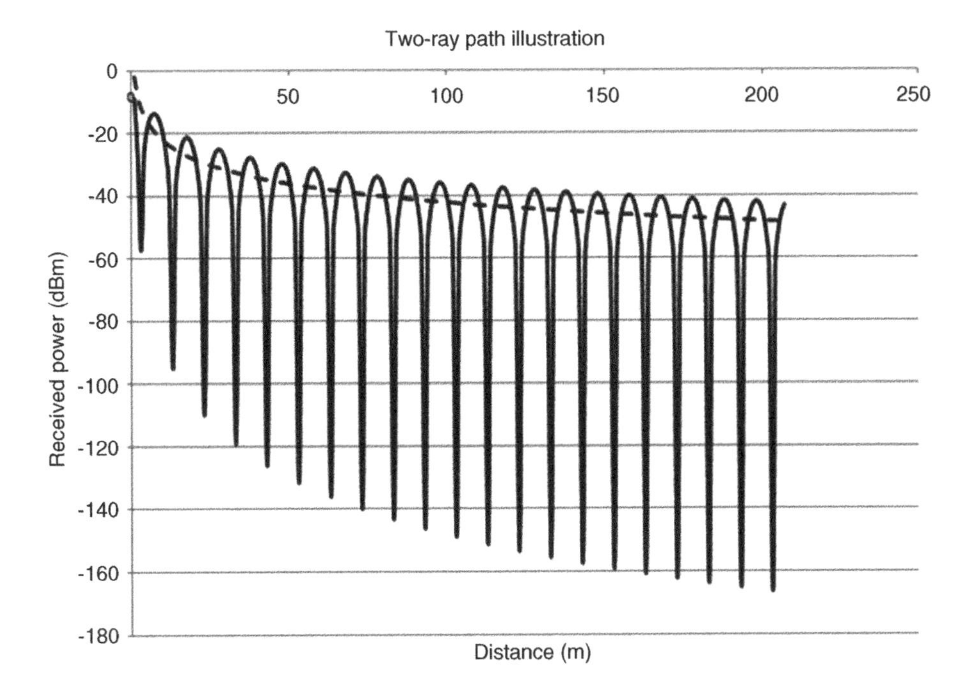

Figure 10.17 An illustration showing the form of received signal with distance due to a perfect ground reflector.

The reflection coefficient is normally near -1 in most practical scenarios. Figure 10.18 shows a graph for the reflection coefficient for two antennas at 10 metres above flat terrain. This is only for short ranges, but even so the value is between -0.9 and -1.0.

The received electric field strength can then be determined for two isotropic antennas by:

$$E = \left[\frac{300}{4\pi fd}\right]^2 \cdot \left\langle 1 + \rho\, exp\left(1 - j\frac{2\pi}{\lambda}\Delta R\right)\right\rangle$$

where

$$\Delta R = \frac{2 \cdot h_t \cdot h_r}{d}$$

h_t = Transmit antenna height (m).
h_r = Receive antenna height (m).
λ = Wavelength in metres.

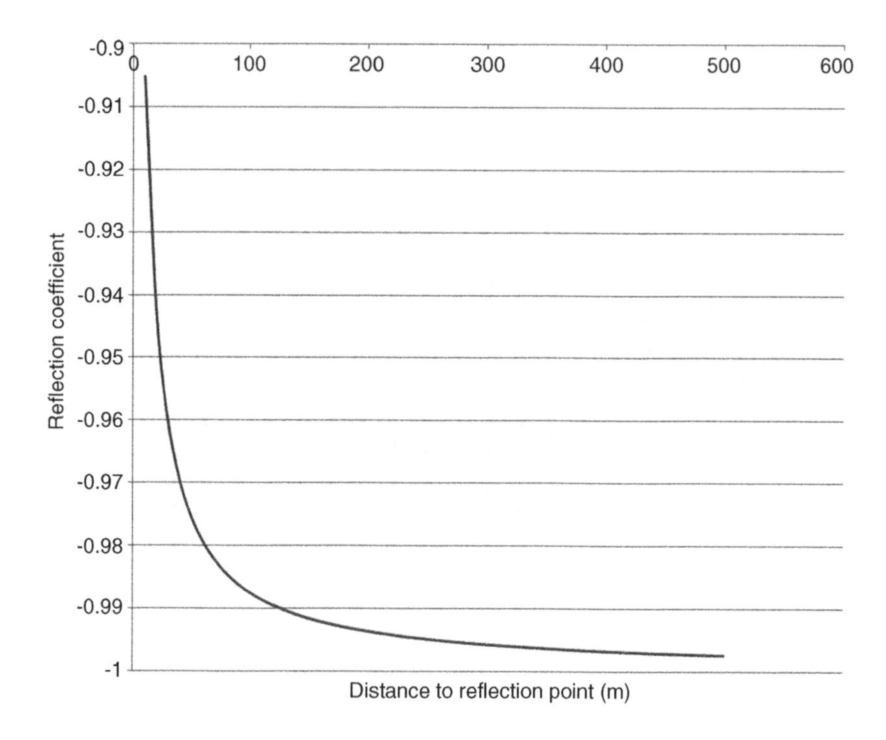

Figure 10.18 Reflection coefficient for two antennas at 10 metres height above flat terrain.

The form of the result is similar to that of Figure 10.17. The importance of the two ray model is that is explains the behaviour of real-life signals. Deep fades due to ground reflections may occur even at short ranges, confirming that 100% coverage from a radio system is impossible in practice.

10.3.3 Common Deterministic Models

10.3.3.1 ITU-R P.526

The ITU 526 model is applicable to a wide range of applications and can form the core model for composite models capable of working from about 20 MHz up to hundreds of gigahertz. The model includes:

- free space loss;
- sub-path losses, which are due to intrusions into the Fresnel ellipse;
- a number of diffraction models, based on single and multiple diffraction edges and including corrections for rounded obstructions;
- a Unified Theory of Diffraction (UTD) model for wedge shaped obstructions such as rooftops.

Without additional correction, the model is suitable for applications clear of ground clutter up to about 6 GHz. Beyond that, additional correction factors are needed.

10.3.3.2 The APM Model

It is worth considering the Advanced Propagation Model (APM) used in the AREPS program. This model uses the concept of a number of regimes as illustrated in Figure 10.19. This is derived from Barrios (2003) who writes about implementation of the APM.

The APM model, unlike the others discussed so far, allows for non-homogenous atmospheric conditions, so it can model surface, evaporation and elevated ducts and other atmospheric effects. It also includes the effects of terrain and multipath interference due to both reflections from the ground and energy arriving at the receiver from refraction regions. Sea conditions are modelled to improve the

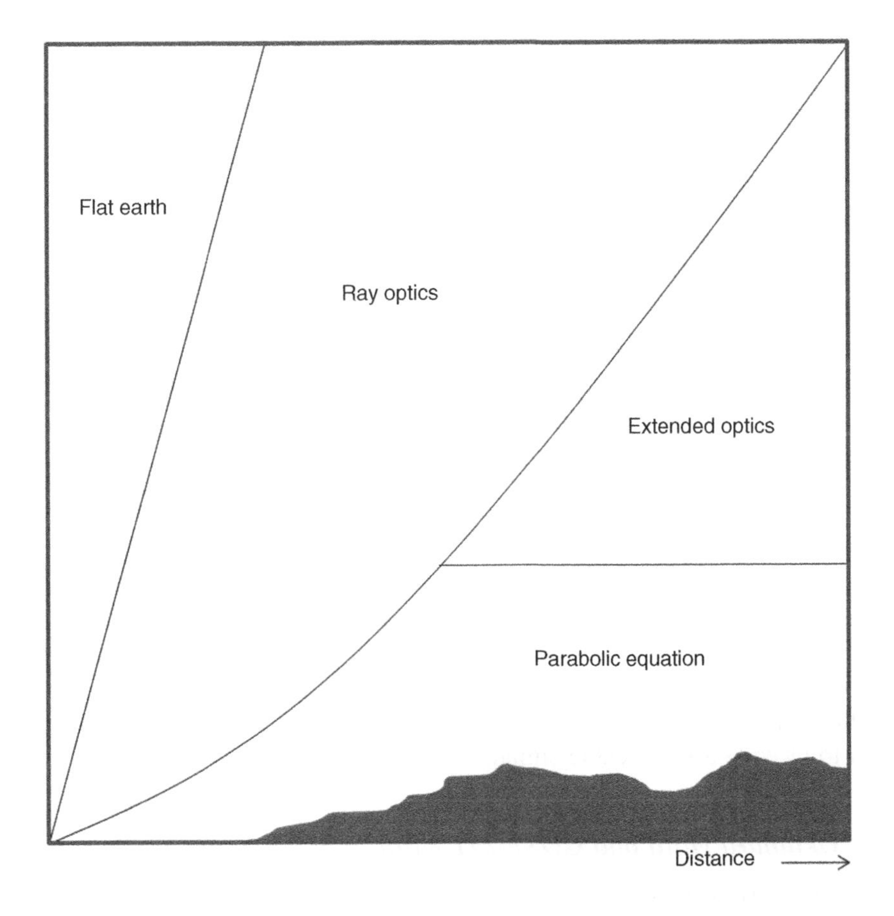

Figure 10.19 The propagation regimes of the APM model, derived from Barrios (2003).

accuracy of reflections from over sea paths. Diffraction in the APM model is calculated using the parabolic equation rather than the Fresnel method used by other models. This is effectively a finite element analysis technique rather than calculating diffraction over the most dominant diffraction over edges. It also includes Tropospheric scatter for paths beyond the diffraction region. The current version of APM does not however account for rainfall and other precipitation or vegetation and urbanisation clutter.

10.4 Empirical Models

10.4.1 Basic Form of Empirical Models

Empirically derived models are generated using data gathered from measurements rather than being based on application of physical models to the scenario under test. The basic form of an empirical model is:

$$E_r = -\gamma \cdot \log(10) + K(P_{BS}, f, h_{BS}, h_{MS})$$

where
E_r is the received field strength in dBμV/m.
d is the path length in km.
γ is the inverse exponent used with path length.
f is the frequency in MHz.
K is a coefficient based on power (PBS, normally in dBW or dBm).
h_{BS} is the height of the base station in metres.
h_{MS} is the height of the mobile station in metres.

The received power in dBm can be calculated for an isotropic receiver with an impedance of 50 Ω by:

$$P(\text{dBM}) = \textit{Field strength}\,(\text{dB}\mu\text{V/m}) - 20\log f - 77.2$$

Other corrections can be added for clutter and other factors relevant to specific applications.

The form of this type of equation shows that it does not require detailed path profile analysis. Rather, it produces a generic result useful for network dimensioning and where the margin added is used as a mechanism to ensure the received signal is strong enough to provide a serviceable signal.

10.4.2 Okumura Hata and COST 231 Hata

The Okumura-Hata model was introduced to model mobile phone applications, where the base station antenna is clear of clutter but the mobile station is embedded within it. It was principally an urban model, but corrections were added for sub-urban and rural environments. The initial model was based on measurements between 150 and 1500 MHz, but the COST 231 model extended this to 2000 MHz to account for newer mobile phone frequencies. The original loss equation for urban environments is:

$$L = 69.55 + 26.16\log_{10}f - 13.82log_e h_{BS} - a(h_{MS}) + (44.9 - 6.55log_e h_{BS})\log_{10}d$$

where
f is the frequency in MHz.
a is a coefficient for the height of the mobile station.
h_{BS} is the height of the base station.
h_{MS} is the height of the mobile.
d is distance in km.

Again, this model is terrain independent and provides a generic model to assess relatively short-range results. It works out the horizon, but some have added diffraction modelling to extend its range.

As well as mobile phone applications, the model can be used for other applications that closely approximate the conditions for which the model was originally developed. This includes assessing the performance of mobile phones used to trigger RCIEDs and the jammers designed to counter them.

10.4.3 ITU-R P.1546

The ITU-R P.1546 model is an extension of the ITU-R P.370 model, which has been used for years to assess long-range interference including cross-border applications to assess interference into other countries. Both of these models are intended for interference assessment rather than detailed network design, and they are both of the same form with the 1546 model including more data input, including clutter options.

The models are intended for ranges of 10–10 000 km for transmitters between 37.5 and 1200 metres, and for receiving antennas of 10 metres above height although corrections can be added to modify receiver antenna heights. The models were derived from measurements taken over long periods and as such, they take into account not only normal propagation conditions but also variations around the median loss values. The loss values are presented in graphical format that can be digitised for computer calculations. These graphs describe median loss for 50% of locations and time percentages of 50, 10 and 1% of the time, with losses lower for the smaller time percentages.

Some reviewers have suggested that 1546 is in error in some circumstances; however, it is in wide use.

10.4.4 ITU-528 Aeronautical Model

The ITU-528 aeronautical model is a subset of the IF-77 model (also known as the Johnson-Gerhardt model). Both models provide loss values for aeronautical applications with the antennas being between 0.5 and 20 000 metres in altitude. The receiving antenna is always higher than the transmitting antenna, which is assumed to be a ground terminal or lower altitude platform. They cover frequency ranges of between 100 MHz and 20 GHz and ranges up to 1800 km. The typical form of the 528 model curves are shown in Figure 10.20.

The IF-77 model is a computer algorithm that produces outputs similar to the 528 curves but it also includes effects such as surface roughness and reflections, climactic variations and the presence of storm cells (a storm cell is a region of high rainfall).

10.4.5 Clutter and Absorption Modelling

Radio clutter is the presence of obstructions and radio noise generators typically found in urban environments.

Clutter can be included in modelling programs by means of an additional loss value above that experienced due to other propagation factors. The method of dealing for this

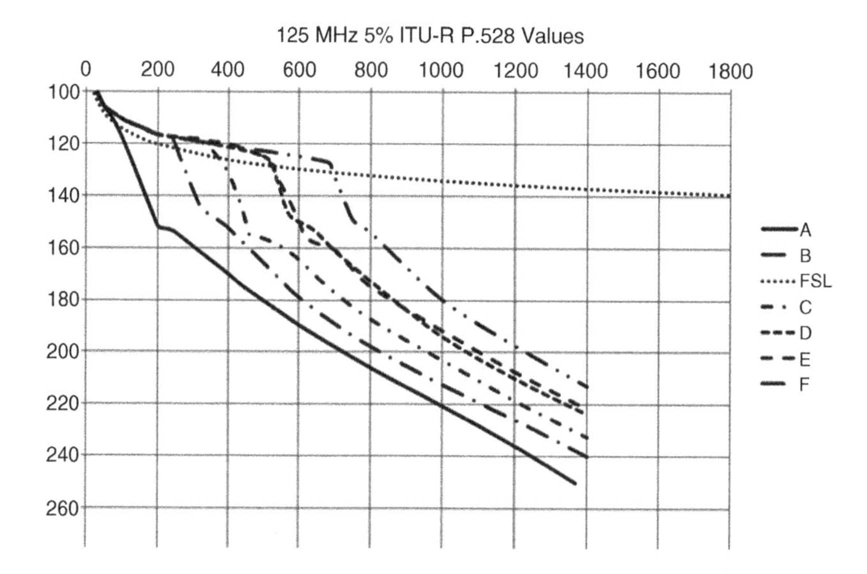

Figure 10.20 A representation of part of the ITU-528 model, showing the form of the curves.

is described in greater detail in Section 11.10, but it is important to recognise that clutter must be included in terrestrial modelling.

Absorption modelling requires understanding of the absorbing material. It is most often used for in-building predictions, where one or more terminals are in a building or other absorbing structure.

10.4.6 Fading Characteristics

The propagation models described in this chapter all report the median field strength received within a short-sector. By default, they do not include the effects of fading due to small, nearby obstructions. This means that fading must normally be applied to the results of any propagation model unless it is specified that this is not necessary.

10.5 Combined Models

10.5.1 Combining Propagation Models

It is possible to combine radio prediction models in some circumstances to improve modelling accuracy. For example, clutter models can be combined with models that do not include their effects such as ITU-R P.526. It is also possible to add other effects such as atmospheric absorption for frequencies above 6 GHz, and rainfall to determine link reliability over a protracted period. It is also possible to combine deterministic, empirical and statistical models in some cases.

As an illustration of this, we will next look at the ITM model to see how deterministic modelling techniques are applied to non-deterministically described propagation paths.

10.5.2 The ITM Model

The Irregular Terrain Model (ITM), also known as the Longley-Rice model, is used is also applicable for radio transmissions of 20 MHz and above, up to 40 GHz. Like the ITU-526 model, this produces the median loss value to the resolution of the short sector. It uses a two-ray model within line of sight and a diffraction model for non-line of sight conditions. It is applicable for antenna heights between 0.5 and 3000 metres and distances of 1 to 2000 km. As such, it is a longer range model than the ITU-526 model. It, like the ITU-526 model, can be used for different surface refractivity, in this case between 250 and 400 N-units, assuming normal vertical characteristics, although these can be accounted for by empirical corrections. Unlike the ITU-526 model, the ITM model can work with fragmentary terrain data, expressed by the variability of terrain height, termed Δh. The form of this can be derived from Table 10.2.

The ITM model also includes the effects of ground conductivity to account for the reflection coefficient used in the two-ray model. It also accounts for antenna location

Table 10.2 Delta-h values for different types of terrain

Terrain Type	Typical Δh
Water or very smooth plains	0–5
Smooth plains	5–20
Slightly rolling plains	20–40
Rolling plains	40–80
Hills	80–150
Mountains	150–300
Rugged mountains	300–700
Very rugged mountains	>700

selection, with models for tactical systems sited randomly and fixed links where the antenna location is more carefully selected. This is used to assess the horizon between terminals. Clutter is taken into account where is known. All of these are used to compute the excess loss over the free space loss condition. Although the model reports the median loss, variations due to fading and time variations are also taken into account. This accounts for short- and long-term variations.

The basic ITM model is deterministic in its handling of diffraction loss and the two-ray model, but since it relies on path approximations rather than explicit path profile analysis, it is a combined model rather than either being deterministic or empirical in nature.

10.5.3 Combination Checklist

So when can models be combined and when can they not? The basic rules are that the models must cover the same frequency band, including the frequency to be modelled. They do not need to be equally applicable, for example, it is possible to use clutter values appropriate for 30–200 MHz with the ITU-R P.526 model, which covers 30 MHz to over 50 GHz – but this combination is only valid over the range 30–200 MHz. Also, the models must be applicable to the same scenarios. So, for example, combining the ITU-R P.528 aeronautical model with a clutter model would not normally be appropriate. Additionally, it is vital to ensure that the models do not both account for the same mechanism, otherwise the effects will be double-counted.

A simple checklist is shown in Table 10.3.

10.6 Link Budgets

Link budgets are used to determine the maximum allowable loss to provide a given level of service. The method of creating a link budget always follows the same process, but there are differences according to the type of system. This section describes a number of different versions of link budgets for different applications and

Table 10.3 A simple checklist for model combination

Factor	Considerations
Frequency range	Both models must be applicable to the frequency range to be modelled
Applicability	Both models must be applicable to the type of link being modelled
Double-counting	Each propagation mechanism must be handled by only one of the models used
Validity	The models must be valid for type of prediction being carried out
Prediction mechanism	The models must work in complimentary methods, so that, for example, they both produce answers for a 50 metre location. If one produces a result for 50 metre points and the other produces results for 500 metres, they may not be compatible
Commonality of units, results	In addition to the above, they must provide answers in the same type of units, or be readily convertible from one to the other. Also, if one model produces a 50% median value and the other does not, they may not be compatible
Added value	Both models must add some value to the accuracy of the combined result, otherwise combining them is pointless

technologies. The link budgets shown can be used as templates, but it is always important to consider whether the link budgets shown are ideal for your application or need tweaking for your particular scenario.

Link budgets are relatively straightforward. The basic principle is to determine the power transmitted in the direction of the receiver, the signal level needed at the receive antenna and the losses calculated by the prediction model. Alternatively, the link budget can be used to determine the maximum allowable losses between two points in the system (see Figure 1.7).

The only major source of confusion in link budget design is that of receiver sensitivity. This arises because sometimes the sensitivity quoted is for an unfaded signal, and sometimes for a faded signal. If the sensitivity figure is unfaded, then fading must be accounted for in the link budget separately. If it is faded, then additional fading should not be added unless the application differs from the scenario for which the faded sensitivity has been designed. Designers should identify the circumstances for which the receiver sensitivity has been derived before building the link budget.

Another critical aspect is to ensure that all the parameters are in consistent units. This means in general that power in Watts or dBW must be converted to dBm before putting them in the link budget to avoid potential confusion later. In all the link budget examples that follow, the units have been converted into compatible units before entering them in the link budget.

10.6.1 Generic Analog Link Budgets

A simple link budget is shown in Table 10.4. The antenna gain is shown for the gain in the direction of the receiver. For directional antennas, this will change over azimuth and/or elevation angles.

Table 10.4 A simple generic analog link budget

System element	Sample value	Typical units
Transmitter side components		
Transmitter output power	30	dBm
Amplifier	10	dB
Feeder losses	2.0	dB
Connector losses	1.0	dB
Antenna gain	3.7	dBi
EIRP	***40.7***	***dBi***
Receiver side components		
Antenna Gain	2.3	dBi
Feeder losses	1.0	dB
Connector losses	1.0	dB
Receiver sensitivity (faded) for 12 dB SINAD	– 100	dBm
Minimum required input signal @ antenna	***– 100.3***	***dBi***
Maximum tolerable loss	**141.0**	**dB**

Table 10.4 has been designed to determine the maximum allowable loss between the two antennas in the faded condition. It can be re-designed as shown in Table 10.5 to determine nominal receive level if the path loss between the two antennas is known.

Table 10.5 Link budget for a particular link where path loss is known

System element	Sample value	Typical units
Transmitter side components		
Transmitter output power	30	dBm
Amplifier	10	dB
Feeder losses	2.0	dB
Connector losses	1.0	dB
Antenna gain	3.7	dBi
EIRP	***40.7***	***dBi***
Path loss	129	dB
Receiver side components		
Received signal at antenna	– 88.3	dBmi
Antenna Gain	2.3	dBi
Feeder losses	1.0	dB
Connector losses	1.0	dB
Received nominal signal power @ antenna	***– 88.6***	***dBm***

Table 10.6 A simple generic digital link budget

System element	Sample value	Typical units
Transmitter side components		
Transmitter output power	44.0	dBm
Feeder losses	2.0	dB
Connector losses	1.0	dB
Antenna gain	4.5	dBi
EIRP	***45.5***	***dBi***
Receiver side components		
Body loss (head level)	4.5	dB
Antenna Gain	– 2.0	dBi
Feeder losses	0.0	dB
Connector losses	0.0	dB
Receiver sensitivity (faded) for 4% BER	– 112	dBm
Minimum required input signal @ antenna	***– 105.5***	***dBi***
Maximum tolerable loss	**151**	**dB**

10.6.2 Generic Digital Link Budgets

Digital link budgets for technologies such as PMR, digital combat net radio and other systems that do not use processing gain are fairly similar to analog link budgets, except that the receiver sensitivity is defined for a given Bit Error Rate (BER). The sensitivity can also be expressed as energy per bit over noise, but again this is referred back to a specific BER.

Table 10.6 shows a link budget for a TETRA system. In this case, typical helical antenna losses are included as a negative gain, and the effect of body loss due to the antenna being held at head height.

10.6.3 Spread-Spectrum Link Budgets

Spread spectrum systems have two major differences between other digital systems. The first is that the processing gain due to the spreading code needs to be included. This is the logarithmic ratio of the spread bandwidth compared to the signal baseband bandwidth. In the example shown in Table 10.6, this is:

$$Processing\ Gain\ (dB) = \frac{Spreading\ Bandwidth}{Signal\ Bandwidth} = 10\,Log\left(\frac{3840\ \text{kHz}}{12.2\ \text{kHz}}\right) = 25.0\ dB$$

The second is that all signals in the system share the same frequency, and that each signal behaves like noise to each other. This means that when the more concurrent calls there are, the higher the noise that the system has to overcome to successfully pass one signal. In general, these signals will be de-correlated and so the power

Table 10.7 Example spread-spectrum link budget, based on an UMTS example

Mobile Transmitter		
Mobile maximum Power	21	dBm
Body loss/antenna loss	1.8	dBm
EIRP	**19.2**	**dBm**
BTS Receiver		
BTS Noise Density	– 168	dBm/Hz
Rx Noise Power (bandwidth correction)	– 102.2	dBm/3.84 MHz
Interference Margin	3	dB
Rx Interference Power	– 102.2	dBm
Noise + Interference (power sum)	– 99.2	dBm
Process Gain for 12.2 kHz Voice	25	dB
E_b/N_o for Speech	5	dB
BTS Antenna Gain	15	dBi
Cable Losses	2	dB
Connector Losses	3	dB
Fast Fading Margin (slow mobile)	4	dB
Rx Sensitivity	**– 125**	**dBm**
–	–	–
Maximum Path Loss	**144.2**	**dB**

sum method can be used to determine their composite effects. To be strictly accurate, the basic noise floor must also be included in this calculation although if there are many over users, the noise will be smaller than the noise energy added by the other calls in the system. Once the composite noise energy has been calculated, the interference margin above which the wanted signal must be can be added. In Table 10.7, the system has no interfering calls in order to determine the maximum loss the system can accept for a single call.

In the table, the interference power is the background noise power alone. However, if noise from other calls is present, then the noise value changes, as per Section 5.2.5.2. If for example, the additional interference plus noise contribution is 4.5 dB, then from Table 10.6 the value of – 99.2 would be increased to – 94.7 dBm. The processing gain still applies as do all of the other values following it in the table. The effect is that the receiver sensitivity is increased by 4.5 dB to – 120.5 dBm and the maximum allowable loss is reduced to 139.7 dB. Since calls in the network will vary continuously, this value also changes continuously and the maximum effective range of a particular cell covered by a BTS will also vary. This is called ‘cell breathing’.

10.6.4 Radar Primary Radar Link Budgets

Radar performance can be handled in exactly the same way as for communications systems by developing similar link budgets. The system can be considered in both directions, or can be split into the transmit and receive paths. The radar range equation

Table 10.8 Example (simple) primary radar link budget

Radar Transmitter		
Radar average output power	0	dBW
Feeder losses	3	dB
Antenna Gain	34.5	dBi
EIRP	**31.5**	**dBW**
Target Data		
Target strength[a]	10	dB
Radar Receiver		
Antenna Gain	34.5	dBi
Feeder losses	3	dB
Bandwidth	1	MHz
kTBF assuming T = 290 K and F = 6 dB[b]	– 132.7	dBW
S/N for given number of pulses and POD[c]	10	dB
Required Signal Strength @ antenna	– 154.2	dBW
Maximum loss inc. TS	185.7	dB
Maximum loss without TS	195.7	dB
$F_r + F_t$, where $F_r = F_t = 6$ dB[d]	12	dB
Equivalent FSL[e]	183.7	dB
Max range for f = 3400 MHz[f]	**107**	**km**

[a] Assumed nominal target strength.
[b] F is the noise factor; a value of 6 dB has been chosen in this case.
[c] This will vary for the given number of pulses that hit the target during a sweep and the desired probability of detection.
[d] This is the propagation factor, which is a reduction from free space loss in dB. It is the same for transmit and receive paths.
[e] This is one way of correcting for the propagation factor. It converts the result into an equivalent value that can be solved using free space loss.
[f] By assuming the frequency used, we can then solve for the range using the free space loss formula, but remembering that there are two paths, hence we need to solve for d^4 in km.

can be used to determine the maximum allowable loss, and this will depend on the type of radar and its application. The example shown in Table 10.8 is therefore only one way of determining this.

Link budgets can be used to solve for different values. For example, the illustration shown in Table 10.8 could be solved for the minimum target strength that is required to ensure a particular range, of for the maximum frequency that can be used.

10.6.5 HF Link Budgets

HF links can be calculated in exactly the same way as for VHF and above links. The only difference is that to ensure that a link works over the entire time it is to be established, it is necessary to calculate the links budgets for each frequency to be used.

References and Further Reading

COST 231, http://www.lx.it.pt/cost231/final_report.htm.

Graham, A.W.; Kirkman, N.C.; Paul, P.M. (2007), *Mobile Radio Networks Design in the VHF and UHF Bands: A Practical Approach,* John Wiley & Sons ISBN 0-470-02980-3.

Recommendation ITU-R P.453: The Radio Refractive Index its Formula and Refractivity Data.

Recommendation ITU-R P.525: Calculation of Free-Space Attenuation.

Recommendation ITU-R P.526: Propagation by Diffraction.

Recommendation ITU-R P.528: Propagation Curves for Aeronautical Mobile and Radionavigation Services using the VHF, UHF and SHF Bands.

Recommendation ITU-R P.676: Attenuation by Atmospheric Gases.

Recommendation ITU-R P.1406: Propagation Effects Relating to Terrestrial Land Mobile Services in the VHF and UHF Bands.

Recommendation ITU-R P.1812: A Path-Specific Propagation Prediction Method For Point-to-Area Terrestrial Services in the VHF and UHF Bands.

Skolnik, 2008 Skolnik, M. (2008), *Radar Handbook,* McGraw-Hill, USA, ISBN 978-0-07-148547-0.

11

Data Requirements for Radio Prediction

11.1 Why Consider Modelling Requirements?

While it is possible to design radio links manually, in most cases the design of radio frequency links and networks is achieved using special software network planning tools. These allow the planner to design and optimise the system prior to deployment with a reasonable expectation of predicting performance in advance. This is far more efficient and cost-effective than traditional manual methods.

In order to perform such simulations, it is necessary to model three types of information:

- the radio transmitters and receivers;
- the radio propagation between transmitters and receivers;
- the environmental factors that influence radio propagation.

This is illustrated in Figure 11.1, which shows the factors involved in radio prediction.

For the radio equipment, we need to model those factors that influence the link budget all the way from output of the transmitter to the input to the receiver. Radio propagation modelling is achieved using mathematical algorithms for the frequency band and application. The environmental factors to be modelled depend on the frequency band and the link topology. For HF skywave, this will be the Ionospheric characteristics and the receiver environment, and for VHF and above, both atmospheric, terrain and terrain clutter are important.

These factors are described in this chapter.

Communications, Radar and Electronic Warfare Adrian Graham

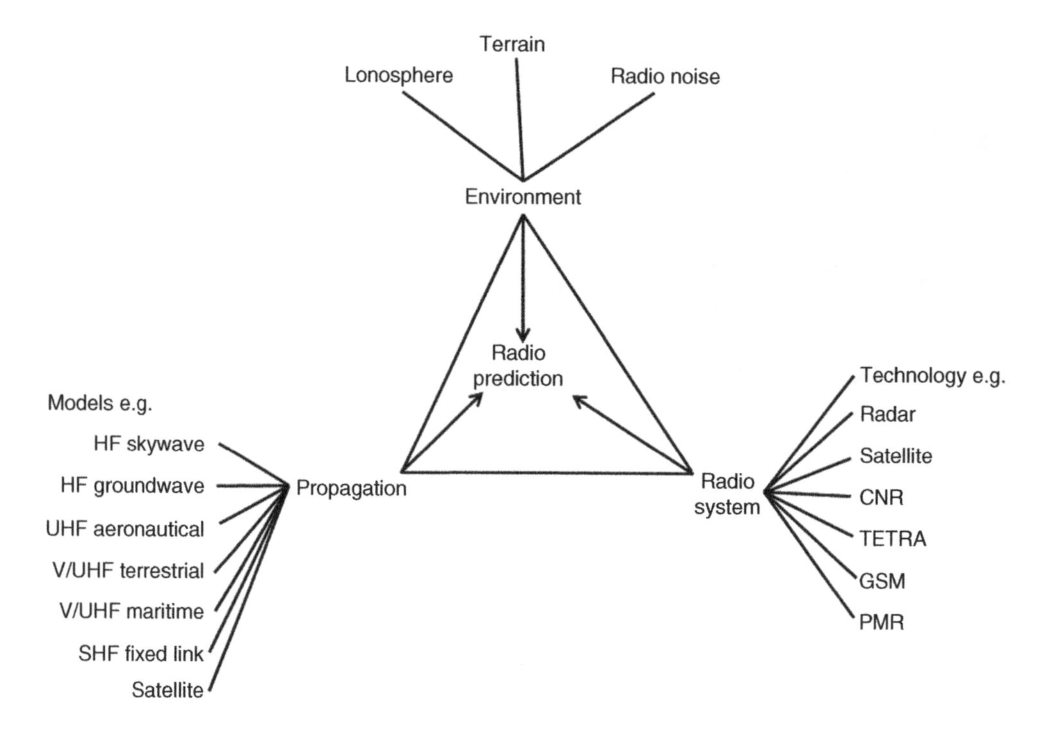

Figure 11.1 Some of the key factors necessary for radio prediction.

11.2 Communications System Parameters

Table 11.1 shows a list of important characteristics of radio transceivers. In this case, the antenna characteristics of the installation are included. The detailed technical characteristics of the antenna polar patterns are described in greater detail in Section 11.3. To carry out uplink and downlink calculations for bi-directional systems, these need to be known for both directions. The table includes a lot of parameters. Not all are necessary for specific predictions, such as coverage, but the combination provides the ability to model most important aspects of radio system performance.

The factors shown in Table 11.1 must be determined for the specific mode and settings used for a particular link. Many radios support different modes of operation, such as voice/encrypted voice, low data rate, high data rate and adaptive modulation. In such cases, it is essential to choose the relevant values used for a particular application.

For radio prediction work, the modulation used to encode the data is useful to know, because from that the Bit Error Rate (BER) for a given value of E_b/N_o can be calculated. However, if the sensitivity is known for a given value of target performance, then this value can be used and calculation is not necessary.

Table 11.1 Communications parameters necessary for radio prediction

Characteristic	Technology	Comments
Tx Output power	All	Important to understand the type of power and the units used. This often varies with the technology and the normal method of expressing the power, e.g. peak power, continuous etc in dBW, dBm, W, kW etc For some technologies, power may be dynamic (i.e. variable)
Tx Frequency	All	Centre or reference frequency in MHz or GHz
Tx Bandwidth	All	The bandwidth can be expressed as a single figure or it can be expressed as part of the power spectral density
Tx Feeder losses	All	Losses due to transmission lines between the radio output and the antenna input. Expressed in dB
Tx Connector losses	All	Losses due to transmission line connectors. Expressed in dB
Tx Amplifier gain	All	Gain due to any amplifiers in the Tx system. Expressed in dB
Combiner loss	Systems with combiners/ splitters	Losses due to combining or splitting the RF power before sending it to antennas. Expressed in dB
Tx Antenna gain/loss	All	Gain or loss of transmit antenna(s), normally reference either an isotropic (dBi) or dipole antenna (dBd)
Tx antenna polar pattern	Directional systems	Directional performance of antenna (see next section)
Tx antenna gain response	All	Frequency gain response of the antenna. Normally expressed in dB down on the highest gain response
Tx Antenna height/altitude	All	Height of the centre of the radiation pattern above local ground or altitude above sea level. Usually expressed in metres or feet
Tx Antenna azimuth	Directional systems	For directional antennas, the direction of the main beam in degrees or mills reference grid or magnetic north
Tx Antenna tilt	All	The electrical or mechanical vertical tilt of the main beam of the antenna. Usually expressed in degrees or mills
Power spectral density	All	The spectral shape of the transmitted energy. Can be used for interference analysis between dissimilar systems. Normally expressed as dB down from the main power
Tx antenna Polarisation	All	The polarity of the transmitted signal. Has an effect on propagation and also on interference
Time slots	TDMA, frequency hoppers	For time based systems, the slot structure of the transmitted signal. Can be used for interference analysis. Normally expressed in microseconds, milliseconds or seconds

(continued)

Table 11.1 (*Continued*)

Characteristic	Technology	Comments
Activity ratio	All	The ratio of activity when the system is transmitting. Again, used for interference analysis. Expressed as a percentage or in a range 0–1
Modulation scheme	All	The encoding scheme used to modulate the baseband signal into the carrier. Has an effect on receiver sensitivity
Diversity gain	Diversity systems	Gain due to the use of diversity antennas. Expressed in dB
Rx antenna height/altitude	All	Height of the centre of the radiation pattern above local ground or altitude above sea level. Expressed in metres or feet
Rx antenna azimuth	Directional systems	For directional antennas, the direction of the main beam in degrees or mills reference grid or magnetic north
Rx antenna tilt	All	The electrical or mechanical vertical tilt of the main beam of the antenna. Usually expressed in degrees or mills
Rx antenna pattern	Directional systems	Directional performance of antenna (see next section)
Rx antenna gain response	All	Frequency gain response of the antenna. Expressed in dB down from the maximum gain
Rx antenna gain/loss	All	Gain or loss of receive antenna(s), normally reference either an isotropic (dBi) or dipole antenna (dBd)
Rx antenna polarisation	All	The polarity of the receive antenna(s)
Rx amplifier	All	Gain of any amplifier used prior to the receiver. Expressed in dB
Time slots	TDMA, frequency hoppers	For time based systems, the slot structure of the received signal. Can be used for interference analysis. Expressed in microseconds, milliseconds or seconds
Activity ratio	All	The ratio of activity when the system is receiving. Again, used for interference analysis. Expressed as a percentage or in a range 0–1
Modulation scheme	All	The encoding scheme used to modulate the baseband signal into the carrier. Has an effect on receiver sensitivity
Processing gain	CDMA	Gain due to the use of a spreading code in a CDMA system. Expressed in dB
Connector losses	All	Losses due to transmission line connectors. Expressed in dB
Feeder losses	All	Losses due transmission lines. Expressed in dB
Receiver sensitivity	All	Sensitivity of the receiver for a given performance level. Can be expressed in a number of ways such as S/N, SINAD, E_b/N_o, BER etc
Noise	All	The environmental noise present at the receiver location
Interference rejection	All	The receiver's ability to reject a particular type of interference signal. Expressed in dB down from the carrier

11.3 ES Specific Parameters

The parameters required for ES systems such as detection, intercept and direction finding are similar to those for communications. However, the characteristics are normally required to examine the transmission of the wanted target into the ES system are slightly different. This is explained in Table 11.2.

11.4 EA Specific Parameters

For electronic attack, the jammer parameters need to be included, as shown in Table 11.3.

11.5 Radar Specific Parameters

A large amount of data is necessary for modelling complex radar systems as shown in Table 11.4.

11.6 Third-Party Characteristics

For radar systems, it is also essential to have knowledge of the type of targets the system is being designed to detect. Reflections of radar energy from realistic targets are complex in nature and it is not usually possible to pre-determine the aspect of the target from the radar transmitter and receiver. In this case, it is normal to use nominal values for planning purposes although complex target models are often used in system performance simulations during the design phase. Thus, a nominal target strength value will be used.

In some cases, microwave systems can use passive reflectors to reflect energy from one microwave terminal to another. In this case, the characteristics of the reflector need to be taken into account.

11.7 General Antenna Characteristics

An antenna is a device used to convert electrical energy into radio energy and vice versa. There are many types of antennas for different applications, and their design is influenced by a range of factors, including:

- frequency of operation;
- bandwidth of operation;
- required directivity and gain;
- size and weight (and wind loading);
- radiated power;
- impedance;
- required polarization.

Table 11.2 Parameters necessary for ES radio prediction

Characteristic	Technology	Comments
Target Output power	All	May need to be estimated. The EIRP of the system may be known or estimated instead, in which case all of the other transmit equipment parameters need not be explicitly known
Target Frequency	All	Centre or reference frequency in MHz or GHz. For analysis of wide bands, it may be necessary to model the top, middle and bottom of the band
Target Bandwidth	All	Bandwidth of target transmission
Target Feeder losses	All	Losses due to transmission lines between the radio output and the antenna input. Expressed in dB. May well not be known. If EIRP is used, will not be required
Target Connector losses	All	Losses due to transmission line connectors. Expressed in dB. May well not be known. If EIRP is used, will not be required
Target Amplifier gain	All	Gain due to any amplifiers in the Tx system. Expressed in dB. May well not be known. If EIRP is used, will not be required
Target Combiner loss	Systems with combiners/ splitters	Losses due to combining or splitting the RF power before sending it to antennas. Expressed in dB. May well not be known. If EIRP is used, will not be required
Target Antenna gain/loss	All	Gain or loss of transmit antenna(s), normally reference either an isotropic (dBi) or dipole antenna (dBd). If EIRP is used, will not be required
Target antenna polar pattern	Directional systems	Directional performance of antenna (see next section)
Target antenna gain response	All	Frequency gain response of the antenna. Normally expressed in dB down on the highest gain response. If EIRP is used, will not be required
Target Antenna height/ altitude	All	Height of the centre of the radiation pattern above local ground or altitude above sea level. Usually expressed in metres or feet. May have to be estimated
Target Antenna azimuth	Directional systems	For directional antennas, the direction of the main beam in degrees or mills reference grid or magnetic north
Target Antenna tilt	All	The electrical or mechanical vertical tilt of the main beam of the antenna. Usually expressed in degrees or mills. May have to be estimated
Target antenna Polarisation	All	The polarity of the transmitted signal. Has an effect on propagation and also on interference
Modulation scheme	All	Most likely detected by ES system. Used for intercept

ES antenna height/altitude	All	Height of the centre of the radiation pattern above local ground or altitude above sea level. Expressed in metres or feet
ES antenna azimuth	Directional systems	For directional antennas, the direction of the main beam in degrees or mills reference grid or magnetic north
ES antenna tilt	All	The electrical or mechanical vertical tilt of the main beam of the antenna. Usually expressed in degrees or mills
ES antenna pattern	Directional systems	Directional performance of antenna (see next section)
ES antenna gain response	All	Frequency gain response of the antenna. Expressed in dB down from the maximum gain
ES antenna gain/loss	All	Gain or loss of receive antenna(s), normally reference either an isotropic (dBi) or dipole antenna (dBd)
ES antenna polarisation	All	The polarity of the receive antenna(s)
DF angular resolution	DF	Angular resolution of DF antenna. May be expressed for different scenarios such as line of sight or diffraction path
ES amplifier	All	Gain of any amplifier used prior to the receiver. Expressed in dB
Connector losses	All	Losses due to transmission line connectors. Expressed in dB
Feeder losses	All	Losses due transmission lines. Expressed in dB
ES Receiver sensitivity	All	Sensitivity of the receiver for a given performance level
Noise	All	The environmental noise present at the receiver location

Table 11.3 Parameters necessary for EA radio prediction

Characteristic	Technology	Comments
Target Output power	All	May need to be estimated. The EIRP of the system may be known or estimated instead, in which case all of the other transmit equipment parameters need not be explicitly known
Target Frequency	All	Centre or reference frequency in MHz or GHz. For analysis of wide bands, it may be necessary to model the top, middle and bottom of the band
Target Bandwidth	All	Bandwidth of target transmission. Used to determine bandwidth to be jammed
Target Feeder losses	All	Losses due to transmission lines between the radio output and the antenna input. Expressed in dB. May well not be known. If EIRP is used, will not be required
Target Connector losses	All	Losses due to transmission line connectors. Expressed in dB. May well not be known. If EIRP is used, will not be required
Target Amplifier gain	All	Gain due to any amplifiers in the Tx system. Expressed in dB. May well not be known. If EIRP is used, will not be required
Target Combiner loss	Systems with combiners/ splitters	Losses due to combining or splitting the RF power before sending it to antennas. Expressed in dB. May well not be known. If EIRP is used, will not be required
Target Antenna gain/loss	All	Gain or loss of transmit antenna(s), normally reference either an isotropic (dBi) or dipole antenna (dBd). If EIRP is used, will not be required
Target antenna polar pattern	Directional systems	Directional performance of antenna (see next section)
Target antenna gain response	All	Frequency gain response of the antenna. Normally expressed in dB down on the highest gain response. If EIRP is used, will not be required
Target Antenna height/ altitude	All	Height of the centre of the radiation pattern above local ground or altitude above sea level. Usually expressed in metres or feet. May have to be estimated
Target Antenna azimuth	Directional systems	For directional antennas, the direction of the main beam in degrees or mills reference grid or magnetic north
Target Antenna tilt	All	The electrical or mechanical vertical tilt of the main beam of the antenna. Usually expressed in degrees or mills. May have to be estimated
Target antenna Polarisation	All	The polarity of the transmitted signal. Has an effect on propagation and also on interference
Modulation scheme	All	Most likely detected by ES system. Used to determine bandwidth that needs to be jammed and possibly jamming signal type

Jammer antenna height/altitude	All	Height of the centre of the radiation pattern above local ground or altitude above sea level. Expressed in metres or feet
Jammer antenna azimuth	Directional systems	For directional antennas, the direction of the main beam in degrees or mills reference grid or magnetic north
Jammer antenna tilt	All	The electrical or mechanical vertical tilt of the main beam of the antenna. Usually expressed in degrees or mills
Jammer antenna pattern	Directional systems	Directional performance of antenna (see next section)
Jammer antenna gain response	All	Frequency gain response of the antenna. Expressed in dB down from the maximum gain
Jammer antenna gain/loss	All	Gain or loss of receive antenna(s), normally reference either an isotropic (dBi) or dipole antenna (dBd)
Jammer antenna polarisation	All	The polarity of the receive antenna(s)
Jammer amplifier	All	Gain of any amplifier used prior to the receiver. Expressed in dB
Connector losses	All	Losses due to transmission line connectors. Expressed in dB
Feeder losses	All	Losses due transmission lines. Expressed in dB
Jammer EIRP	Jammer	Used to calculate jammer output
Jammer Receiver sensitivity	All	Used for look-through
Noise	All	The environmental noise present at the receiver location

Table 11.4 Radar characteristics

Characteristic	Comments
Name/Type	Nomenclature of system or name or station
Location	Coordinates of station
Associated systems	Any other systems associated with emitter
Platform	Type of platform to which system is fitted
Mode	Observed mode
List of modes	List of available modes
Output power	Observed output power in dBW, dBm or W
Output device	Type of system used to generate the output power
Tx Frequency	Observed centre or reference frequency in MHz or GHz
Tunability	Type of tuning e.g. continuous, raster etc
Tuning step	Tuning step in kHz or MHz
Frequency Tolerance	Variation around tuned frequency in Hz
Frequency hopset	Hopset for frequency hopping systems
Hop rate	Number of hops per second
Dwell rate	Duration of dwell in milliseconds
Occupied Bandwidth	The bandwidth can be expressed as a single figure or it can be expressed as part of the power spectral density
Antenna type	Category of antenna
Horizontal aperture	Horizontal aperture
Vertical aperture	Vertical aperture
Diameter	Diameter of a dish
Beam type	Type of beam emitted
Horizontal beamwidth	Observed horizontal beamwidth
Minimum horizontal beamwidth	Minimum system horizontal beamwidth
Maximum horizontal beamwidth	Maximum system horizontal beamwidth
Vertical beamwidth	Observed vertical beamwidth
Minimum vertical beamwidth	Minimum vertical beamwidth
Maximum vertical beamwidth	Maximum vertical beamwidth
Antenna gain	Observed gain of transmit antenna(s), normally reference either an isotropic (dBi) or dipole antenna (dBd)
Antenna polar pattern	Directional performance of antenna (see next section)

Antenna gain response	Frequency gain response of the antenna. Normally expressed in dB down on the highest gain response
Antenna height/altitude	Height of the centre of the radiation pattern above local ground or altitude above sea level. Usually expressed in metres or feet
Antenna azimuth	For directional antennas, the direction of the main beam in degrees or mills reference grid or magnetic north
Antenna tilt	The electrical or mechanical vertical tilt of the main beam of the antenna. Usually expressed in degrees or mills
Rotation speed	Observed rotation speed in rotations per second
Horizontal scan speed	Observed horizontal speed
Horizontal scan rate	Observed horizontal scan rate
Horizontal scan type	Scan type
Vertical scan speed	Observed scan speed
Vertical scan rate	Observed scan rate
Vertical scan type	Scan type
Phased array number of main beams	Number of beams in the phased array
Phased array number of elements	Number of elements in phased array in horizontal and vertical directions
Power spectral density	The spectral shape of the transmitted energy. Can be used for interference analysis between dissimilar systems. Normally expressed as dB down from the main power
Tx antenna Polarisation	The polarity of the transmitted signal
Modulation	Observed modulation schemes
Modulation types	System modulation types
Pulse burst rate	Observed number of bursts per second
Pulse burst duration	Observed duration in microseconds
Number of pulses per burst	Observed number of pulses per burst
Pulse burst off time	Observed time in microseconds between pulse bursts
Burst type	The type of pulse format
PRF	Observed Pulse Repetition Frequency
Duty cycle	Observed duty cycle
Power per pulse	Observed power per pulse
Pulse compression method	Type of pulse compression
Pulse compression ratio	Observed pulse compression ratio
Rise time	Observed rise time
Fall time	Observed fall time

Antenna size is heavily influenced by the frequency of operation, with antennas operating at lower frequency and hence longer wavelengths being significantly larger than those designed for higher frequencies. At HF, where the wavelength can be as long as 100 metres, skywave antennas can be large, although the simplest HF antenna is a simple piece of wire (the 'random wire' or 'long wire' antenna) raised as high above the ground as possible, so although it is physically long it is not heavy. However, directional HF antennas can be very large compared to those used in for example mobile UHF communications.

Table 11.5 shows the size of wavelength for different frequency bands. The wavelength column is arranged in the same order as the frequency, so that at 3 MHz the wavelength is 100 metres and at 30 metres the wavelength is 10 metres.

Antennas can be split into two distinct types; those designed to work omni-directionally and those designed to provide gain in one direction, i.e. directional antennas.

Practical omni-directional antennas provide the same energy in each direction, normally horizontally, but are not omni-directional in the vertical plane. This is normally fine because in general communications, even for aeronautical applications, work mostly on or fairly near the horizon. Naturally, satellite antennas work vertically but they are highly directional.

The most commonly used omni-directional antenna type is a simple dipole, normally of either a half-or quarter wave of the transmission wavelength. Different types of omni-directional antenna are shown in Figure 11.2.

There are a large variety of directional antennas. Some common types are shown in Figure 11.3. The gain of directional antennas varies according to design, with the size of the antenna versus the frequency being a very important factor. Also note that although some are shown with horizontal or vertical polarisation, they can be mounted at different angles to provide different radiated polarisation.

Parameters often used to describe antenna characteristics are shown in Figure 11.4. The boresight is the direction of maximum gain. From that angle, the beamwidth can be specified for any angle, and in this case the 3 dB and 10 dB beamwidths are shown. The angles of the sidelobes and nulls can also be expressed, where nulls are the positions on minimum transmitted energy. The backlobe is the gain in the opposite

Table 11.5 Comparison of frequency and wavelength

Frequency band	Frequency range (MHz)	Wavelength (m)
HF	3–30	100–10
VHF	30–300	10–1
UHF	300–3000	1–0.1
SHF	3000–30 000	0.1–0.01
EHF	30 000–300 000	0.01–0.001

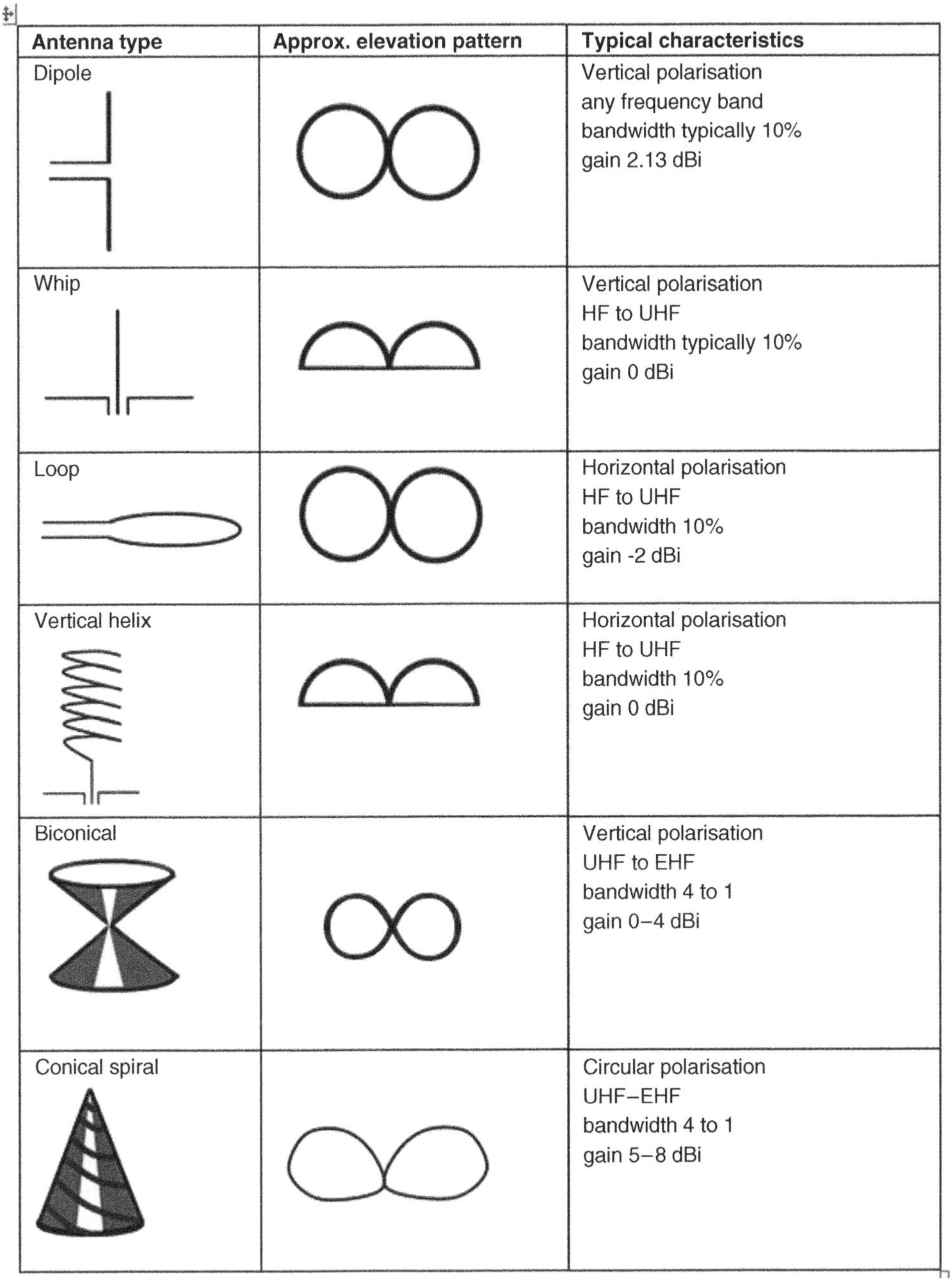

Antenna type	Approx. elevation pattern	Typical characteristics
Dipole		Vertical polarisation any frequency band bandwidth typically 10% gain 2.13 dBi
Whip		Vertical polarisation HF to UHF bandwidth typically 10% gain 0 dBi
Loop		Horizontal polarisation HF to UHF bandwidth 10% gain -2 dBi
Vertical helix		Horizontal polarisation HF to UHF bandwidth 10% gain 0 dBi
Biconical		Vertical polarisation UHF to EHF bandwidth 4 to 1 gain 0–4 dBi
Conical spiral		Circular polarisation UHF–EHF bandwidth 4 to 1 gain 5–8 dBi

Figure 11.2 Non-directional antenna characteristics.

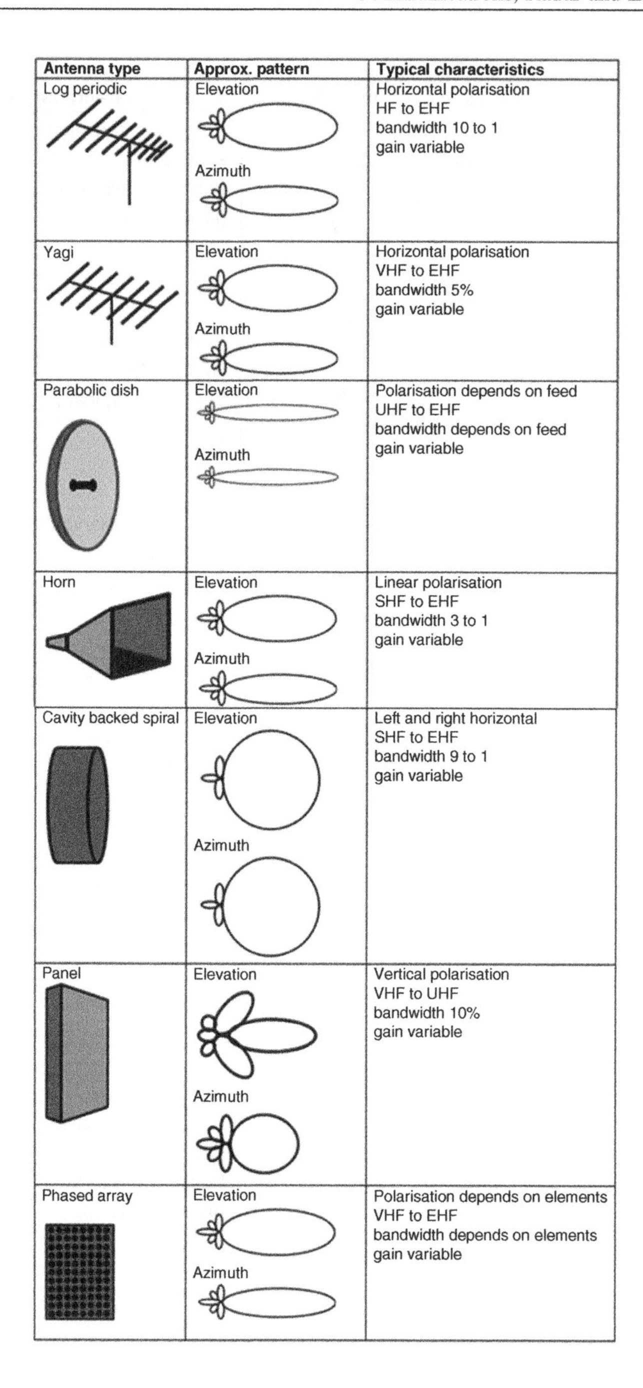

Antenna type	Approx. pattern	Typical characteristics
Log periodic	Elevation Azimuth	Horizontal polarisation HF to EHF bandwidth 10 to 1 gain variable
Yagi	Elevation Azimuth	Horizontal polarisation VHF to EHF bandwidth 5% gain variable
Parabolic dish	Elevation Azimuth	Polarisation depends on feed UHF to EHF bandwidth depends on feed gain variable
Horn	Elevation Azimuth	Linear polarisation SHF to EHF bandwidth 3 to 1 gain variable
Cavity backed spiral	Elevation Azimuth	Left and right horizontal SHF to EHF bandwidth 9 to 1 gain variable
Panel	Elevation Azimuth	Vertical polarisation VHF to UHF bandwidth 10% gain variable
Phased array	Elevation Azimuth	Polarisation depends on elements VHF to EHF bandwidth depends on elements gain variable

Figure 11.3 Some common directional antenna characteristics.

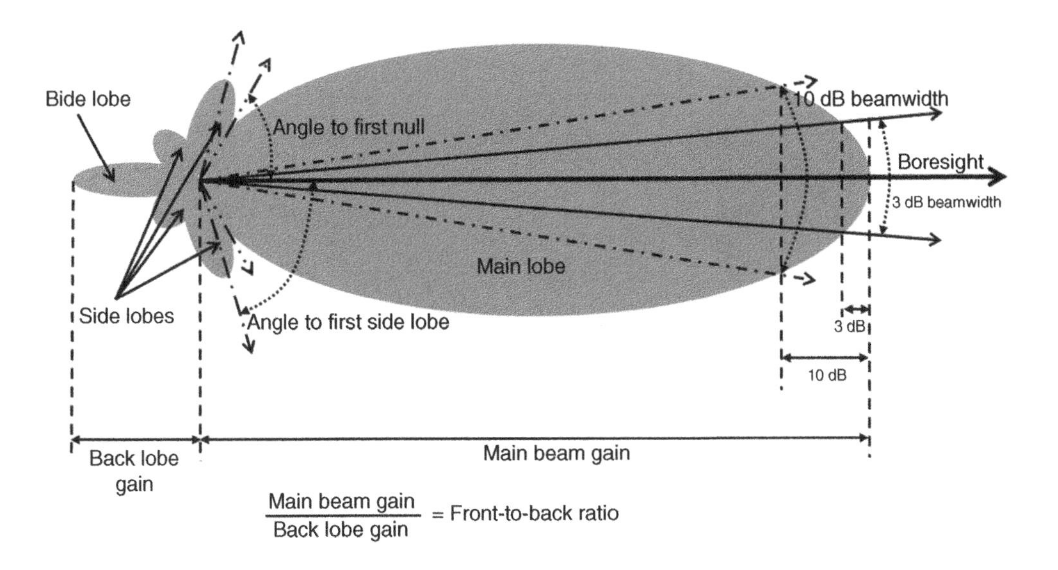

Figure 11.4 Typical antenna characteristics.

direction to the boresight, and the ratio of main beam gain to backlobe gain is referred to as the 'front-to-back' ratio (expressed in dB).

These parameters are useful for a first-cut approximation to antenna characteristics; however, they may not be sufficient to allow accurate predictions for three-dimensional simulation of system performance. Increasingly, antennas are described for many different angles so that the performance in all directions can be determined. A simple, often used method is to create a table of angle versus antenna gain for both the horizontal and vertical plane. Usually, the values are shown in dB down from the main beam. A typical example is shown in Table 11.6. In this case, the gain is expressed in five degree intervals.

This data can be used to generate a graphic display of the horizontal pattern, as shown in Figure 11.5, where the angular units are degrees and the internal units are dB down on the main lobe.

For the type of pattern shown (known as a cardiod), the five degree interval of values is sufficient to allow the total pattern to be described. However, the vertical pattern is more complex as illustrated in Figure 11.6.

In this case, the pattern has to be described in one-degree intervals in order to capture all of the salient points. In highly directional antennas, it may be necessary to describe a pattern in even more detail by expressing the gain at intervals of less than a degree.

As well as expressing the vertical and horizontal patterns, it is common to include a header with additional information as shown in Table 11.7.

Expressing the antenna characteristics allows advanced and accurate prediction of system performance.

Table 11.6 A typical antenna pattern table, in this case in five degree intervals

Horizontal pattern							
Angle	Loss (dB)	Angle	Loss (dB)	Angle	Loss (dB)	Angle	Loss (dB)
0	*0*	90	*14*	180	*40*	270	*13.5*
5	*0*	95	*15.8*	185	*40*	275	*12*
10	*0.1*	100	*17.9*	190	*40*	280	*10.4*
15	*0.3*	105	*20.1*	195	*40*	285	*9*
20	*0.6*	110	*22.5*	200	*40*	290	*7.8*
25	*0.9*	115	*25.4*	205	*40*	295	*6.6*
30	*1.3*	120	*28.4*	210	*40*	300	*5.6*
35	*1.8*	125	*31.5*	215	*40*	305	*4.7*
40	*2.4*	130	*35*	220	*40*	310	*3.8*
45	*3.1*	135	*38.6*	225	*40*	315	*3.1*
50	*3.9*	140	*40*	230	*37.4*	320	*2.4*
55	*4.8*	145	*40*	235	*32.2*	325	*1.8*
60	*5.8*	150	*40*	240	*28.3*	330	*1.3*
65	*6.8*	155	*40*	245	*24.9*	335	*0.9*
70	*8*	160	*40*	250	*22.1*	340	*0.5*
75	*9.3*	165	*40*	255	*19.6*	345	*0.3*
80	*10.7*	170	*40*	260	*17.3*	350	*0.1*
85	*12.2*	175	*40*	265	*15.3*	355	*0*

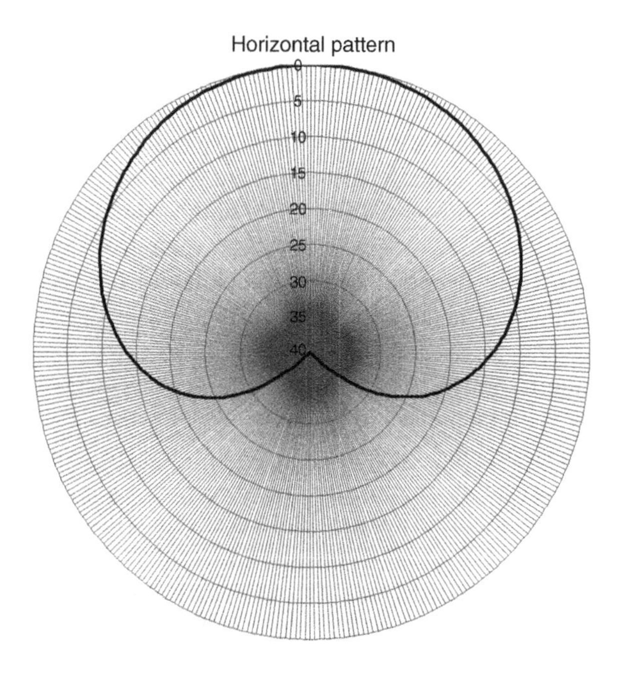

Figure 11.5 Antenna pattern representation.

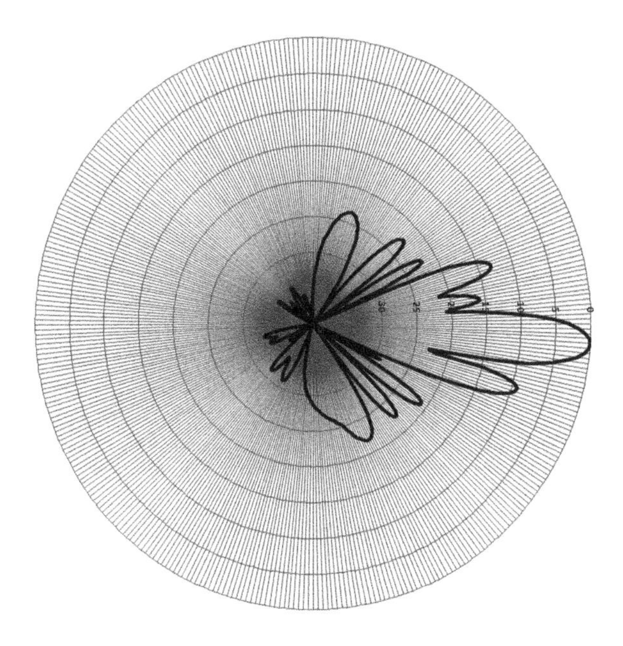

Figure 11.6 Vertical pattern representation.

It must also be noted that the antenna characteristics of HF antennas need to include the effects of the proximity of the antenna to the ground and the ground characteristics. This was described in more depth in Chapter 9.

11.8 Antenna Environment Considerations

Antennas, particularly mobile antennas will be affected by their environment. This can be accounted for by a geospatial model of ground clutter, combined with lookup tables for attenuation values as described in Section 11.10.

Table 11.7 Typical header included in antenna file

Antenna ID	GSM antenna # 01234
Manufacturer	ANDREW
Frequency (MHz)	860
Polarisation	Vertical
Horizontal beamwidth (degrees)	90
Vertical beamwidth (degrees)	7.5
Front-to-back ratio (dB)	40
Gain (dBd)	14.1
Tilt type	ELECTRICAL

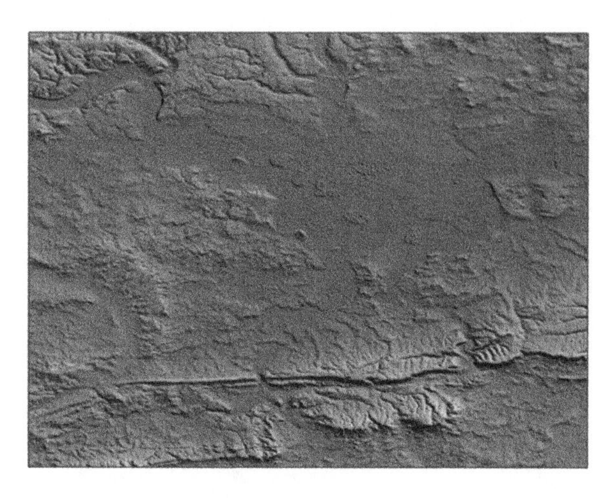

Figure 11.7 Graphic representation of terrain. These are normally shown in colour on most software systems.

11.9 Terrain Data

11.9.1 Introduction to Terrain Data

As well as describing equipment parameters, it is often necessary to include the effects of terrain on propagation. Hills, valleys and ridges can all act to block radio propagation and to produce shadow zones where communications or detection is not possible. A typical plan view of a terrain database is shown in Figure 11.7. In this figure, different altitudes are shown as different shades, with the lighter shades showing higher ground.

This can be more clearly seen in Figure 11.8, which shows a pseudo three-dimensional view of the same terrain. The vertical (altitude) scale is exaggerated to

Figure 11.8 3-D representation of terrain. This is normally shown in colour in most software systems.

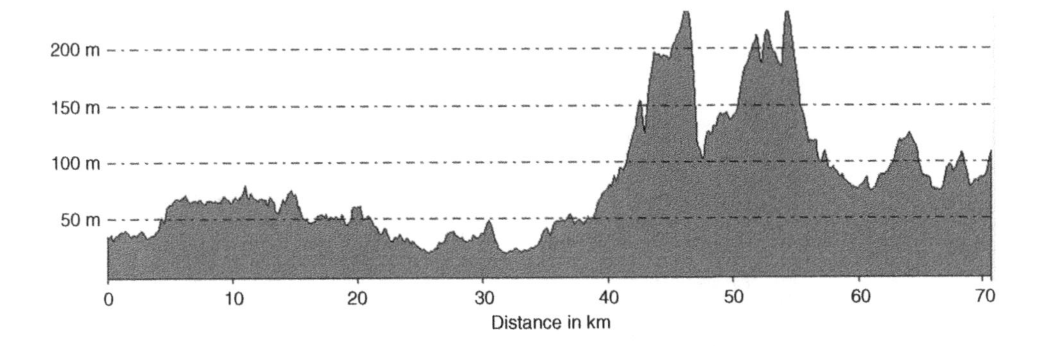

Figure 11.9 A typical path profile representation.

highlight terrain features. It needs to be exaggerated because the horizontal scale is likely to be in kilometres whereas the altitude is normally in metres.

To predict radio propagation between two points, a path profile can be generated on the direct great circle path (the shortest distance between two points on the ground). This is illustrated in Figure 11.9. The path profile values are fed into an appropriate propagation model to calculate the additional loss due to terrain obstructions. Note that for all radio prediction work, the effect of the effective Earth radius must be accounted for, typically by applying a correction to the actual radius of the Earth to account for the refractive index of the atmosphere. In many parts of the world, a value of 4/3 is used, which equates to an equivalent radius of approximately 8500 km compared to the real radius of 6370 km (approximately).

The use of path profile analysis is described in Section 13.1.

11.9.2 Sources of Terrain Data

Terrain data can be generated by a number of means. Early Digital Terrain Models (DTM) were made by interpolation between contour lines from existing paper maps, but these days remote sensing provides a more convenient and accurate method. Space based terrain databases, such as the SRTM (Shuttle Radar Topographical Mission) data shown in the figures above provide wide coverage at a resolution of three seconds of an arc, which can be converted into data with terrain points approximately 90 metres or better. A more recent mission called ASTOR has captured data at approximately three times better resolution and covering over 99% of the world's land mass. At the time of writing, the raw data is available and processing to improve the data is underway. Such post-processing is necessary to overcome data artefacts caused by the capture method. An example of this is shown in Figure 11.10.

The spots shown on the sea are artefacts of the data capture process. A small number may be caused by temporary structures such as large ships, but more are caused by random effects or are a by-product of the methods used. Post-processing removes these

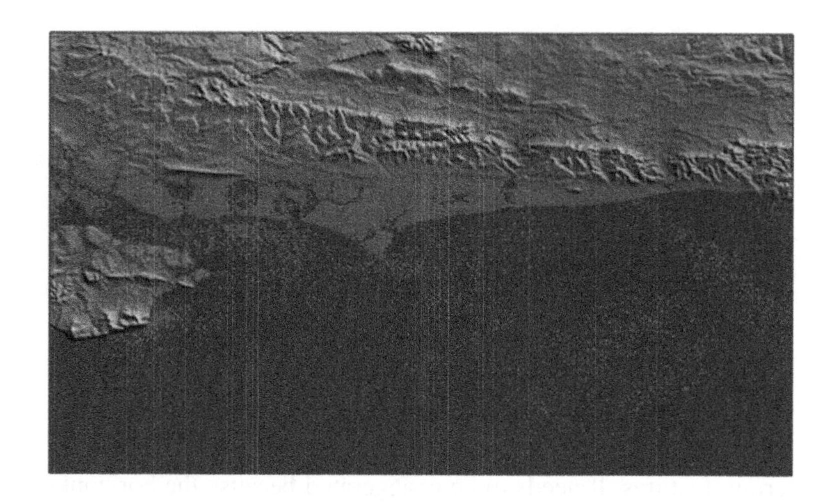

Figure 11.10 Raw SRTM data, showing collection artifacts.

artefacts as far as possible although some data voids may still persist after this. Fortunately, the most severe artefacts are visible to the eye when viewed. However, some may remain undetected.

The difference between the pre- and post-processed data can be seen by comparing Figure 11.10 to Figure 11.11, which shows the same data after having been processed to improve accuracy. The artefacts have now been removed.

Apart from space-based collection, aircraft using synthetic aperture radar or LIDAR (Light Detection and Ranging) can collect higher resolution data. Such data can be of resolution in the range of a few centimetres. However, it is expensive and time

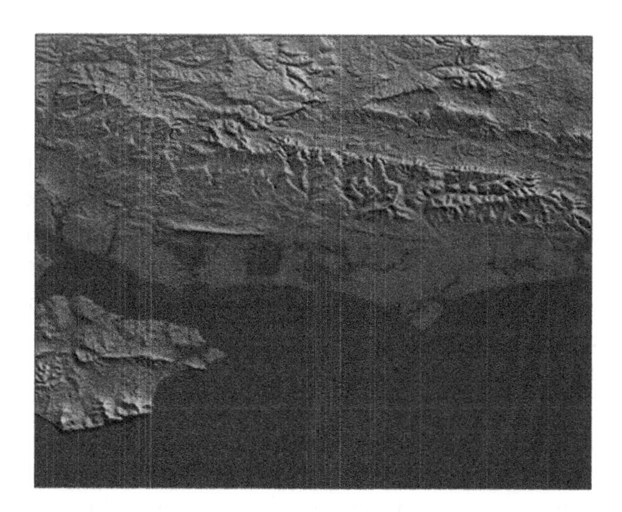

Figure 11.11 Post-processed SRTM data with artefacts removed.

consuming to collect and there is the risk that the data latency (the time during which the data is a valid representation of the measured area) is low. This is particularly true when there is a great deal of building in the area, or when there is a rapid change, e.g. during earthquakes, heavy bombing or other incidents. This means, for example, that high resolution data captured even recently may be useless to emergency relief planners after the event. At the time of writing in early 2010, the earthquake in Haiti has just happened and no doubt part of the emergency relief is a rapid re-surveying of the island to assist in planning for rescue, aid and reconstruction.

Some people believe that higher resolution data is always better than lower resolution, however for radio and EW planning this is not always the case, for the reasons discussed in Section 11.5.4.

11.9.3 Geographic Projections and Datums

It is worth at this point discussing geographic data projections, at least at a high level. This is a complex subject, and some recommended reading is included at the end of the chapter. However, it is worth knowing that data for the surface of the Earth can be expressed in a number of ways.

Typical formats include the following:

- 'Lat-long', e.g. 50 degrees, 30 minutes, 25 seconds North; 3 degrees, 5 minutes, 50 seconds West, or 50.3025, −3.0550 in common numeric format.
- Military Grid Reference System (MGRS), e.g. 29SNC018630.
- Universal Transverse Mercator (UTM), e.g. RHMF4658
- National grid reference systems, which depend on individual countries. For example in the UK, UK National Grid Reference system (NGR), e.g. TQ 123 456 or its numerical equivalent 512300, 145600.

Each of these systems is built on a geographic datum. This is the mathematical description of the Earth, which rather than being a perfect sphere is an oblate spheroid, and is slightly flattened at the poles compared to the equator. The most commonly used datum currently is WGS84 (World Geodetic System, 1984), which is the system used by the Global Positioning System (GPS). In most radio and EW planning tools, the datum will be expressed and if necessary, conversions between both geographic projection and datums will be available.

11.9.4 Terrain Data Resolution

A commonly asked question is what data resolution is necessary for realistic radio prediction work? This depends on a number of factors, the most critical of which is the frequency to be predicted. A good working assumption is based on the length of the 'short sector'; the distance over which shadowing is modelled but fast fading is not.

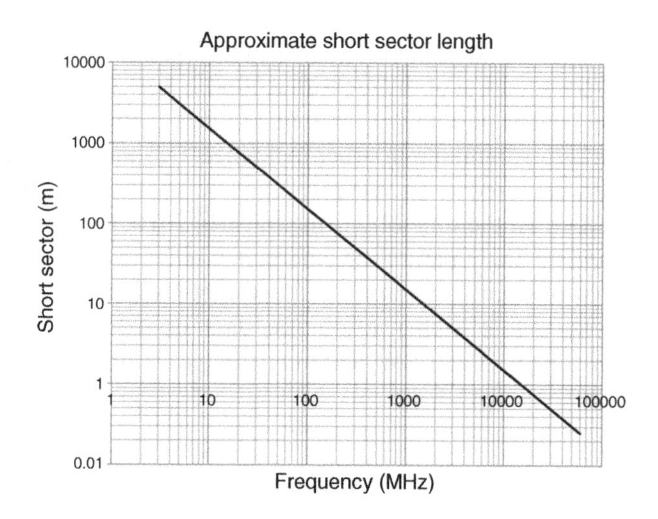

Figure 11.12 Approximate short sector length based on frequency, using 50 wavelengths.

This is typically held to be about 40–50 wavelengths of the transmission frequency. This is illustrated in Figure 11.12 for 50 wavelengths.

It can be seen from Figure 11.12 that the size of the short sector ranges from over a kilometre at the low end of HF, to less than 5 millimetres at 60 GHz. Realistically, at higher frequencies it is highly unlikely that data at the resolution of the short sector will be available. However, useful modelling can still be performed at these high frequencies for the reasons discussed in the next section.

11.9.5 Required Precision

The resolution discussed in the previous section is important, but there are other factors that influence the applicability of lower resolution data to propagation modelling. These include:

- The type of prediction being performed. Long range predictions at VHF and above depend as much on atmospheric conditions as terrain effects, and for these predictions small terrain features are of less importance. Also, so-called 'site-general' predictions do not require terrain models at all, and in other cases, a nominal model of terrain variability can be used.
- How close the terrain is to the two radio terminals. Communications between two aircraft in flight, between a ground terminal and an aircraft or between two high towers may not require accurate terrain modelling, with only the highest intrusions being significant.
- Where terrain variations are relatively small over the link ranges, a coarser terrain model can be entirely acceptable.

- Very short range paths, such as those encountered for RCIED prediction may depend more on factors such as ground conductivity than terrain, which usually will not vary substantially over the link range.
- Where other mechanisms are more dominant, the importance of terrain is less. This is true for many paths between mobile radio base stations and mobiles in urban environments, where radio clutter may be of higher significance.

These are not the only conditions where terrain need not be represented in lower resolution than indicated in Figure 11.12. It must be remembered that radio prediction is an engineering activity, and the experience of the engineer can account for risks associated with using less than ideal data.

There are situations where high resolution of even higher than short sector lengths may be required. Some examples are:

- Rooftop to rooftop or street predictions at UHF and above where buildings are treated as solid objects through which radio energy is not transmitted. Since building outlines and heights change over very small distances, a high resolution model is needed to model them. A digital model of terrain plus buildings and other manmade obstructions is known as a DEM (Digital Elevation Model).
- Where the propagation model requires very high precision, such as ray tracing methods.

Often, the question of which data to use is a question for system designers and has already been dealt with before operators get to use resulting planning tools; however, it is important that it is addressed and understood in order to ensure that the system is valid.

11.10 Ground and Radio Clutter Data

11.10.1 Ground Conductivity and Permittivity

For antennas near the ground, conductivity and permittivity affect propagation. This is particularly true for HF groundwave, but is also true for higher frequencies where the antennas are less than a few wavelengths above the ground. Maps of median ground conductivity and permittivity can be found from the ITU and other sources. Tools designed for modelling of HF groundwave may include this data internally. Conductivity and permittivity vary according to ground type and the amount of water present; loamy soil may have a high conductivity when wet but significantly less when completely dry. Table 11.8 shows some figures produced by a Professor Duncan Baker and Peter Saveskie for different ground types. These figures should be regarded as indicative of the relative variations between and within particular categories. There will be variations from these figures; for example, salt content of sea water (which is

Table 11.8 Conductivity and permittivity of common materials

Earth Type	Conductivity (S/m)	Permittivity (ε_r)
Poor	0.001	4.0–5.0
Moderate	0.003	4.0
Average	0.005–0.01	10.0–15.0
Good	0.01–0.02	4.0–30.0
Dry, sandy, flat (typical of coastal land)	0.002	10.0
Pastoral Hills, rich soil	0.003–0.01	14.0–20.0
Pastoral medium hills and forestation	0.004–0.006	13.0
Fertile land	0.002	10.0
Rich agricultural land (low hills)	0.01	15.0
Rocky land, steep hills	0.002	10.0–15.0
Marshy land, densely wooded	0.0075	12.0
Marshy, forested, flat	0.008	12.0
Mountainous/hilly (to about 1000 m)	0.001	5.0
Highly moist ground	0.005–0.02	30.0
City Industrial area of average attenuation	0.001	5.0
City industrial area of maximal attenuation	0.0004	3.0
City industrial area	0.0001	3.0
Fresh water	0.002–0.01	80.0–81.0
Fresh water at 10.0 deg C (At 100 MHz)	0.001–0.01	84.0
Fresh water at 20.0 deg C (At 100 MHz)	0.001–0.01	80.0
Sea water	4.0–5.0	80.0–81.0
Sea water at 10.0 deg C (to 1.0 GHz)	4.0–5.0	80.0
Sea water at 20.0 deg C (to 1.0 GHz)	4.0–5.0	73.0
Sea ice	0.001	4.0
Polar ice	0.00025	3.0
Polar Ice Cap	0.0001	1.0
Arctic land	0.0005	3.0

not constant over the world or at all times in particular locations) will change the values.

These figures can then be used in propagation models that accept conductivity and permittivity as inputs.

11.10.2 Radio Clutter

Particularly at frequencies at VHF and above, the question of radio clutter arises. Radio clutter can be defined as ground-based obstructions that affect the propagation of radiowaves near the ground. The two main categories of clutter are:

- built up areas. These can range from the heavily cluttered dense urban environments, ports, airports and sports stadia to low-rise spread out suburban environments;
- vegetation, including dense jungle, forestry, agriculture and so on.

These can be further split into different sub-categories, such as:

- dense urban (50 metre + average building heights);
- urban (30 metres average building height);
- urban (15 metres average building height);
- suburban (10 metres average building height) and so on.

Figure 11.13 shows some estimated measured median values for excess loss (F_a). The graph includes the values from ITU-R P.372 plus some more recent figures generated by Mass Consultants on behalf of the UK regulator Ofcom.

The values for F_a are used to determine the minimum signal level that has to be present at the antenna in order for the link to work to a given degree of performance.

Along with values for excess loss, it is also necessary to have some kind of geospatial map so that individual locations can be categorised according to the clutter on the ground. Such maps can be derived from digital social maps that show the outlines of the categories based on urbanisation. Many commercial companies offer such maps, although at present there is no freely available worldwide clutter map, unlike the case for terrain maps.

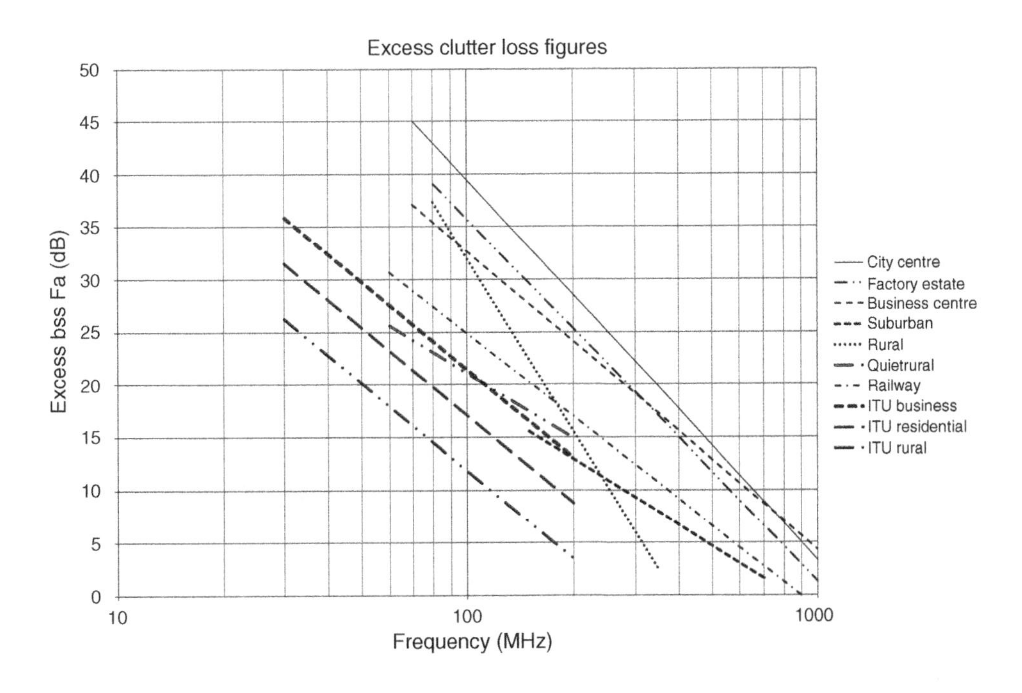

Figure 11.13 Measured median loss values for different environments.

11.11 Sunspots, Ionospheric and Atmospheric Data

11.11.1 Sunspots

Sunspot numbers are predicted into the future using smoothed curves based on analysis of historical data. The predictions are known as smoothed sunspot numbers. The predictions to 2019 are illustrated in Figure 11.14.

The data up to the end of 2015 is shown in tabular format in Table 11.9.

11.11.2 Ionospheric Conditions

Ionospheric data is also available on the internet from many sources. One such source is http://www.swpc.noaa.gov/Data/index.html. This gives values for many different measured values such as the maximum usable frequency and the heights of various layers such as the F2 layer on an hourly basis.

11.11.3 Gaseous Absorption

Gaseous absorption is described in ITU-R P.676. It provides methods of calculating the values and also some graphs from which wanted figures can be read. One of these is reproduced in Figure 11.15 in approximate form.

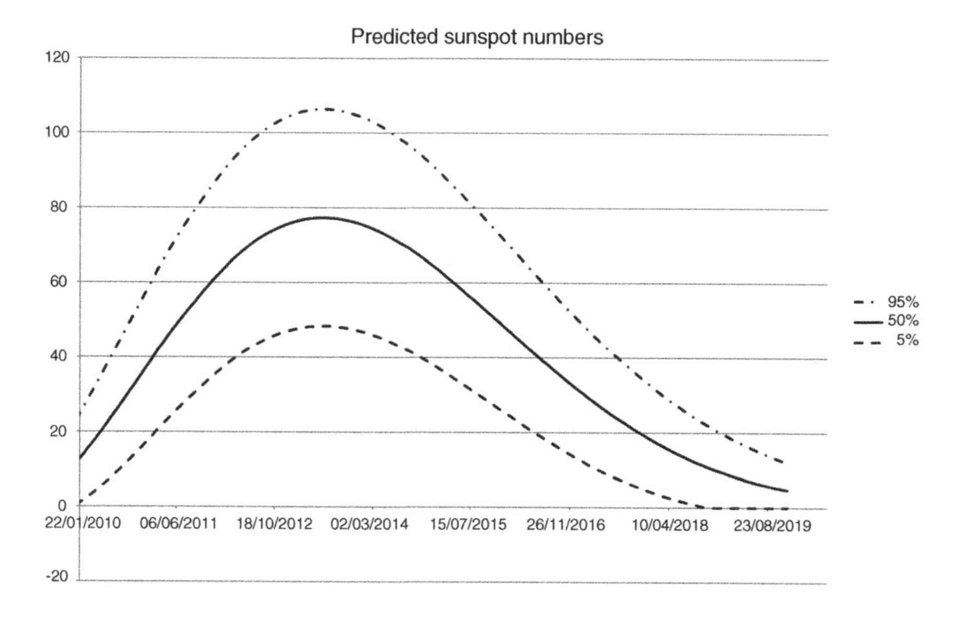

Figure 11.14 Predicted smoothed sunspot numbers to 2019. Obtained from http://solarscience.msfc.nasa.gov/SunspotCycle.shtml.

Table 11.9 Predicted smoothed sunspot numbers to 2015, obtained from http://solarscience.msfc.nasa.gov/SunspotCycle.shtml

Date	95%	50%	5%		95%	50%	5%
Jan-10	23	11.7	0.4	**Jan-11**	57.4	37.3	17.1
Feb-10	25.5	13.4	1.3	**Feb-11**	60.3	39.6	18.8
Mar-10	28.2	15.3	2.4	**Mar-11**	63.2	41.9	20.5
Apr-10	30.9	17.2	3.5	**Apr-11**	66.1	44.1	22.2
May-10	33.7	19.2	4.8	**May-11**	68.8	46.4	23.9
Jun-10	36.6	21.3	6.1	**Jun-11**	71.6	48.6	25.6
Jul-10	39.5	23.5	7.5	**Jul-11**	74.2	50.7	27.2
Aug-10	42.5	25.7	9	**Aug-11**	76.8	52.8	28.8
Sep-10	45.5	28	10.5	**Sep-11**	79.3	54.8	30.4
Oct-10	48.5	30.3	12.1	**Ort-11**	81.6	56.8	31.9
Nov-10	51.5	32.6	13.8	**Nov-11**	83.9	58.7	33.4
Dec-10	54.4	34.9	15.4	**Dec-11**	86.1	60.5	34.8
Jan-12	88.2	62.2	36.2	**Jan-13**	104.4	75.7	47
Feb-12	90.2	63.8	37.5	**Feb-13**	105	76.2	47.4
Mar-12	92.1	65.4	38.7	**Mar-13**	105.5	76.6	47.7
Apr-12	93.8	66.9	39.9	**Apr-13**	105.8	76.9	48
May-12	95.5	68.2	4	**May-13**	106.1	77.1	48.1
Jun-12	97	69.5	42	**Jun-13**	106.2	77.2	48.2
Jul-12	98.4	70.7	42.9	**Jul-13**	106.2	77.2	48.2
Aug-12	99.7	71.8	43.8	**Aug-13**	106.1	77.1	48.2
Sep-12	100.9	72.8	44.6	**Sep-13**	105.9	77	48
Oct-12	102	73.6	45.3	**Oct-13**	105.6	76.7	47.8
Nov-12	102.9	74.4	46	**Nov-13**	105.2	76.4	47.6
Dec-12	103.7	75.1	46.5	**Dec-13**	104.7	76	47.2
Jan-14	104.2	75.5	46.8	**Jan-15**	91	64.5	38
Feb-14	103.5	74.9	46.4	**Feb-15**	89.5	63.3	37
Mar-14	102.7	74.3	45.8	**Mar-15**	88	62	36
Apr-14	101.9	73.6	45.3	**Apr-15**	86.4	60.7	35
May-14	100.9	72.8	44.6	**May-15**	84.8	59.4	33.9
Jun-14	99.9	71.9	43.9	**Jun-15**	83.2	58	32.9
Jul-14	98.8	71	43.2	**Jul-15**	81.5	56.6	31.8
Aug-14	97.7	70.1	42.4	**Aug-15**	79.8	55.3	30.7
Sep-14	96.5	69.1	41.6	**Sep-15**	78.1	53.8	29.6
Oct-14	95.2	68	40.8	**Qct-15**	76.3	52.4	28.5
Nov-14	93.9	66.9	39.9	**Nov-15**	74.6	51	27.4
Dec-14	92.5	65.7	39	**Dec-15**	72.8	49.6	26.3

11.11.4 Precipitation

Rainfall varies widely across the world. ITU-R P.838 provides a method of calculating rainfall attenuation per kilometre, and ITU-R P.837 contains worldwide maps for rainfall throughout the world. A graph of rainfall rate against attenuation per kilometre is shown in Figure 11.16 for different frequencies and polarisations.

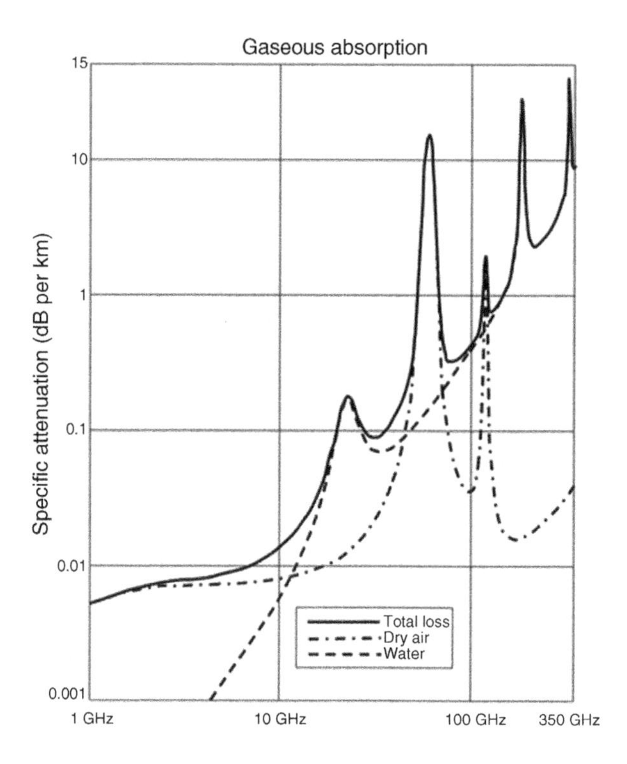

Figure 11.15 Approximation of gaseous absorption based on ITU-R P676.

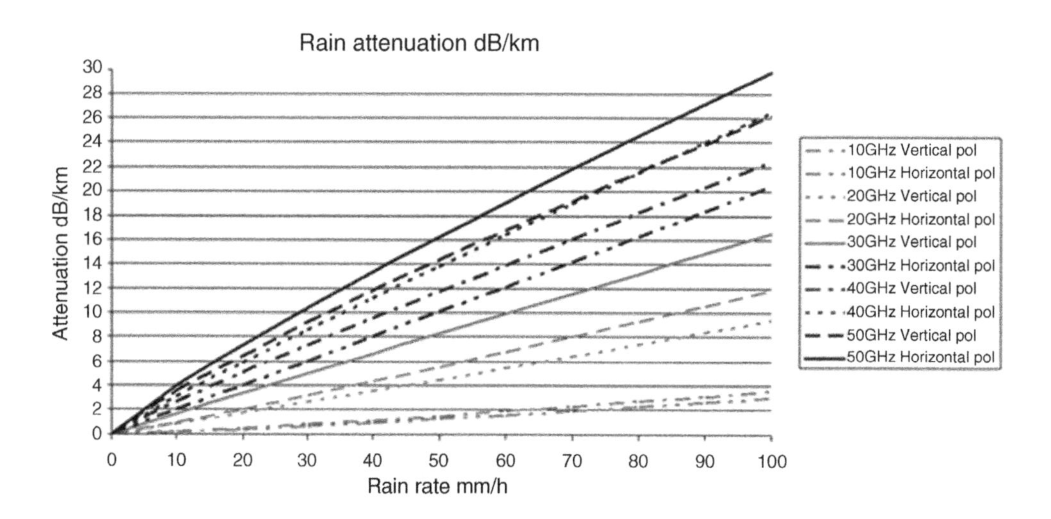

Figure 11.16 Rainfall attenuation for different frequencies and polarizations.

References and Further Reading

All internet references correct at time of writing.

Ademy, D. (2001), *EW101: A First Course in Electronic Warfare,* Artech House, MA, USA, ISBN 1-58053-169-5.

Ademy, D. (2004), *EW102: A Second Course in Electronic Warfare,* Horizon House Publications, USA, ISBN 1-58053-686-7.

Graham, A.W.; Kirkman; N.C.; Paul, P.M. (2007), *Mobile Radio Networks Design in the VHF and UHF Bands: A Practical Approach,* John Wiley & Sons ISBN 0-470-02980-3.

Recommendation ITU-R P.372: Radio Noise.

Recommendation ITU-R P.676: Attenuation by Atmospheric Gasses.

Recommendation ITU-R P.837: Characteristics of Precipitation for Propagation Modelling.

Recommendation ITU-R P.838: Specific Attenuation Model for Rain for Use in Prediction Methods.

Recommendation ITU-R P.1058: Digital Topographic Databases for Propagation Studies.

Skolnik, M. (2008), *Radar Handbook,* McGraw-Hill, USA, ISBN 978-0-07-148547-0.

12

Planning and Optimising Radio Links

12.1 Path Profile Prediction

The most basic tool available to the radio planner is the path profile prediction. Most of the more sophisticated tools used by the planner are extensions of this; for example, coverage predictions are only path profile predictions carried out for a large number of paths.

The basic path profile is shown in Figure 12.1. The terrain is shown. In this case, no clutter is shown on the profile, but some tools will show clutter as a coloured overlay on top of the terrain. In all cases, the profile should be along the great circle path between the two points and the effective Earth radius bulge must also be included for the path under consideration.

To this path profile, we can apply a suitable propagation model, in this case ITU-R P.526, to determine the additional loss due to diffraction. This is illustrated in Figure 12.2, where a transmit antenna has been added at the beginning of the link and a receiver added at the end. The black line shows the result of applying string theory to the path. In this case, there is double diffraction over two separate obstructions.

We need to consider the factors that need to be considered when computing the power leaving the transmit antenna and arriving at the receiver. A typical process is shown in Table 12.1.

An example of a link budget created to determine the minimum working input power to the antenna is shown in Table 12.2.

Without including the additional loss factors, a planner may conclude erroneously that it is acceptable to simply subtract the EIRP from the required input power into the receiver, giving a value of 144 dB maximum allowable path loss. This would of course be wildly inaccurate and would lead to major problems on the ground. The difference

Communications, Radar and Electronic Warfare Adrian Graham

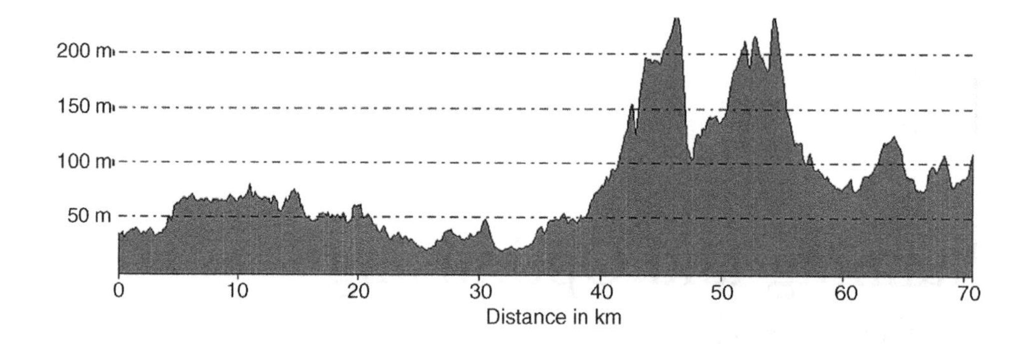

Figure 12.1 A typical path profile.

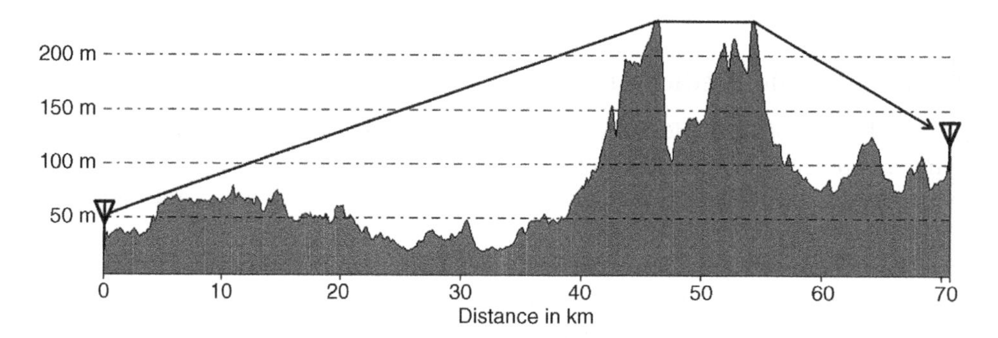

Figure 12.2 String theory used to identify diffraction objects in a path profile.

in maximum system range between a total loss of 144 dB compared to 103 dB is very significant.

Once the maximum allowable loss has been calculated, the predicted loss can be compared to the maximum allowable loss to determine whether the loss is acceptable or not, and by how much.

If, for example, the predicted loss including diffraction loss for a path is 137 dB, we can determine that the link will not be workable; the signal at the receiver antenna is $103 - 137 = 34$ dB too low.

Analysing individual path profiles is useful but time-intensive for large systems. Often, computer planning tools will be able to analyse many links in a single simulation to determine overall performance of point-to-point networks and to examine interference, detection or jamming from one or multiple sources.

12.2 Optimising a Link

If the path profile shows that the wanted link is viable, then all well and good. But what actions can be taken if the link is not viable. There are a number of options that can be carried out to improve performance.

Table 12.1 Considerations in determining parameters to be used in a path prediction

Aspect	Considerations
Radiated power	Determine power radiated in the direction of the receiver. For directional antennas, this may be lower than the main beam if the receiver is not in the direction of the horizontal boresight
Transmitter platform	Determine whether the platform affects the radiated power. For body-mounted antennas, body loss may need to be considered
Transmitter clutter	Assess whether the transmitter is transmitting through clutter. If so, additional losses may need to be added to account for losses close to the transmit antenna
Mid-path clutter	Determine whether clutter in the mid-path needs to be taken into account. In Figure 12.2 only clutter present at the two obstruction points need to be considered. Clutter obstructions may account for in a simplified fashion by adding height to the obstruction heights above local terrain. This is not normally included in the link budget but is included in the propagation loss prediction
Receiver clutter	Determine whether additional loss needs to be added to account for clutter in the location of the antenna
Receive antenna pointing loss	Determine whether there is any loss down from the main beam in the direction from the transmitter
Receiver platform	Consider whether the receiver platform influences the received power, again, factors such as body loss may need to be included. For ships and other mobile platforms, pointing loss may need to be considered
Fade margin	Depending on whether the receiver sensitivity is faded or unfaded, a loss factor may need to be added. Even if the receiver sensitivity is quoted for the fading conditions, there may need to be additional corrections to account for differences between the conditions under which the faded sensitivity was calculated and the actual conditions
Receiver sensitivity	The required input power into the receiver to achieve the required degree of performance must be determined so that the energy needed at the antenna can be calculated

Table 12.3 shows some parameters that can be changed by the operator, and the effects that they can have.

We will now look at the effect of adding a re-broadcast terminal to an otherwise unviable link between two physical locations.

12.3 Re-Broadcast Links

A re-broadcast ('re-bro') or relay terminal is used to overcome problems caused by terrain or distance. The principle is that an additional radio is added to the link to accept a transmission, amplify it and re-direct it to the other terminal. A re-bro can be used for example to route communications around a difficult hill. An example is illustrated in

Table 12.2 Link budget created after consideration of parameters

Parameter	Value	Units
Transmit EIRP	40	dBm
Tx antenna Pointing loss	5	dB
Tx platform loss	0	dB
Transmit clutter loss	8	dB
Effective power (in direction of Rx)	***27***	***dBm***
Receiver clutter loss	12	dB
Receiver antenna loss/gain	1	dBi
Receiver platform loss	5	dB
Fade margin required	10	dB
Receiver sensitivity	−104	dBm
Required power at Rx antenna	***−76***	***dBm***
Maximum allowable path loss	***103***	***dB***

Figure 12.3, where a hill is on the direct path between the transmitter and receiver. A re-bro terminal is placed in a different position to allow viable links between transmitter, re-bro and receiver.

The re-bro can either work on the original transmission frequency by providing isolation between two different antennas, one pointed at the transmitter and one pointed at the receiver, or frequency isolation can be achieved by receiving on one frequency and re-transmitting on another. In this case, the receiver must be tuned to the frequency transmitted out from the re-bro terminal and not the original transmission frequency.

Of course, the question then arises how the location of a successful re-bro can be determined. The key point is that the re-bro must be able to talk to both the transmitter and receiver simultaneously. One way of doing this is to examine the coverage of both transmitter and receiver and to look at the overlap between the two. This is illustrated in Figure 12.4 where the dark, black bounded region shows where transmitter coverage and receiver coverage both overlap.

There is also overlap on the obstructing hill and this would be a viable alternative. However, this may not be desirable; a high re-bro is vulnerable to detection, localisation and jamming and it also may be physically difficult to install equipment on a hill with no road transport.

12.4 Linked Networks

Path profile analysis can be easily extended to multiple links in order to determine the performance of entire point-to-point networks and interference or jamming vulnerability. It is also useful for analysing connectivity and potential vulnerabilities. Figure 12.5 shows a point-to-point network consisting of 13 callsigns. The links can

Table 12.3 Parameters that can be changed by the operator to improve radio performance

Parameter	Effect
Increase transmit power	Increasing the power will increase the EIRP, however, this is limited by the amount of power available, there is an increased risk of interference to other systems and the risk of detection by the enemy is increased
Add an amplifier	This has the same effect as increasing transmit power
Increase antenna heights	For paths with only small mid-path intrusions, increasing the transmit and/or receiver antenna height can help to overcome diffraction loss. However, where intrusions are large, the effect is likely to be minimal
Re-orient antennas	If the antennas are directional and the main beam is not in the direction of the receiver or transmitter respectively, including both azimuth and tilt, then the antennas can be re-oriented to point directly at the other terminal. However, this will be at the expense of other potential receivers in other directions. This cannot be done for omni-directional antennas
Change to a higher gain antennas	If it is possible, changing antennas to one with a higher gain can help. However, this will again be at the expense of other possible receivers in other directions. A higher gain (and hence more directional) antenna can also help to reduce the probability of interference and detection
Reduce Tx frequency	Reducing the transmission frequency will reduce the overall loss in many cases; however this will be limited by equipment minimum frequency and possibly also by its ability to transmit a sufficiently high bandwidth
Change modulation scheme	A simpler modulation scheme has a higher resilience to noise and thus can be transmitted further. However, this will be at the expense of the achievable data rate
Move one or both terminals	If possible, moving either or both terminals to a better location will improve the link, but for operational reasons this may not be possible
Add a re-broadcast terminal	Adding a physically separate re-broadcast antenna in a location that can be reached by both transmitter and receiver has the effect of removing the difficult path and replacing it by two better paths that allow transmission from the transmitter to the re-broadcast terminal and then to the receiver. However, this does require further planning, additional equipment and logistics

been split into three different categories; black solid lines for robust links, long dash lines for marginal links and dotted lines for sub-standard links. In a colour system, a traffic light system of green, yellow and red is often used since it is intuitive.

As well as giving a good visual representation of the state of health of the network, it is also useful when examining potential vulnerabilities. For example, assuming the network is all-informed there is only one link between Callsign 7 and Callsign 8. If this

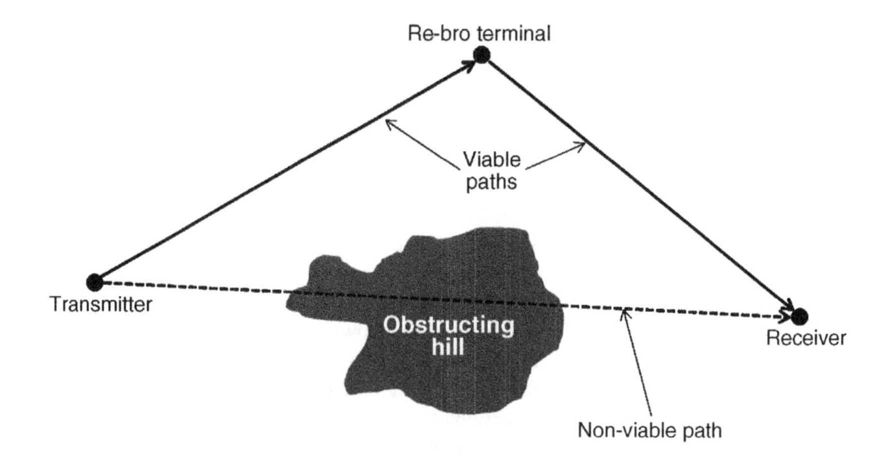

Figure 12.3 Use of a re-bro to link otherwise obstructed terminals.

link suffers failure or is jammed, then the network is split in two. Note that this analysis is useful to help design out vulnerabilities of own networks and is also useful to examine vulnerable points of enemy networks once the terminal locations have been localised.

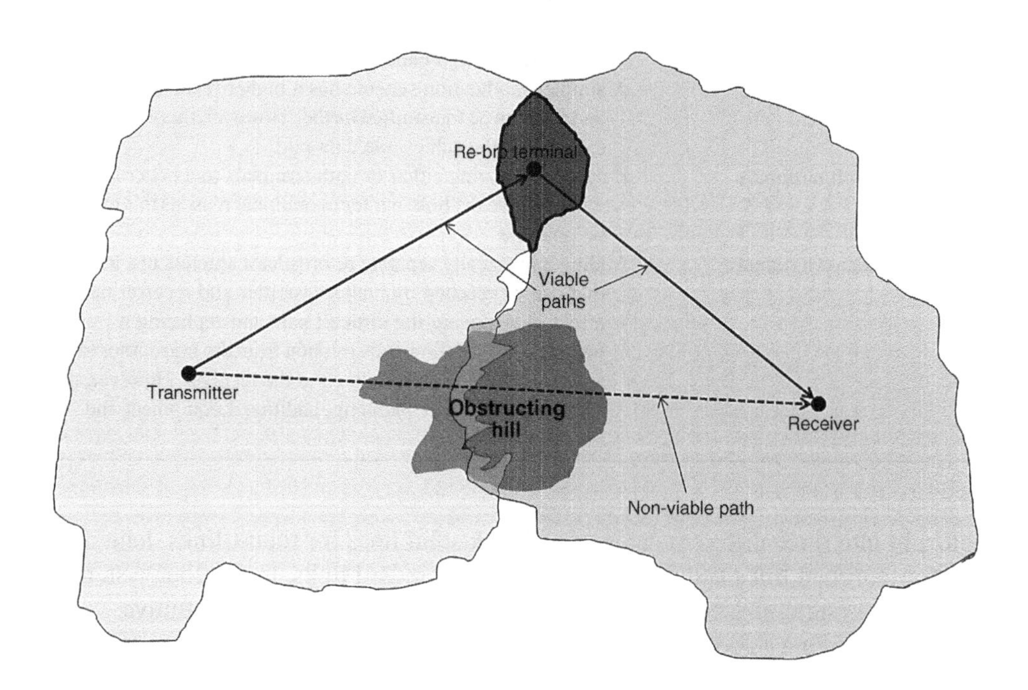

Figure 12.4 Using coverage overlap to identify potential re-bro locations.

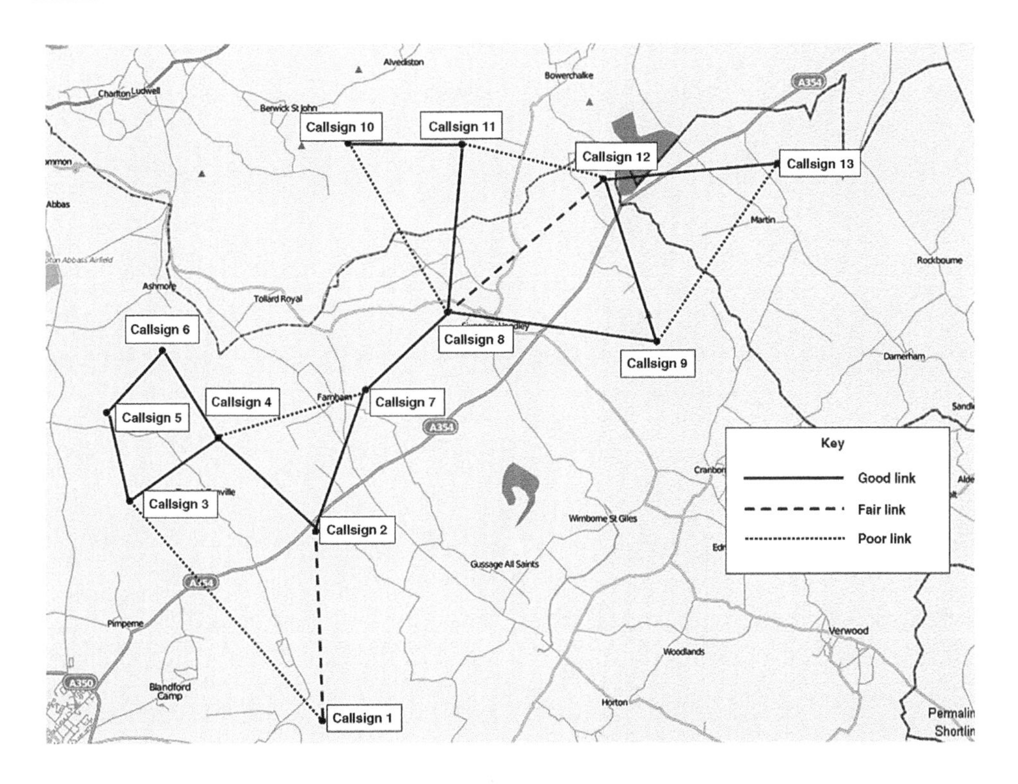

Figure 12.5 Point to point network, showing link predicted performance. Map data © OpenStreetMap contributors, CC-BY-SA.

References and Further Reading

Graham, A.W.; Kirkman, N.C.; Paul, P.M. (2007), *Mobile Radio Networks Design in the VHF and UHF Bands: A Practical Approach*, John Wiley & Sons. ISBN 0-470-02980-3.

For information on the map images used, please visit:
www.openstreetmap.org and www.creativecommons.org.

13

Planning Radio Networks for Coverage

13.1 Coverage Predictions

Coverage predictions are created by a multiplicity of path profiles between a base station (or in different terminology, a fixed station or terminal) and all locations within a given specified maximum range. This is illustrated in Figure 13.1, which shows a fictitious base station and its associated coverage. This is a mock-up of a typical coverage display since almost all planning tools provide coverage in colour and they are not clear in black and white. For this reason, all coverage-type predictions are shown in the same manner for clarity. The best coverage is bounded in the solid black line, close to the base station. Marginal coverage is shown bounded by large black dashes and very marginal coverage is shown bounded by dots.

The number of levels shown can be altered from a simple go-no go analysis right the way through to a multi-coloured display with many different levels.

The interpretation of the coverage prediction is that a mobile element operating in within the area enclosed by the solid black line would be able to communicate clearly with the fixed base station, and those in the other categories would have a lower link probability of success. Beyond the region enclosed by the dotted line, communications are unlikely to work.

Coverage predictions are a vital tool for modern network design and mission planning. The principle is extendible beyond simple communications coverage. By setting appropriate parameters it is also possible to predict:

- Probability of detection against an enemy emitter of known parameters but unknown location. In this case, the 'base station' is NOT transmitting, but the law of reciprocity is used to determine the probability of an enemy transmitter being detected by the passive detector.

Communications, Radar and Electronic Warfare Adrian Graham

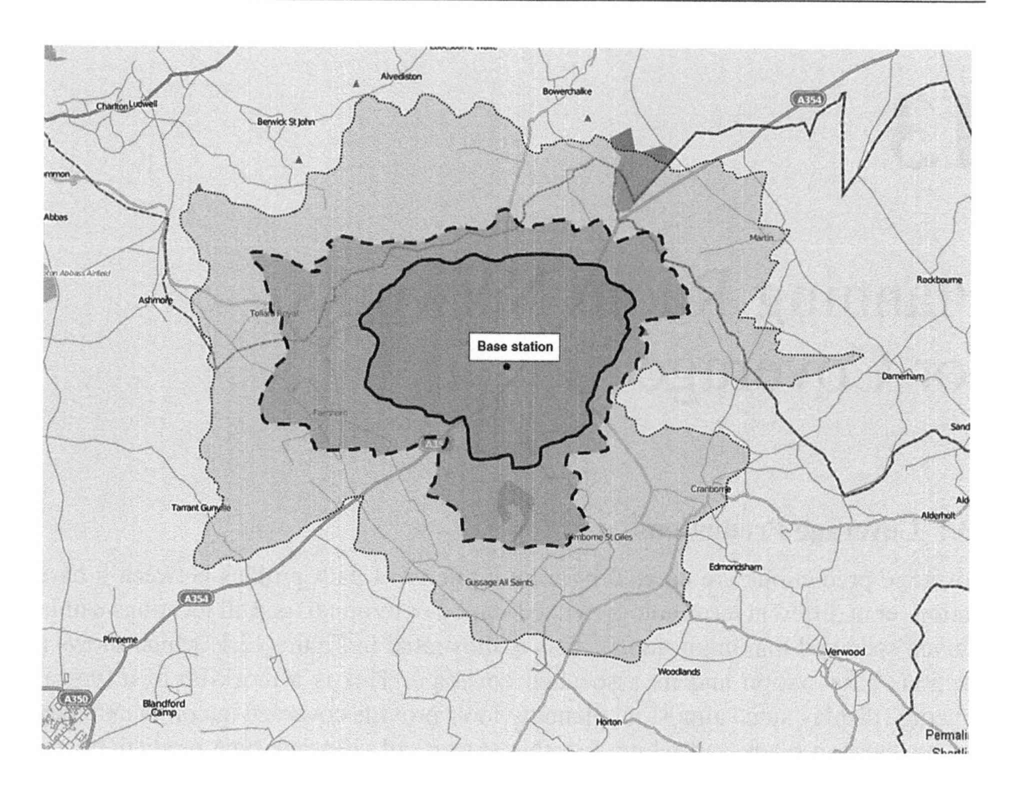

Figure 13.1 A representation of a coverage display. Map data © OpenStreetMap contributors, CC-BY-SA.

- Direction finding probability, in the same way. However, there may be additional factors that limit the quality of signal received even if the signal strength is detectable. However, when comparing potential sites for a DF, this type of analysis is very useful.
- Predicting radar performance. In this case, the 'base station' parameters are those of the radar emitter and the wanted target strength is used to represent the mobile system. The altitude of the target can be set above ground level for tactical aircraft or against sea level for higher level targets. Exactly the same process can be used for radar performance against maritime or ground-based targets, assuming that the ground clutter is also taken account of.

The coverage prediction of radars is explained in Figure 13.2 below. The instantaneous coverage is shown in the diagram on the left. If the circular sweep is integrated over an entire revolution, then the coverage can be shown as on the right. In this case, there are no terrain effects and so the coverage is a circle. However, in many cases, terrain will vary the radar horizon and thus the coverage will not be circular. For radars that do not have a circular sweep such as a fixed phased array, the

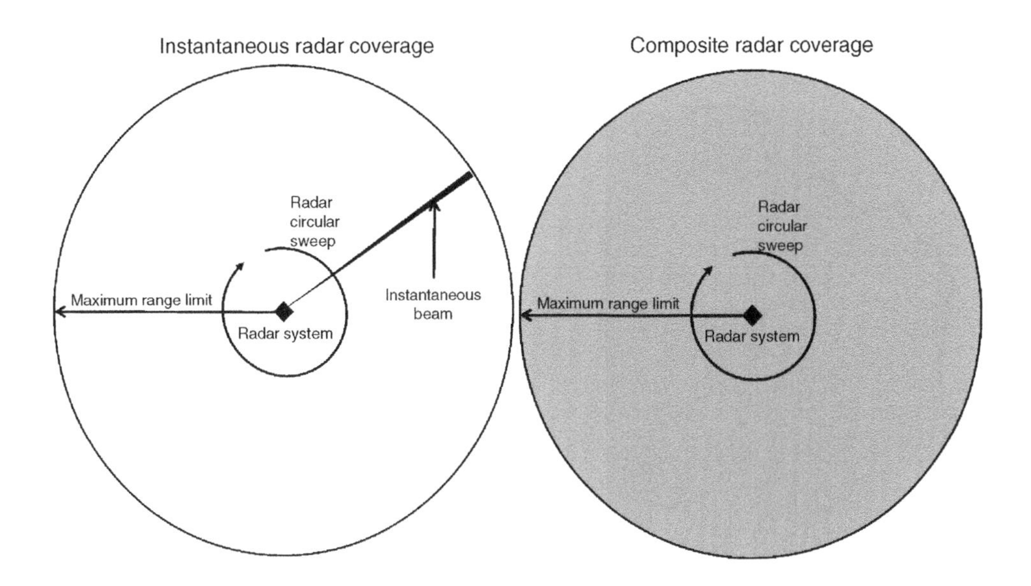

Figure 13.2 Instantaneous radar coverage integrated over the period of a sweep to show total coverage. Map data © OpenStreetMap contributors, CC-BY-SA.

composite pattern generated by integrating each instantaneous beam can be used in the same way. This would lead to a directional coverage pattern covering the horizontal arc of coverage.
- Predicting the raw power produced by a communications jammer.

Coverage predictions are also the basis for more sophisticated analyses, including:

- Composite coverage of a number of systems. This can be used to describe the coverage of an entire network or parts of it. This process can be used to display the predicted performance of radio networks, direction finding base lines, detection and intercept networks, radar network coverage and so on. In the case of radar networks, we examine the integrated coverage over the entire horizontal angle sweep range.

  This is illustrated in Figure 13.3, where the black dots are base stations and the coverage, this time limited to a go-no go threshold (it does not need to be; the coverage levels can be split into as many performance boundaries as necessary as for the coverage of a single system). The coverages of all the base stations are combined to produce this display of total network coverage.

  Similarly, for radar systems, the same type of plot can be created. Figure 13.4 shows an illustration of four tactical air defence systems (black diamonds) set up to protect a base from an air threat. The composite network coverage is shown for small

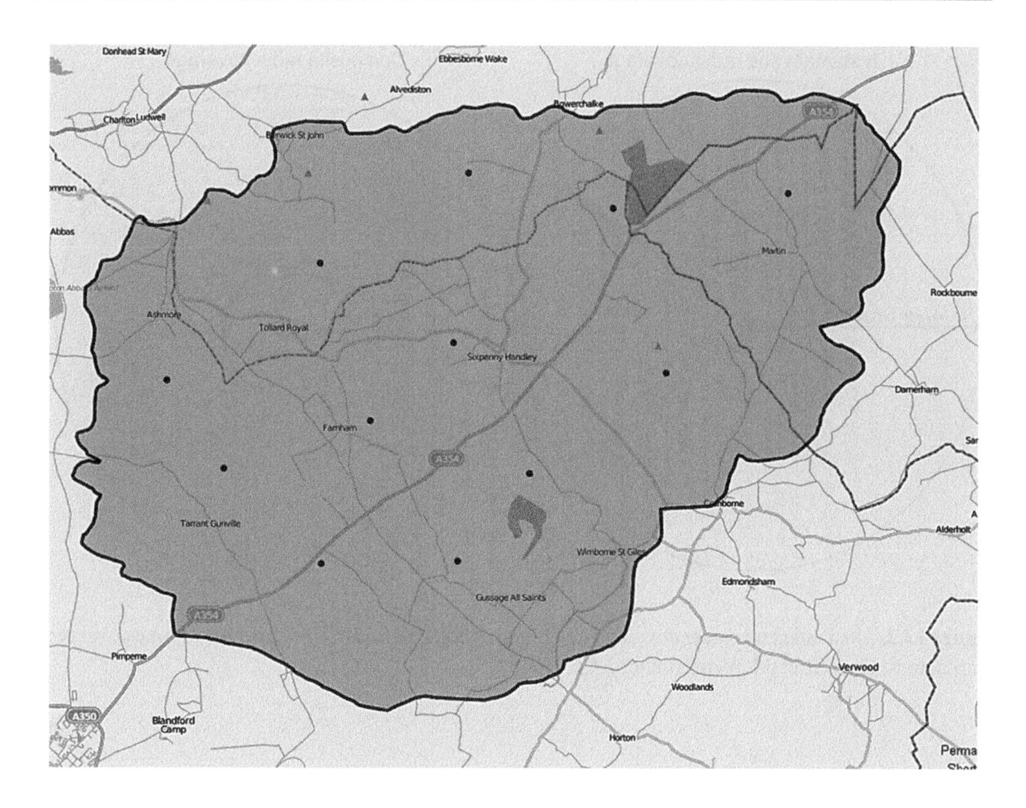

Figure 13.3 Composite coverage representation. Map data © OpenStreetMap contributors, CC-BY-SA.

military jets flying at 250 metres above the terrain or higher. In this case, terrain does have an effect on the coverage.

- Best server predictions. This is an analysis of the relative power of a number of base stations to determine which provides the strongest signal at each location. By again utilising the reciprocity theory, the same principle can be used to determine which station in a detection or DF network provides the highest probability of detecting a target in a specific location. The same is true for radar networks and most other kind of systems. An illustration of a best server display is shown in Figure 13.5.
- Coverage overlap. This is useful for analysing potential redundancy in communications or radar networks and also for determining where a DF baseline will be able to localise targets because at least two (or preferably at least three) DF stations can simultaneously receive a target transmission. This is illustrated in Figure 13.6.
- Communications handover analysis. This occurs in regions where there is a workable signal strength from at least two base stations, that the relative strengths of signals are within a given design value and that network handover rules are met.

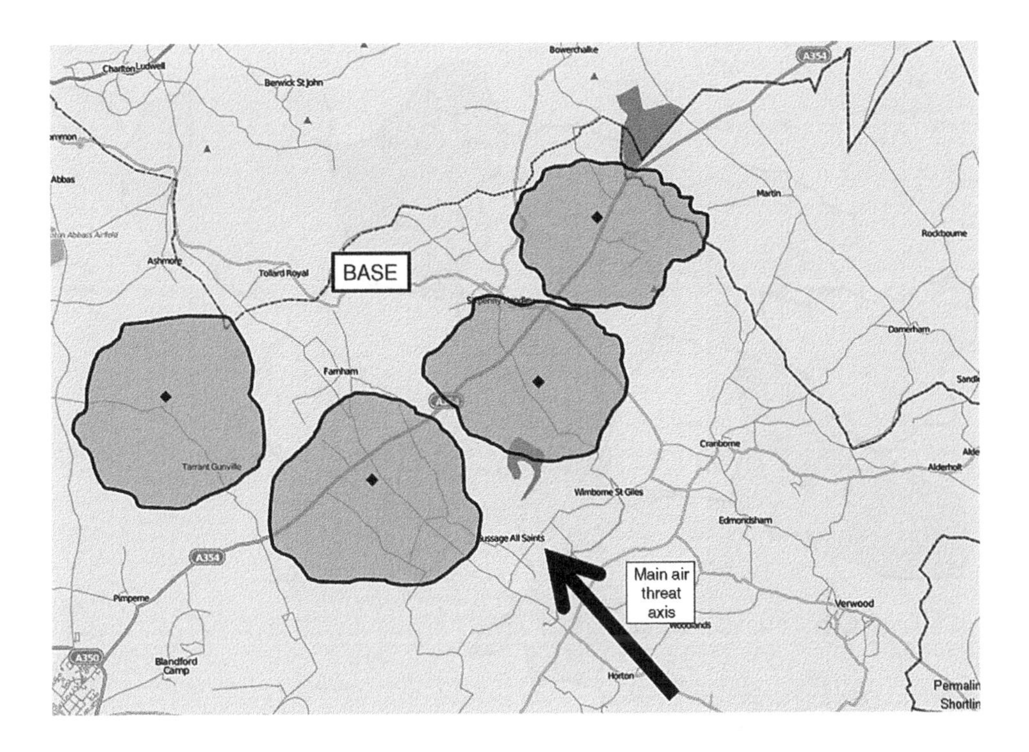

Figure 13.4 Composite radar network coverage for targets at a given minimum height above ground. Map data © OpenStreetMap contributors, CC-BY-SA.

- As we will see in Chapter 15, coverage predictions are also instrumental in mobile interference predictions.

13.2 Optimisation of Radio Networks

Coverage plots are more than just a method of displaying predicted coverage from fixed station locations. They are also the primary means of optimising the coverage from such stations, and by extension from entire networks. Optimisation is the process of improving the wanted characteristics of a radio network and reducing the unwanted effects. Note that this is not always a desire to maximise coverage; often it is more important to attempt to match the coverage to some wanted metric which may or may not be related to making signals go as far as possible. Some typical design metrics are listed below.

- Coverage maximisation may be a wanted characteristic in some cases.
- Coverage maximisation in a particular direction or arc of angle may be important.
- Coverage of a specific target area is often a key design metric for communications and especially communications electronic warfare.

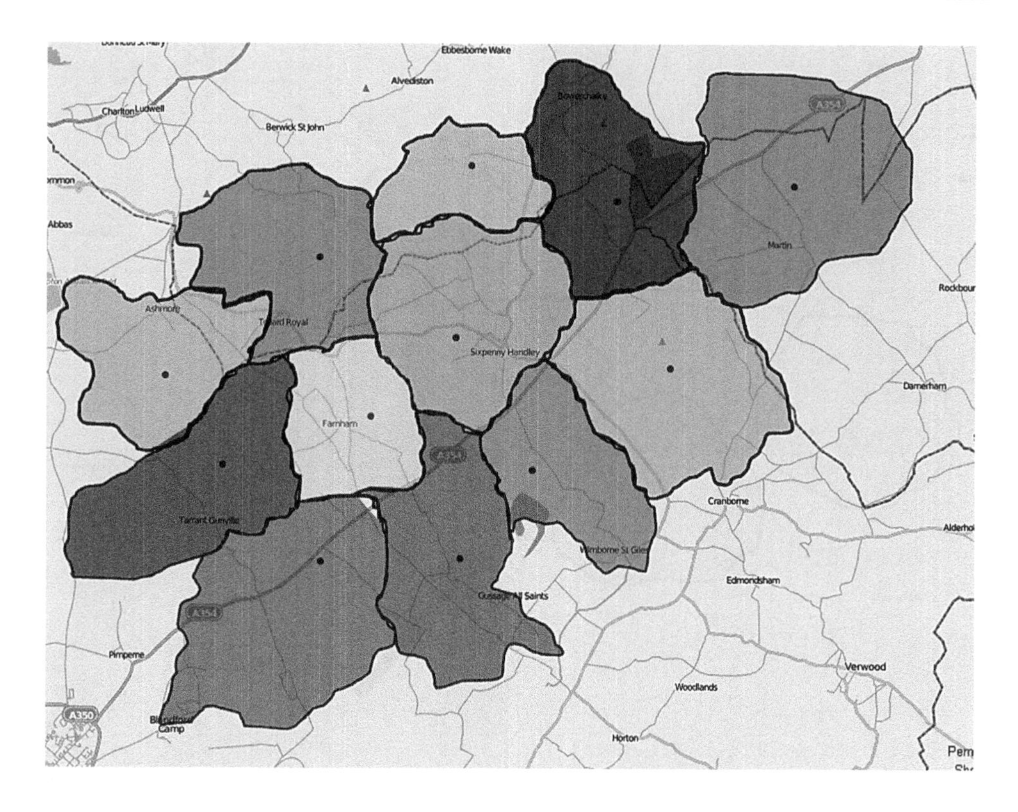

Figure 13.5 Best server coverage representation. Map data © OpenStreetMap contributors, CC-BY-SA.

- Minimisation of coverage in a particular direction or over a specified arc may also be important to avoid interference or to minimise detection vulnerability.
- Coverage design to meet and not exceed a particular station's portion of the service area (the service area is the identified area to be covered by the network). This is often required to balance radio traffic over the required number of channels and to allow orderly handover between cells or sections of the network.
- Coverage design to maximise coverage within the service area and minimise it everywhere else is also a commonly desirable metric.

Some of these desirable metrics require fairly complex description so that there is a reasonable metric to measure optimisation attempts against. Often, polygons describing the geographic boundaries of the wanted performance are used, as illustrated in Figure 13.7. Without such a detailed description, it is not really possible to produce an effective design.

A network or mission planner has to make use of those parameters under his or her control. This generally does not include radio equipment design characteristics but

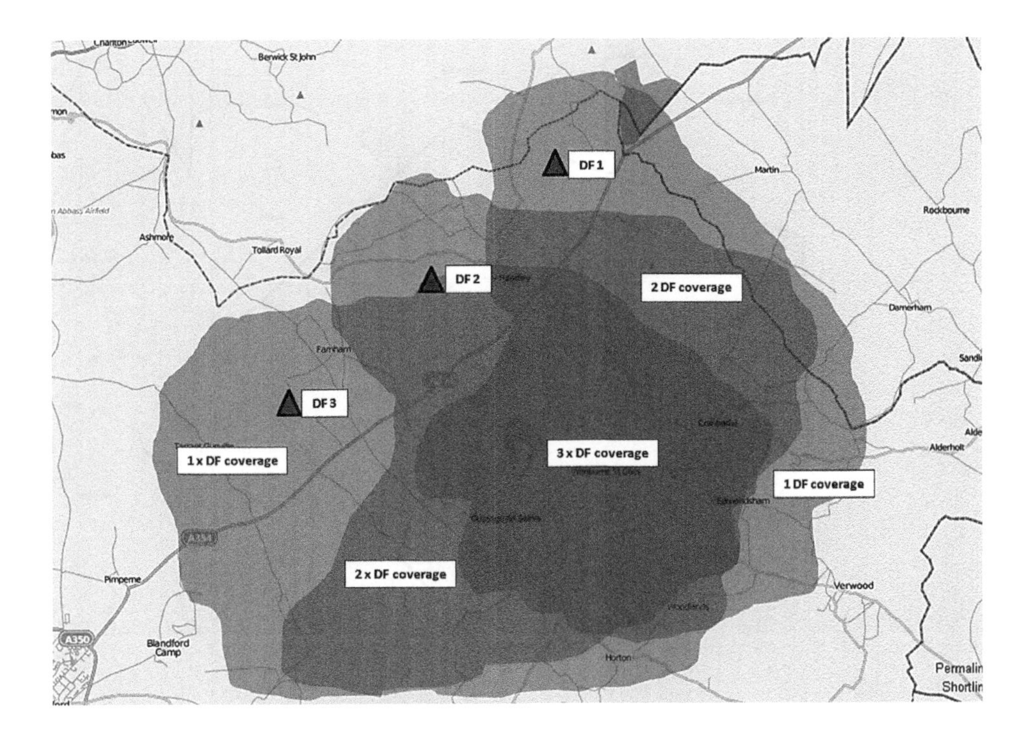

Figure 13.6 Overlapping DF coverage representation. Map data © OpenStreetMap contributors, CC-BY-SA.

rather those to do with the use of equipment already designed and selected for use. The typical characteristics that the planner can control include:

- antenna location(s);
- antenna height above local ground;
- sometimes, selection from between a number of antenna options;
- antenna direction in azimuth;
- antenna vertical tilt;
- antenna polarisation in some cases (this is not an alteration of the antenna polarisation per se but is rather the angle at which the antenna is mounted. For example, a horizontally polarised antenna rotated on its axis can be used to transmit vertically with respect to the ground);
- transmit power;
- the selection and use of amplifiers;
- transmission frequency (although this has to be within the constraints of frequencies available to the planner; this is likely to be significantly less than the system is capable of producing);

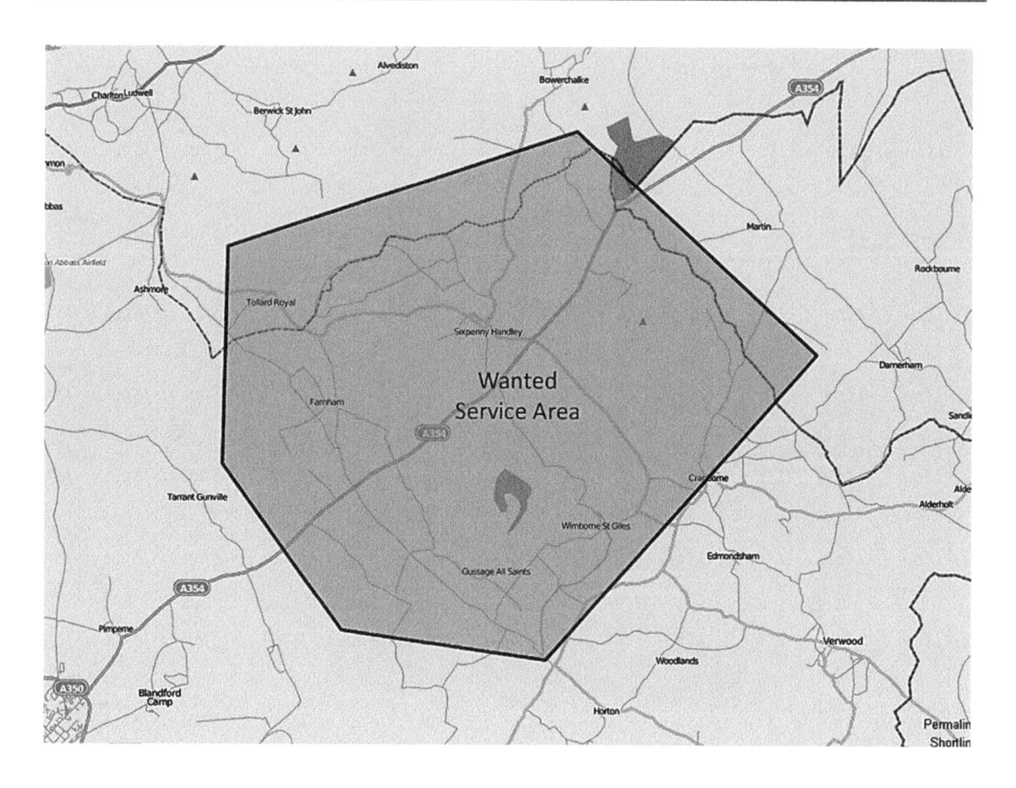

Figure 13.7 Use of a polygon to represent a wanted service area. Map data © OpenStreetMap contributors, CC-BY-SA.

- modulation scheme, for those systems featuring multiple or adaptive modulation. This will affect coverage range at the expense of transmitted data rate;
- time slots, launch delay, spreading code selection and other time-based effects optimised to reduce interference.

We can now look at the effects of changing these parameters, as shown in Table 13.1.

We can now look at some illustrations of changing some of these parameters. Figure 13.8 shows the effect of increasing antenna height in two different conditions. In the first on the left hand side, the terrain is flat and the effect of increasing antenna height is to provide good additional coverage. On the right, there is variable terrain with a hill obstructing coverage to the southwest. In this case, the effect of terrain reduces the effect of increasing antenna height is less pronounced and, where the hill is, virtually no improvement is seen. This is a typical type of effect, showing that changing antenna height is highly dependent on the environment.

The concept of moving an antenna to a more central location is fairly simple, as illustrated in Figure 13.9. In the left hand figure, the base station antenna is on the left hand side of the wanted area and there are gaps to the West. In the second, the antenna

Table 13.1 Effects of changing parameters to optimise site performance

Parameter	Change made	Effect
Antenna location	Move antenna to a higher location	Generally improves coverage; dependent on surrounding terrain
	Move antenna to the top of a hill	Improves coverage in all or many directions. May introduce coverage gaps around the foot of the hill, close to the antenna
	Move antenna closer to service area or more centrally within the service area	Can move more of the coverage into the service area and reduce amount wasted in surrounding areas
	Move antenna to the side of a hill	Provides terrain shielding in the direction of the hill. Good for reducing detection vulnerability
	Move antenna to a lower location (e.g. a valley)	Reduces overall coverage; reduces detection vulnerability
	Move antenna into an urban location	Reduces overall coverage but can optimise it for the urban environment
Antenna height	Increase antenna height	Improves coverage
	Reduce antenna height	Reduces coverage
Antenna selection	Change to a higher gain antenna	Improves gain in direction of main beam, reduces it in other directions. Good for point-to-point links, not so good for coverage-based systems
	Change to a lower gain antenna	Provides lower signal in direction of main beam but increases it in other directions
Antenna direction	Change antenna direction	Optimises main beam of directional antennas in antenna pointing direction at the expense of other directions
Antenna tilt	Increase vertical tilt	Vertical tilt is normally expressed in degrees DOWN from the horizontal hence increasing vertical tilt reduces coverage at a distance
	Reduce vertical tilt	Increases longer range coverage
Antenna polarisation	Change polarisation by rotating antenna along the axis of polarisation	Can be used for interference minimisation or to correct links where terrain modifies polarity between transmitter and receiver
Transmit power	Increase transmit power	Increases range, but typically only by a few dB. E.g. changing from 10W to 15W only increases power by 1.76 dB
	Reduce transmit power	Reduces range inverse to increasing it. Can be used to manage transmit distance

(*continued*)

Table 13.1 (*Continued*)

Parameter	Change made	Effect
Amplifiers	Add a transmit amplifier	Improves radiated power out of the antenna, however introduces a risk that bi-directional links will be unbalanced and that the return link will not work as well as the transmit path
Frequency	Increase frequency	Generally reduces range, all other factors being constant
	Reduce frequency	Generally increases range, but may not support higher modulation schemes
	Select specific frequency	Used in frequency assignment to minimise interference
Modulation	Move to higher modulation scheme	Reduces range but increases data rate
	Move to lower modulation scheme	Increases range at the expense of data rate
Time slot	Select particular time slot	Used to minimise interference
Spreading code	Select particular spreading code	Used to minimise interference
Hop set parameters	Select starting point in hop set	Used to minimise interference
Launch delay	Increase launch delay	Can be used in simulcast networks to move interference to less important areas. Increasing launch delay generally mean interference occurs further away from station
	Reduce launch delay	As above, but moves interference closer to station. Also, increased signal power in these locations may counteract interference

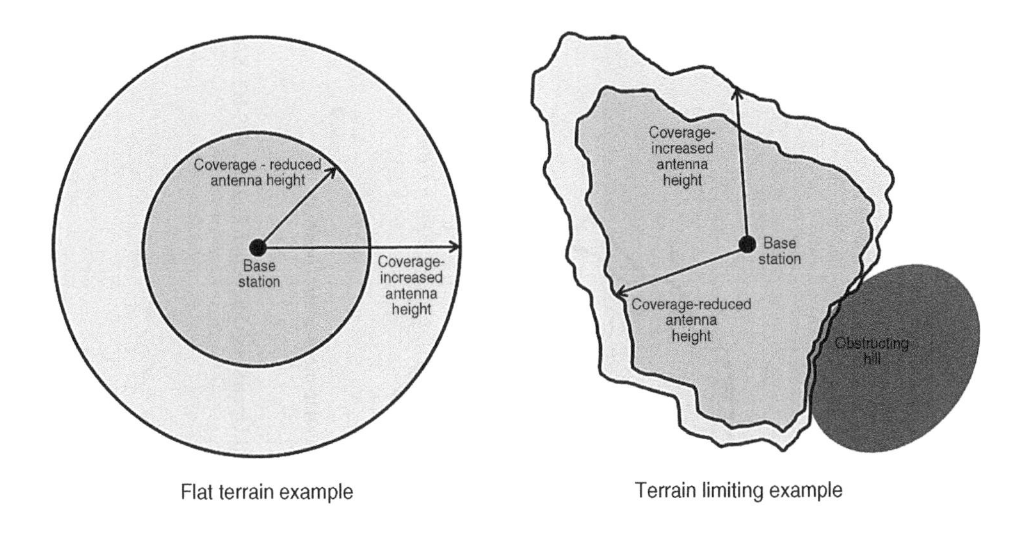

Figure 13.8 Effects of changing antenna heights in non-terrain limited and terrain limited conditions.

has been moved to a more central location. However, in practice, such things are rarely so straightforward and the situation shown where the wanted service area is fairly close to the coverage area is rare. In general, the design will be far more complex.

The effect of antenna tilt is illustrated in Figure 13.10 for a directional antenna on flat terrain. When the antenna tilt is increased, the coverage in the direction of the main beam is foreshortened in this case. This will depend on the vertical pattern grazing angle, which must be taken into account allow with the antenna vertical polar diagram. When correctly engineered, the coverage can be adjusted to provide coverage in the wanted area and to limit coverage beyond it. This helps both with interference (and jamming) vulnerability and anti-detection.

The effects of adaptive modulation on range are illustrated in Figure 13.11 for a WiMax system, with all data rates representative of typical rates (but not exact). The key parameter that affects the range performance is the Signal to Noise Ratio (SNR). The relative ranges are also indicative only, but show the principle that accepting a lower modulation scheme, with concomitant reduction in required SNR provides longer coverage at the expense of data rate.

The principles of modifying parameters to optimise system performance are fairly straightforward, but applying them can require either planner experience or careful design of automated tools. However, such optimisation is a necessary part of modern network design or mission planning.

13.3 Limiting Coverage

We have identified some methods of reducing coverage, but now we will look at some reasons to do so and in a little more depth at how this can be achieved in practice, with

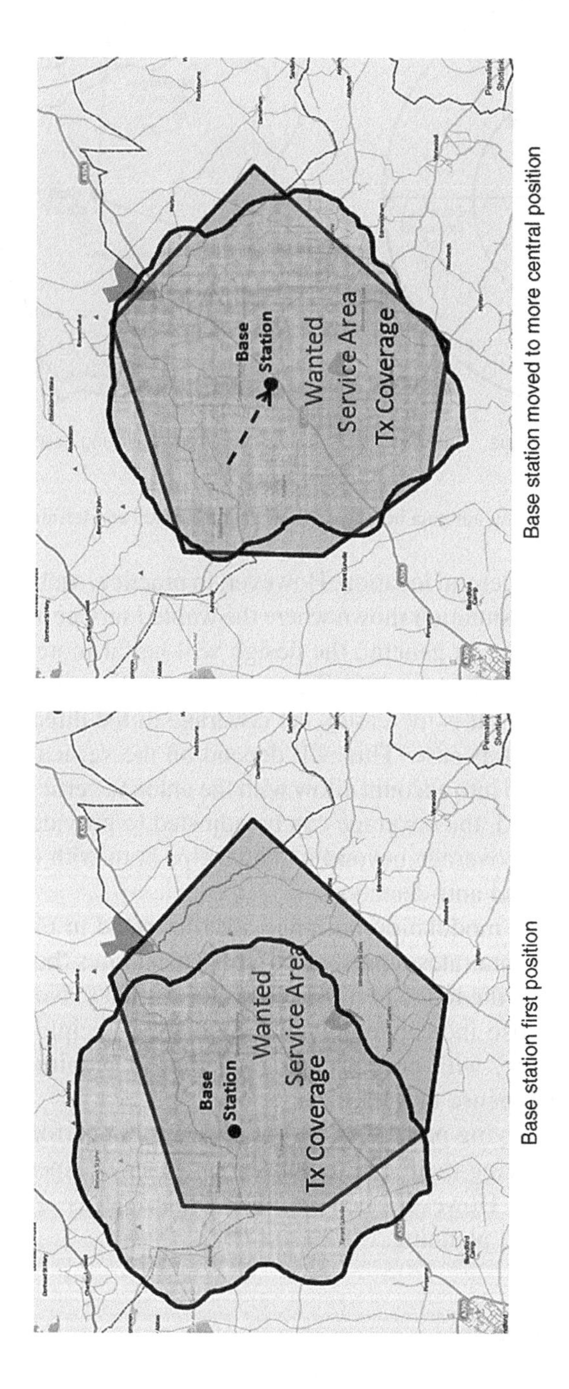

Figure 13.9 Effect of moving a base station towards the centre of the wanted service area. Map data © OpenStreetMap contributors, CC-BY-SA.

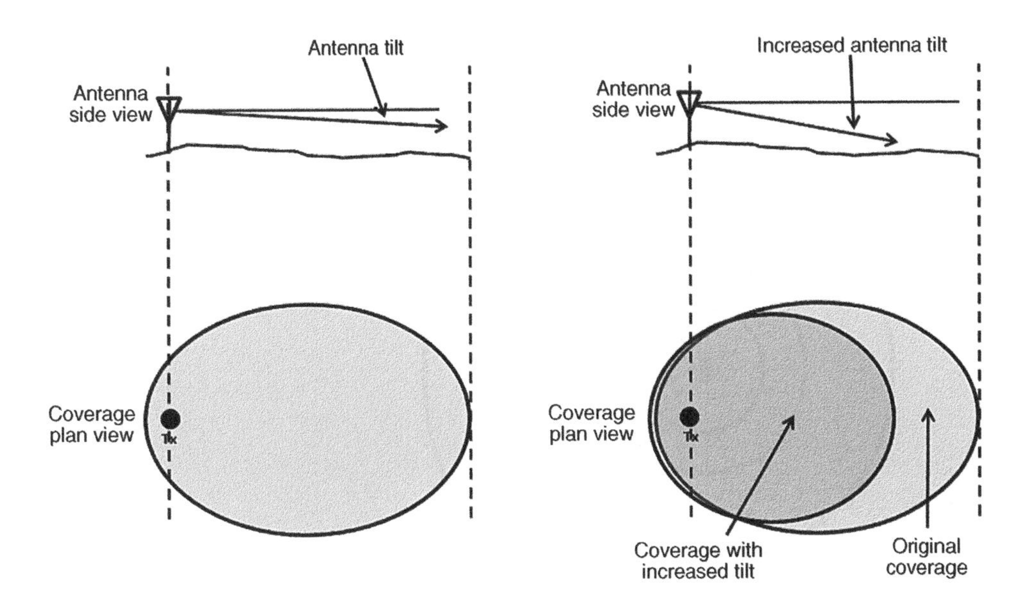

Figure 13.10 Effects of antenna tilt changes.

the emphasis on systems working in an electronic warfare environment where there is a need to avoid detection, interception, localisation and jamming. In essence, we wish to be able to use a communications system while minimising vulnerability by reducing the energy available to the enemy and reducing incoming jamming energy from enemy jammers. We can do this by a number of different means, all of which require an understanding of the operational picture and the relative positions of our own communications and enemy assets. However, it is not necessary to have localised enemy assets, although if possible this is desirable.

The first potential method is to use terrain as a deliberate shield against the enemy. Figure 13.12 shows an illustration of a scenario where the enemy is to the North, with a line of demarcation (Forward Edge of the Battle Area [FEBA]) shown as a straight line running across the terrain. In this case, we can see a range of hills which we can use for terrain shielding. These are shown towards the East side, with the dark areas showing steep south-facing slopes. These are facing directly away from the enemy side, wherever enemy assets are placed in this area.

If we examine a path from our own side to the enemy side as illustrated in Figure 13.13, we can look at the terrain along the path.

This is shown in Figure 13.14. Here we have positioned a transmitter on the south-facing slope fairly close to the FEBA. The intention of this transmitter is to provide coverage to the West, East and South but little to the North. The two hills provide shadowing to the north, including a dip within our own territory. However, from the perspective of the enemy side, the energy radiated from the transmitter is shielded by the terrain.

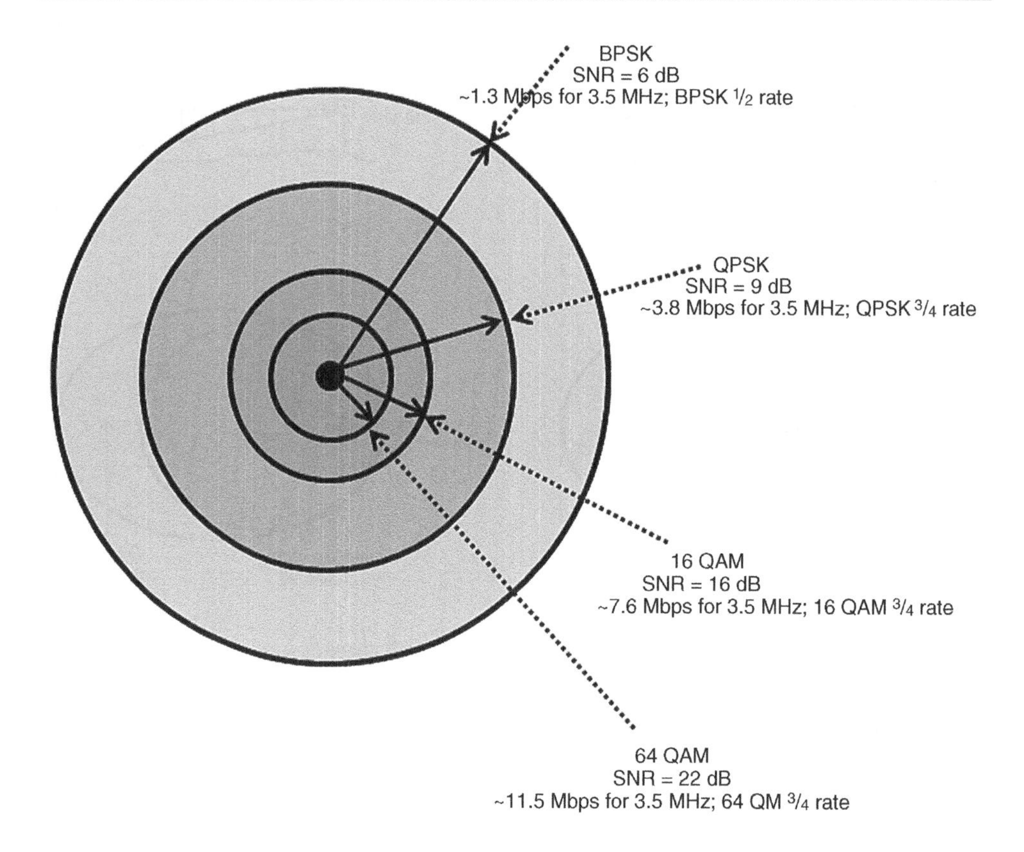

Figure 13.11 Adaptive modulation example.

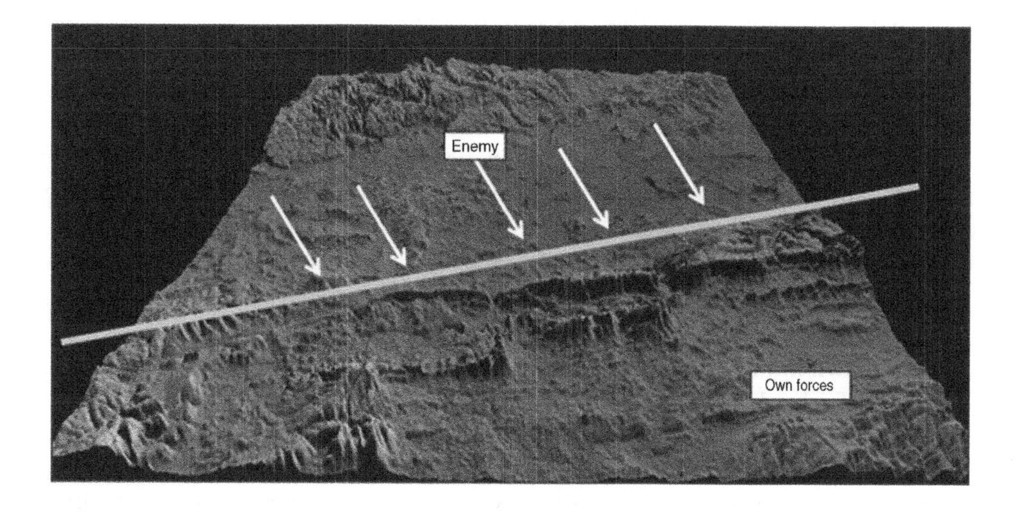

Figure 13.12 A scenario where coverage limiting would be useful.

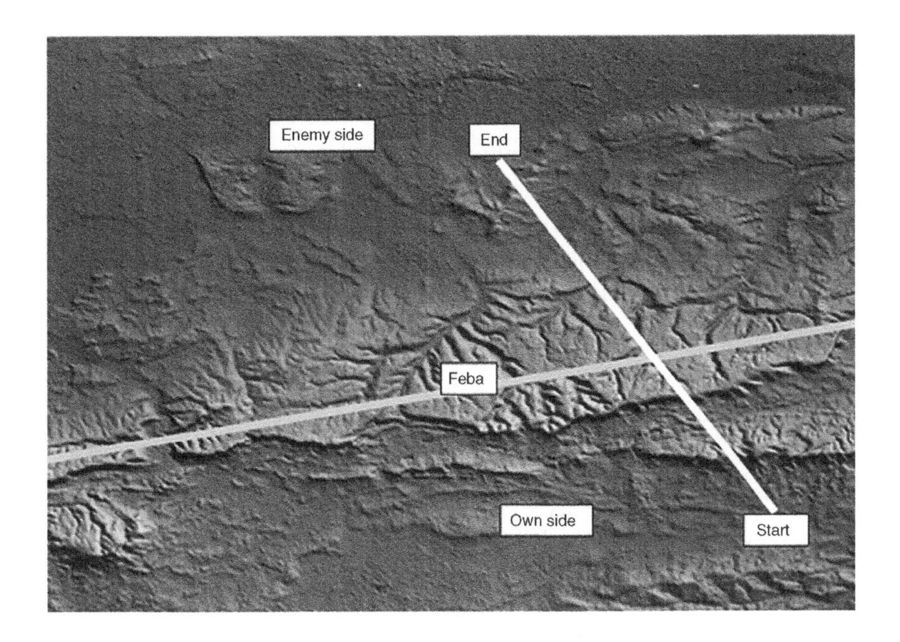

Figure 13.13 A path chosen to examine the terrain towards enemy ES assets.

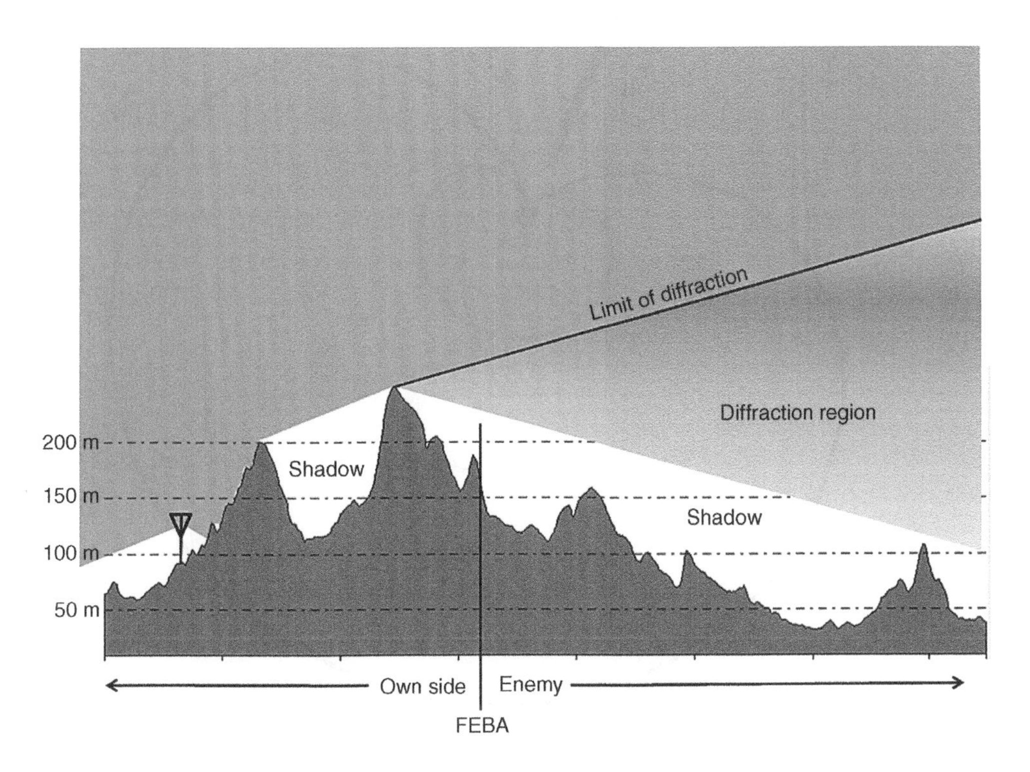

Figure 13.14 Examination of the selected path.

We could use the same method to add another base station in the leftmost shadow region to provide communications in the dip without providing detection opportunities to the enemy.

Note from the figure that terrain shielding is not only effective against ground-based sensors and jammers. The terrain shielding extends high into the atmosphere over the enemy area, meaning that airborne sensors or jammers need to fly higher over their own territory, or fly over ours in order to be in effective range of our transmitter. In both cases, the airborne platform is more vulnerable to ground or air fire.

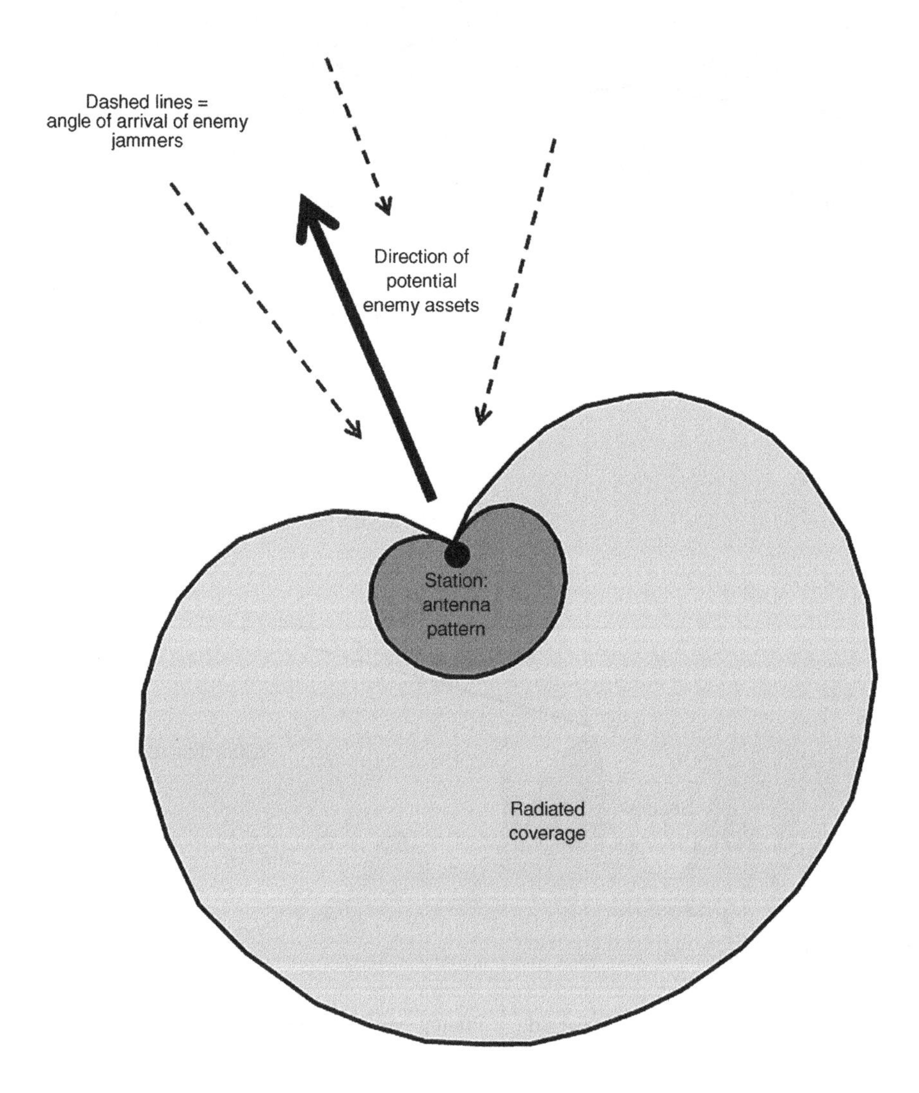

Figure 13.15 Use of a directional antenna to limit signal energy towards the enemy.

Also, note that although we are examining an individual path profile, the lie of the land means that what is true for this profile is also generally true for the rest of the operational area under consideration.

As well as terrain shielding, we can also use large clutter obstructions in the same way. This can be dense urbanisation or high trees or jungle. The skilled planner will use whatever factors work best in his or her advantage in the particular circumstances of the situation.

A second effective method of managing transmitted energy is to make good use of available antennas in order to direct energy in the wanted directions and away from vulnerable regions.

Figure 13.15 shows an example of using a cardiod antenna, which has up to 40 dB of null in the back lobe direction. The pattern is shown on the horizontal plane and terrain is ignored. In this case, where the direction of the enemy is as shown, energy arriving from enemy jammers will have to overcome up to 40 dB of energy rejection before being able to influence the communications station. This adds a great deal of difficulty for enemy jammers to achieve.

Use of antennas with narrow vertical patterns can also be effective against the threat of airborne enemy systems. This is shown by a slightly simplified diagram in Figure 13.16. The top image shows an antenna with a relatively wide vertical beam

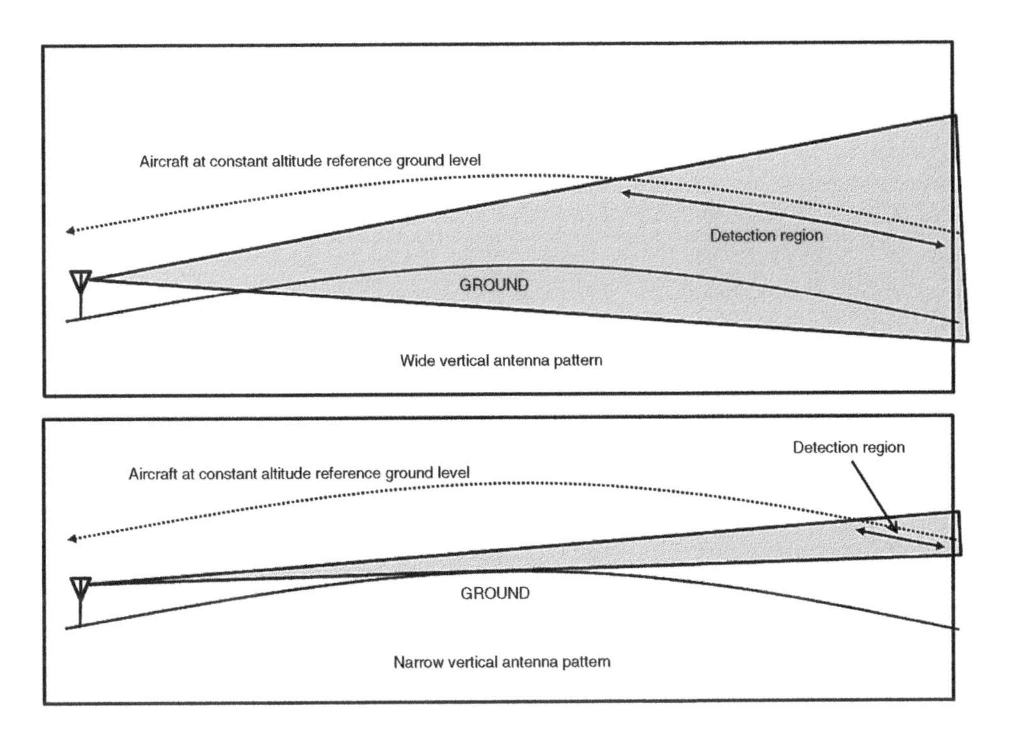

Figure 13.16 Effects of reducing antenna vertical beamwidth.

compared to the one on the bottom. An aircraft travelling at the same constant height above the terrain in both diagrams will have a wider regime in which to detect transmissions from the radio antenna at the left of both diagrams when the vertical pattern is wider than when it is narrower. Also, with a correctly tilted antenna, the vulnerable area can be made to be further away than for the wider vertical pattern, with consequential drop in detectable signal power.

References and Further Reading

Freeman, L. (2007), *Radio System Design for Telecommunications*, John Wiley & Sons, NJ, USA, ISBN 978-0-471-75713-9.

Graham, A.W.; Kirkman, N.C.; Paul, P.M. (2007), *Mobile Radio Networks Design in the VHF and UHF Bands: A Practical Approach*, John Wiley & Sons ISBN 0-470-02980-3.

Hess, G.C. (1993), *Land Mobile Radio System Engineering*, Artech House, ISBN 978-0890066805.

For information on the map images used, please visit:
www.openstreetmap.org and www.creativecommons.org.

14

Interference Analysis

14.1 Introduction to Radio Interference Analysis

Figure 14.1, reproduced from an earlier diagram for convenience, shows the basic effect of a radio interferer.

The basic influence of interference is to raise the noise floor of a radio or EW receiver. It also affects radars in the same way, if the interferer raises a radar receiver noise floor. In fact, interferers have the same effect as a noise jammer. The effect on radio system performance is to reduce the range from which a transmission can be detected or received at a given level of link performance. Figure 14.2 shows the received signal power for a high mobile phone base station operating in an urban environment. Terrain and diffraction are not considered, but this is based on the typical inverse exponential fall-off rate found in such scenarios.

The effect of an interferer is to increase the power necessary at the base station receiver to counter the increase in noise caused by the interferer or jammer. Figure 14.3 shows the effects of 10, 20 and 30 dB of jamming (or interference) present at the receiver.

From Figure 14.3, if the non-interfered base station has a maximum range of just over 19 km, 10 dB of jamming or interference reduces this to approximately 10.5 km. Adding another 10 dB of jamming reduces the range further to less than 6 km, and another 10 dB reduces this to just over 3 km. Notice that the reduction in range is less the closer the mobile is to the base station. This shows that at some point, communications will be possible despite the presence of the jammer. In jamming scenarios, this is known as burn-through. The purpose of jamming therefore has to be to reduce the effective range of communications to the point that it is useless to the enemy, rather than absolutely denying the enemy communications at all. Of course, this is not the objective in the case of inadvertent interference, but the effect is the same.

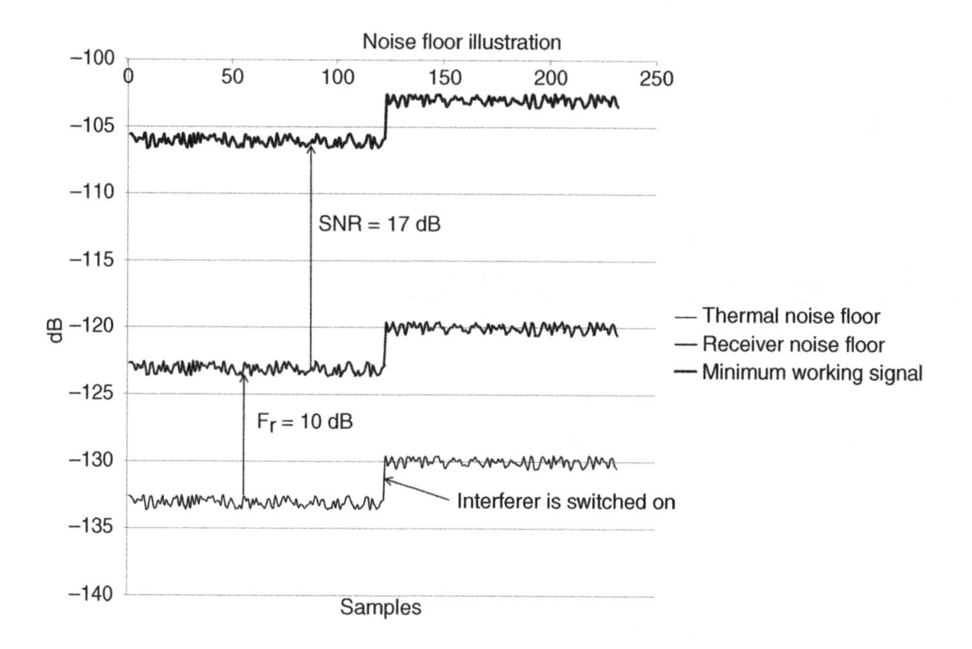

Figure 14.1 System performance in the presence of an interferer.

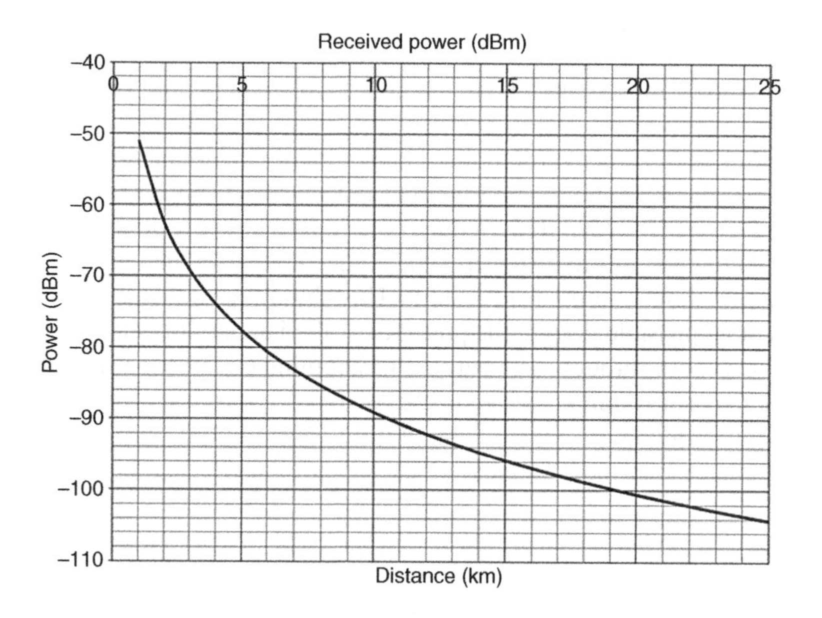

Figure 14.2 Inverse exponent method used to assess nominal range of a radio system.

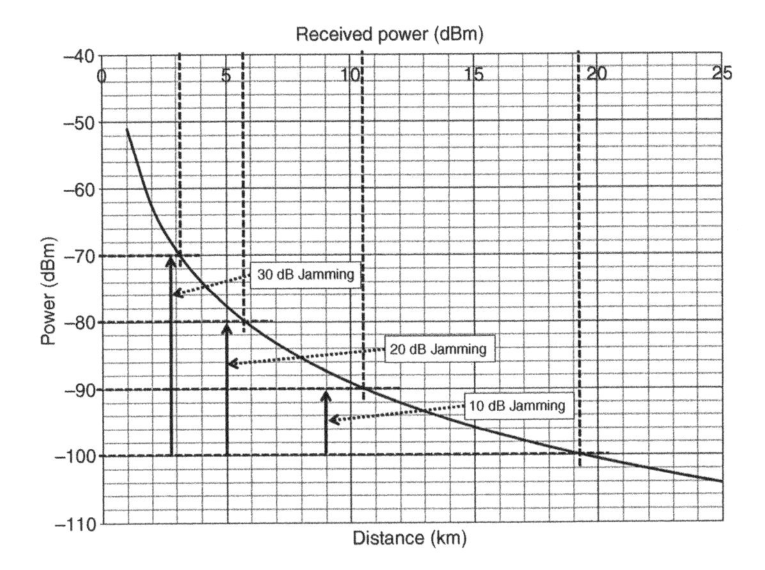

Figure 14.3 Effect of jamming on the range of a mobile phone system. The effect is to reduce the range that the receiver can successfully receive signals from mobile systems. Noise jamming and interference have exactly the same effect.

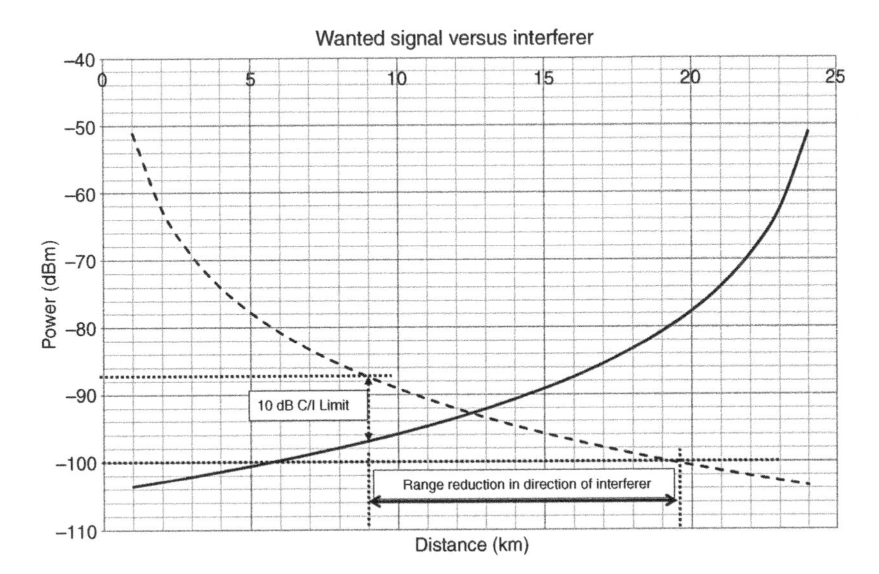

Figure 14.4 A wanted base station and an interferer of equal power 25 km away. The power of both is the same 12.5 km away. Assuming that the wanted signal has a C/I of 10 dB then range reduction occurs where the interfering signal is within 10 dB of the wanted power.

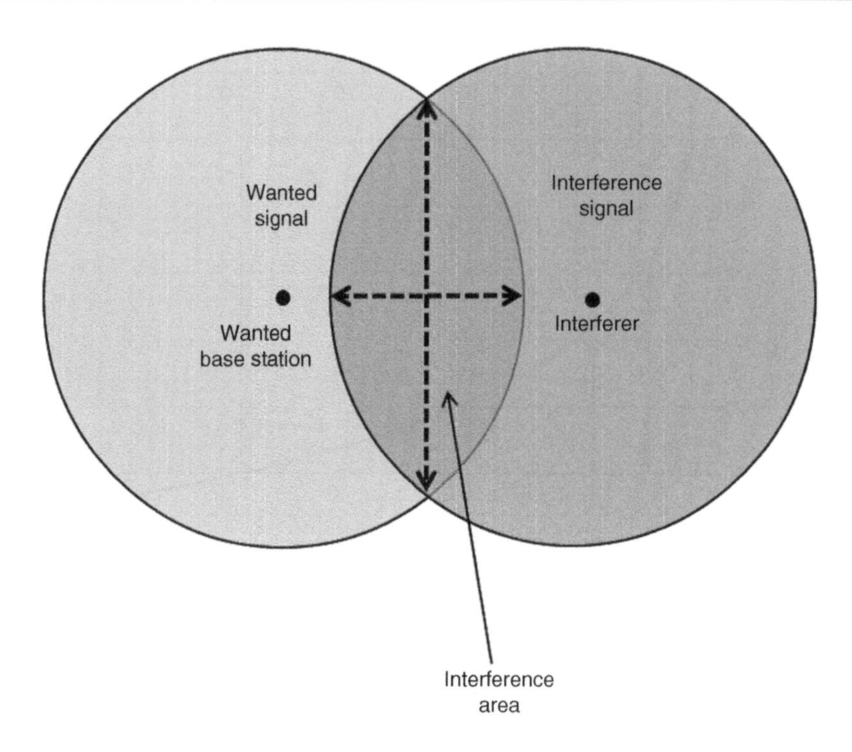

Figure 14.5 An interference scenario for mobiles in a mobile network. The original wanted service area is circular, but the area closest to the interferer suffers from interference.

The analysis just described looks at interference on a base station, but it is also possible to jam mobiles in a network, even when their location is not explicitly known. Consider the system shown in Figure 14.2. Now, if an interferer of equal power is placed 25 km away. Assuming the power of both systems fall off at the same rate, their power will be equal at half the distance, i.e. 12.5 km. This is illustrated in Figure 14.4.

The interference to mobiles will not be equal over the whole service area. An idealised case is shown in Figure 14.5. This ignores terrain and clutter. The wanted service area is shown on the left, with circular coverage. Coverage to the same power level for the interferer is shown on the right hand side. The area of interference to mobiles is the centre region. The interferer has 'taken a bite' out of the wanted coverage area closest to the interferer.

The same principles can be applied to multiple interferers and to terrain and clutter limited coverage.

14.2 Fading Considerations

So far we have looked at median field strength and have not considered the effects of fading. As we have seen in previous parts of the book, fading is almost always present

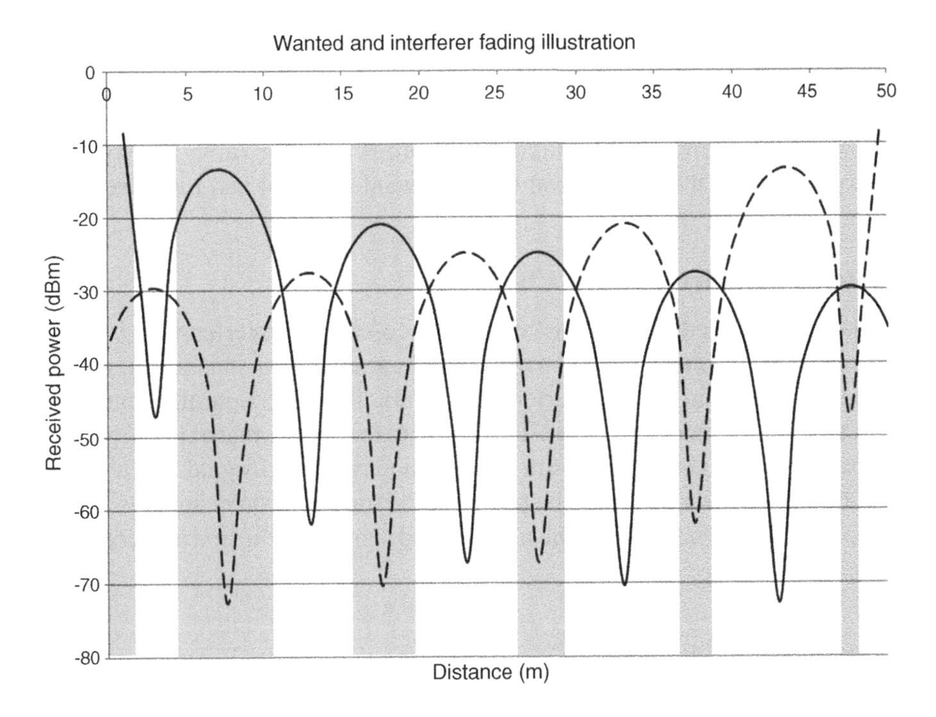

Figure 14.6 Fading overlaps of a wanted signal (solid black line) and an interferer (dashed line). The grey areas shown where the faded wanted signal exceeds the faded interferer by the minimum C/I of 10 dB, and thus communications would work despite the interferer. In all other areas, the interference will block successful communications.

to some degree, and this is the case when considering interference as well. This means that for radio systems subject to interference, both wanted signal and interferer are both subject to fading. Figure 14.6 shows an example of fading for a wanted signal and an interferer of similar power levels. We would not expect the fading between the wanted and interferer to be correlated unless they happen to be placed in exactly the same location. This is highly unlikely in practice. The figure shows the wanted signal variation due to fading in black solid lines, and the interferer in dashed lines. If we assume that the minimum C/I ratio the wanted signal can tolerate is 10 dB, this is true at the micro scale as well as the macro scale. In the figure, the grey areas show where the wanted signal exceeds the interferer by at least 10 dB and communications would work. As the level of the interferer rises towards the right of the diagram and as the strength of the wanted signal decreases, then the amount of grey diminishes, showing that interference is more prevalent. However, even at the left hand side of the diagram, interference is present as the faded wanted signal is reduced in strength.

Fading is not normally modelled explicitly in most radio prediction systems. Instead, as we have seen, it is more often modelled by applying statistical methods to determine signal variation. Exactly the same process can be applied to interference systems. The easiest method is to determine the median value at each point, as is normally returned and then determine the margin that needs to be applied to correct for availability. Using Rayleigh fading as an example, a Rayleigh correct of 0 dB provides the median value, whereas 90% availability requires a margin of 9.7 dB. This can be taken into account in the comparison between wanted signal and interferer.

14.3 Interference from other Channels

The scenarios discussed so far have considered co-channel interference, where both the wanted signal and interferer are on the same channel. But co-channel interference is not the only situation in which interference occurs. Interferers on different channels can also cause interference problems. This is due to the energy that extends beyond the wanted bandwidth as shown in Figure 14.7 for a narrowband signal and a wideband one. The nominal power down on the power in the transmission channel is shown for the first four adjacent channels for the narrowband signal and the first adjacent channel

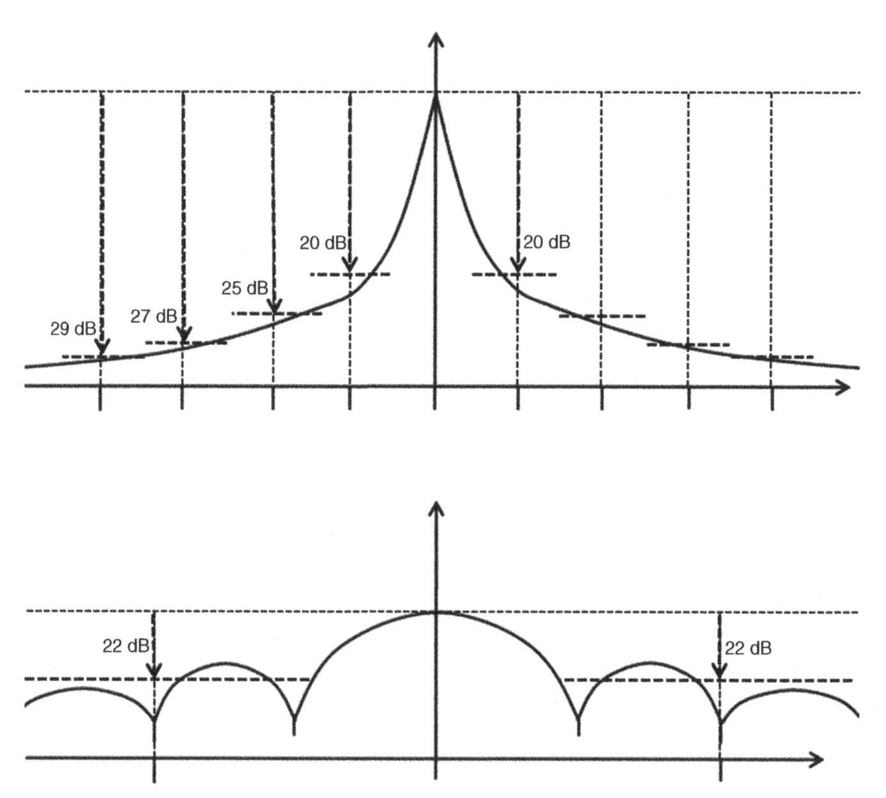

Figure 14.7 Power reductions in a narrowband signal (top diagram) and a broadband signal (bottom diagram). This shows that energy is present outside of the channel of transmission. This can cause interference to systems on other frequencies that are close to the frequency of the interferer.

Table 14.1 Corrections to out of band response to provide corrected C/I figure for interference assessment. A value of 7 dB C/I has been used to produce this table

Channel offset	Power down on interferer channel (dB)	Corrected C/I value (dB)
0	0	7
1	20	−13
2	25	−18
3	27	−20
4	29	−22

in the broadband case. It is normally assumed that the reduction is symmetrical above and below the wanted channel.

The signals shown in the figure are not well-filtered. In practice, we would expect lower out of channel responses. However, the figure demonstrates the principle, which we can now build on this to look at the effects of adjacent channel interference. Assume that the narrowband signal requires 7 dB C/I to function acceptably. In this case, we can use the figures in the figure to calculate the equivalent power in the adjacent channels required to cause interference to a wanted signal. This is illustrated in Table 14.1.

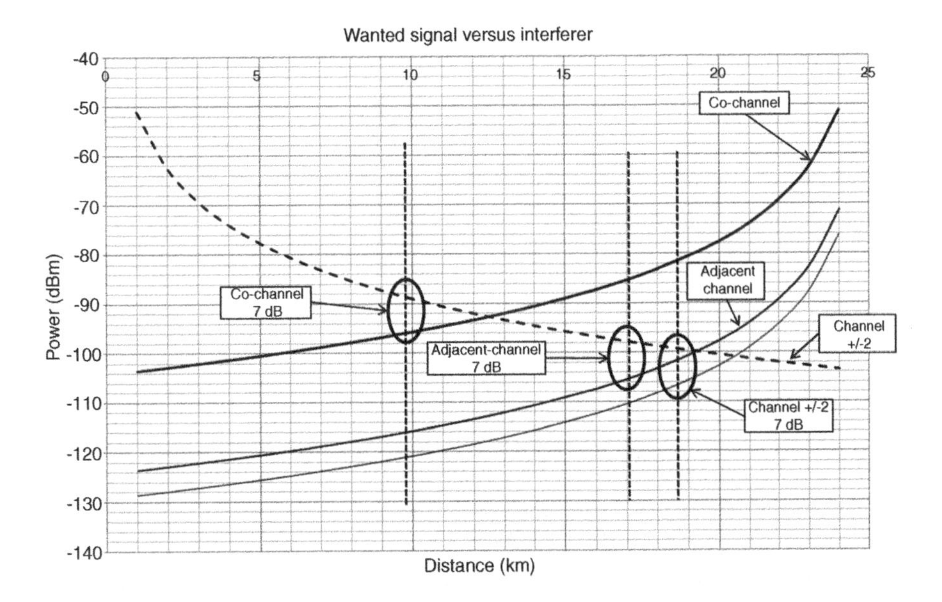

Figure 14.8 Co-channel, adjacent channel and the next channel (+/− 2 times the channel spacing) interference. Although these channels have less power than the channel the interferer is tuned to, excess energy can still cause interference outside the band. The point at which interference occurs in distance is shown. At the edge of coverage, the interference power may be below the receiver sensitivity.

The interpretation of Table 14.1 is that if an interferer is on-channel with a victim, the victim power at any point has to be 7 dB above the interferer. If the interferer is one channel away from the victim, then the power in the interferer channel has to be 13 dB above the level of the victim in its own band to cause interference, because the signal in the victim channel is 13 dB down on the power in the transmission channel. The effects of interferers are shown in Figure 14.8. This shows the co-channel, adjacent channel and +/−2 channels, based on the reduction in power due to offset from the primary transmission channel. Although the power is reduced, it can still cause interference. The figure shows where this will occur for interferers tuned to different channels from the wanted signal as well as the co-channel case.

The figure shows the case for an interferer of the same power as the wanted signal, but it could be far higher in which case the interference would be worse and even interference from several channels away could still be troublesome. As described in Section 14.2, fading should still be considered.

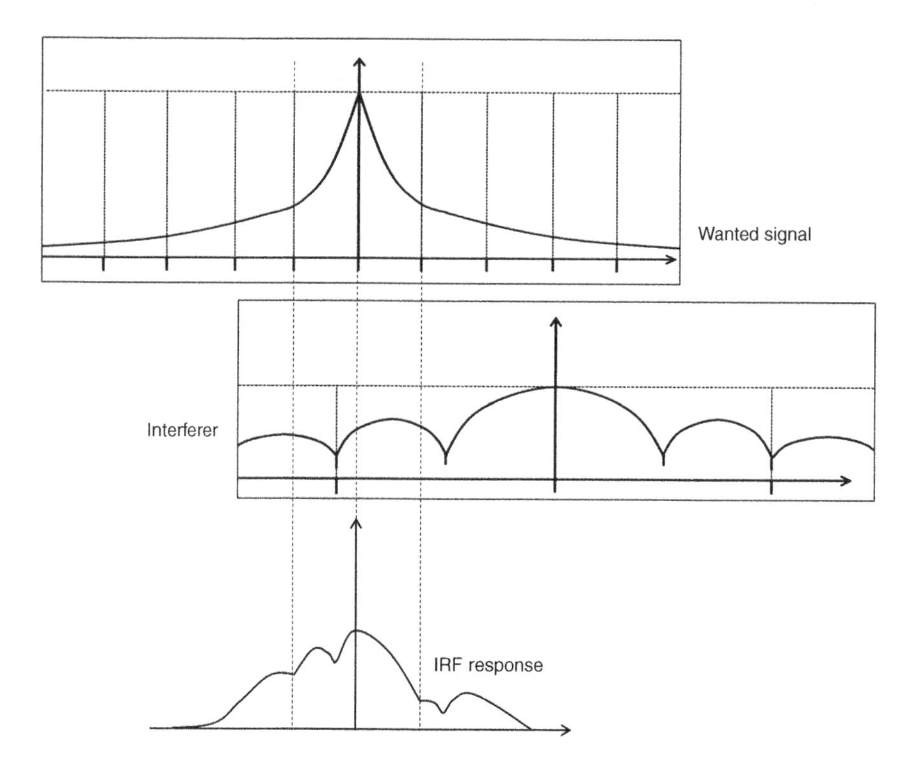

Figure 14.9 IRF used to assess the interference into the wanted receiver by a dissimilar and offset interferer. The receiver response is assumed to be matched to the spectral characteristics of the transmitted wanted signal. The IRF response is a composite of the receiver response and the spectral characteristics of the interferer. This is a slight simplification; in fact, the response is the mathematical convolution of the receiver response and the interferer spectral characteristics. However, the diagram is useful in understanding the general approach used.

14.4 Different Ways of Representing Co-existing Signals

The adjacent channel interference can be represented well for similar systems using the C/I method. However, it is not comprehensive enough for expressing dissimilar systems. In that case, the Interference Rejection Factor (IRF) method is more flexible and is a better method of calculating the effects of interference. The IRF method uses the transmission spectral power density and the receiver rejection characteristics to determine the effect on the receiver. This is illustrated in Figure 14.9. The same signals as used in the previous example are used. The narrowband signal is the wanted one, and the wideband signal, which is dissimilar to the wanted signal, is the interferer. The IRF response is dependent on both of these.

The data shown in Figure 14.9 can be used to derive the equivalent C/I figures for use in interference analysis. But the approach taken to generate these figures is more robust and flexible than the simple C/I method described previously.

The IRF approach is difficult to apply in practice because often the data required is not available. It may seem strange, but in both military and commercial environments, the required data is simply not available for legacy systems of even a few years old. However, with modern spectrum management requirements it is to be expected that such data will become available during the next few years.

References and Further Reading

Graham, A.W.; Kirkman, N. C.; Paul, P.M. (2007), *Mobile Radio Networks Design in the VHF and UHF Bands: A Practical Approach*, John Wiley & Sons ISBN 0-470-02980-3.

Hess, G.C. (1993), *Land Mobile Radio System Engineering*, Artech House, ISBN 978-0890066805.

14.4 Different Ways of Representing Co-existing Signals

15

Management Techniques for Interference

15.1 Preventing Interference

The first defence in preventing interference is the adoption of appropriate spectrum management methods. These are administrative methods used to prevent interference before it becomes necessary to adopt technical measures. This starts with administration at United Nations level via the International Telecommunications Union (ITU) as described in Chapter 2. They produce allocations that link radio services with particular bands of the radio spectrum. In doing so, they necessarily deny other, potentially interfering systems from operating in those bands. There is a clause that overrides this in the case of national defence; however, if other services are already using their allocated bands, then military services using this opt-out will be subject to uncontrolled interference in many cases. The allocations process at the international level is backed up by allotments at the national level. This is managed by the national regulator and can be devolved to other agencies by agreement. When national spectrum management is based on the command and control method, de-confliction can be avoided by these administrative methods. However, in other cases, technical methods must be used. This includes modern de-regulated spectrum management approaches and military spectrum management. In these cases, interference must be avoided by lower-level administrative methods and by technical de-confliction.

De-confliction can be effected in a number of ways, including:

- spatial methods;
- spectral methods;
- time-based methods.

Communications, Radar and Electronic Warfare Adrian Graham

These are investigated in the following sub-sections.

15.1.1 Spatial Methods

Spatial methods simply mean separating networks operating on the same frequency by distance. Clearly, two networks operating on the same UHF channel for terrestrial applications separated by several hundred or thousand kilometres apart are not going to interfere with each other, simply because any signal from the other network will be of such a low level that it cannot possible cause interference. However, this is an extreme case. In order to maximise spectral efficiency it is necessary to re-use frequencies as optimally as possible. This can be achieved by considering the following factors:

- interfering signal minimum potential interference level;
- wanted signal minimum signal level to achieve wanted level of performance;
- ability of wanted signal to reject interference from co-channel sources;
- extent of wanted service area(s);
- shielding provided by terrain or clutter features;
- use of directional antennas.

It is possible to start by considering the minimum faded signal required to achieve the wanted level of performance. For example, if this is −100 dBm, then with an approximate 10 dB margin to account for the required level of availability, then the minimum acceptable median signal level is −90 dBm. If the C/I required is 12 dB, then the maximum interference level that can be tolerated is −102 dBm. This can be considered either for the unfaded condition, or for a given level of faded availability. The first case would ensure that interference does not occur at all, but it is sub-optimal in terms of critical engineering. Instead, it may be possible to consider the wanted to unwanted interference level under different levels of fading. Thus, if we were to consider 90% availability for the wanted signal and 5% availability for the interferer as being acceptable, then the combination of the two calculated values to achieve 12 dB minimum C/I can be determined. This is illustrated in Figure 15.1. This shows two signals, a wanted signal and an unwanted signal. They have been spaced vertically to provide more room, but the horizontal axis shows the difference in received signal power. The interference difference is the value that corresponds to 12 dB difference, based on 90% availability for the wanted signal and 5% for the interferer. Also shown is the difference in median power received. This is shown because in many mission planning and radio planning tools the median value only is calculated. The difference, based on the statistics of the faded signal can then be used in such tools to allow direct comparison based on the median received power.

If, in the previous example, the minimum wanted value is −90 dBm for 90% availability and the correction from the 90% value to the median value is 10 dB,

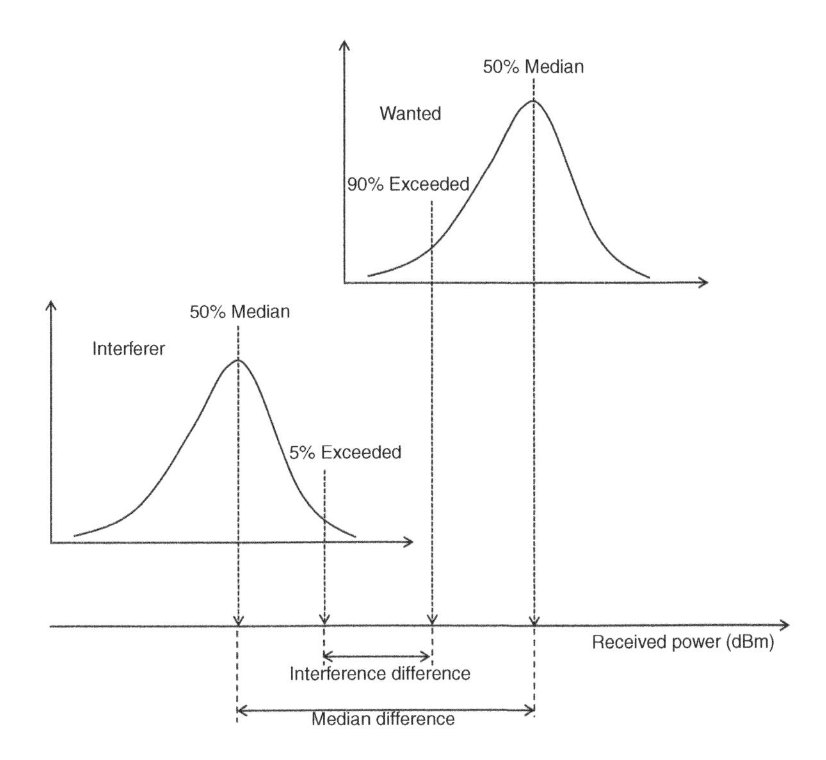

Figure 15.1 Diagram showing two signals (vertical difference between the two signals has no relevance), a wanted and unwanted signal. The difference in the signal level between the 90% exceeded (wanted) and 5% (unwanted) is shown on the horizontal axis, along with the median difference.

then an additional 10 dB needs to be added, giving −80 dBm. If the correction for the interferer between the median value and the 5% value is 5 dB, then the maximum signal that can be present for 5% availability is 12 dB down on this and the median value must be 17 dB down. Thus, the minimum median signal level that can be used is −80 dBm and the maximum median signal level of the interferer is −97 dBm.

Once the minimum acceptable signal level for the wanted level of performance is known, three different approaches can be taken to determine where this signal level must be protected. This is shown in Figure 15.2. The outer area uses a very simple prediction model to determine the maximum possible extent of the wanted level of signal. This produces a circular area to be protected. The middle area is based on the wanted service area, which may be smaller than the area calculated by the first method. This produces a smaller area to be protected, which is more efficient. The inner most area shows the actual predicted coverage. This is the area that really needs to be protected and is the most efficient way of describing the protection area. However, it is the most computationally intensive.

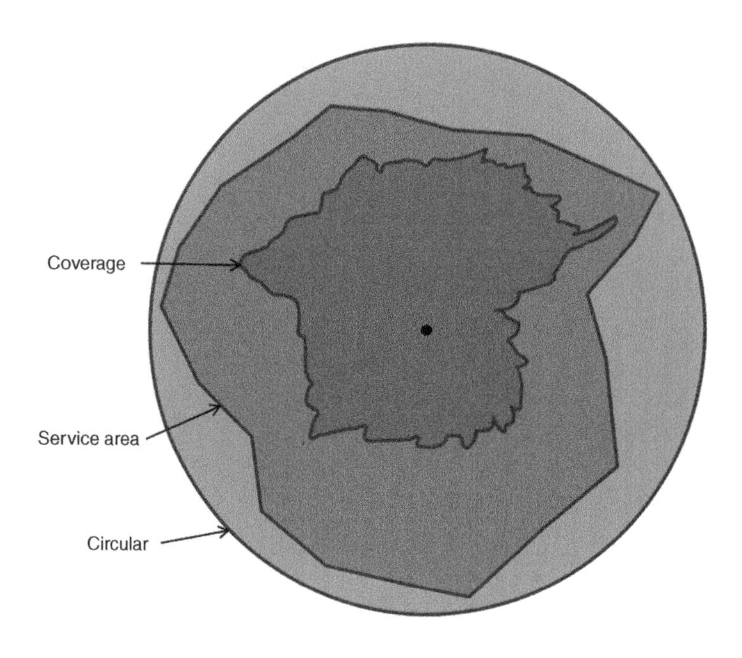

Figure 15.2 Different ways of describing a protection area.

15.1.2 Spectral Methods

We have already discussed spectral methods in some depth. However, in this section we will look at the role that filters can play in reducing interference to other services. Filters can be added to both transmitters and receivers. Filters added to transmitters will reduce the unwanted energy radiated by the transmitter. Filters added to receivers will help to reduce the unwanted energy from entering into the downstream parts of the receiver. This is illustrated in Figure 15.3 for a transmitter.

Filters fitted to receivers work in exactly the same way, in this case reducing the energy entered into the receiver from the sidelobes or any other energy not within the bandwidth of the wanted signal.

Filters are important to reduce co-site interference, as described in Chapter 16.

15.1.3 Time-Based Methods

Time-based methods can also be used to de-conflict radio systems that do not need to operate concurrently. This can be based on allotments when two radio systems do not need to be available at the same time. This could be a network that functions only during the night and another that functions only during the day. It can also be achieved between systems that are operating at the same time if the timing of messages can be planned. This is only typically possible for systems with low activity ratios.

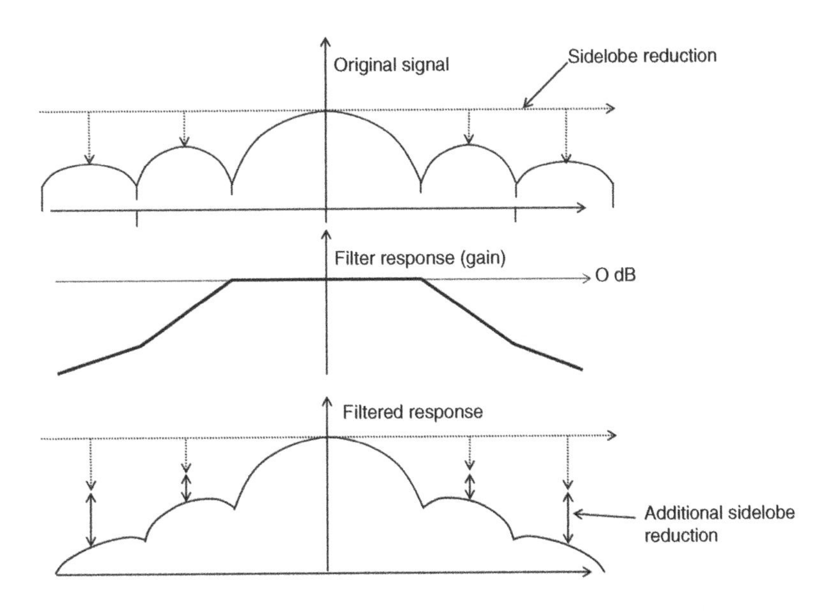

Figure 15.3 Effect of a filter on a transmission. It reduces the energy present in the sidelobes.

Highly-directional, rotating systems such as radars can also be assigned the same frequencies if they can be coordinated such that the level of interference from the sidebands will not affect the main beam. For this to work the systems must be rotating at exactly the same speed, otherwise there will be interference during some rotations. This can make the process difficult to achieve in practice.

15.2 Managing Interference

The process of interference can be managed at a number of levels. This illustrated in Figure 15.4.

The allocation process is used to separate systems and services according to the ITU regional and national table of allocations. This may change in the future with the development of more flexible spectrum management systems. The allotment process is used at national and sub-national level to allot groups of channels to particular users. Within their designated service area, the allotments can only be used by the organisation to which the channels are provided. It is up to that organisation to assign individual frequencies to specific channels. Within the structure shown, this is handled by the same spectrum manager who assigns individual channels. This is commonly the case for military systems. In this definition, the spectrum manager is the individual responsible for managing spectrum across the operational theatre. Individual spectrum management tasks may be devolved to others, either co-located

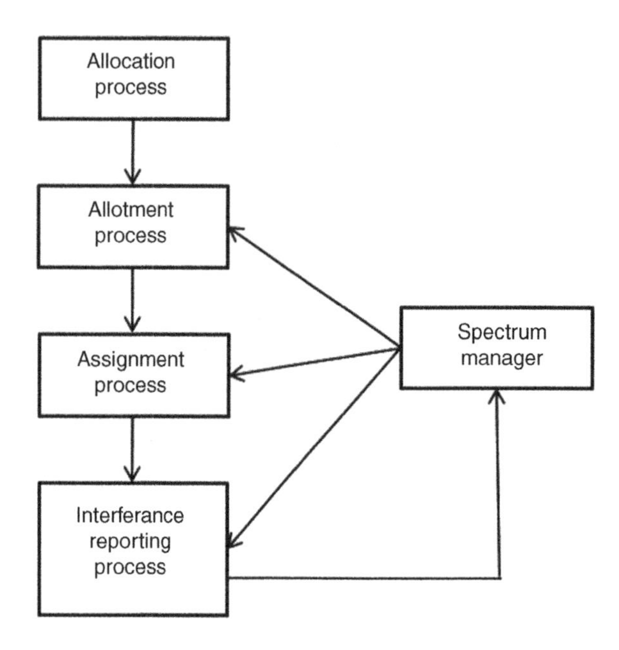

Figure 15.4 Interference management occurs at a number of levels.

with the main spectrum management team or may be separated and responsible for a specific sub-area.

Authorised spectrum users may be the first to identify interference happening to their systems. When this occurs, there is an interference reporting process. The user will inform the spectrum manager. The spectrum manager will try to identify the cause of the interference. This could be an unauthorised user, or interference from another user assigned the same channel.

The spectrum manager can investigate the interference and identify the best method to deal with the situation. This could be to switch the interferer to a different channel, switch the victim to a different channel, or determine that the interference is unfortunate but that nothing can be done. If the last is the case, then the user has to live with the interference and look at other alternatives to overcome the problem.

The EW management process is also invoked to determine whether the interference is due to enemy jamming. If this is the case, the actions to be taken against the jammer will be determined according to the rules in place for the operation.

15.3 Interference Reports

Interference reports are used by authorised spectrum users when they encounter interference. Since users only see the effects of interference, usually they cannot determine whether the problem is unintentional interference or intentional jamming. This is determined during the investigation prompted by the interference report

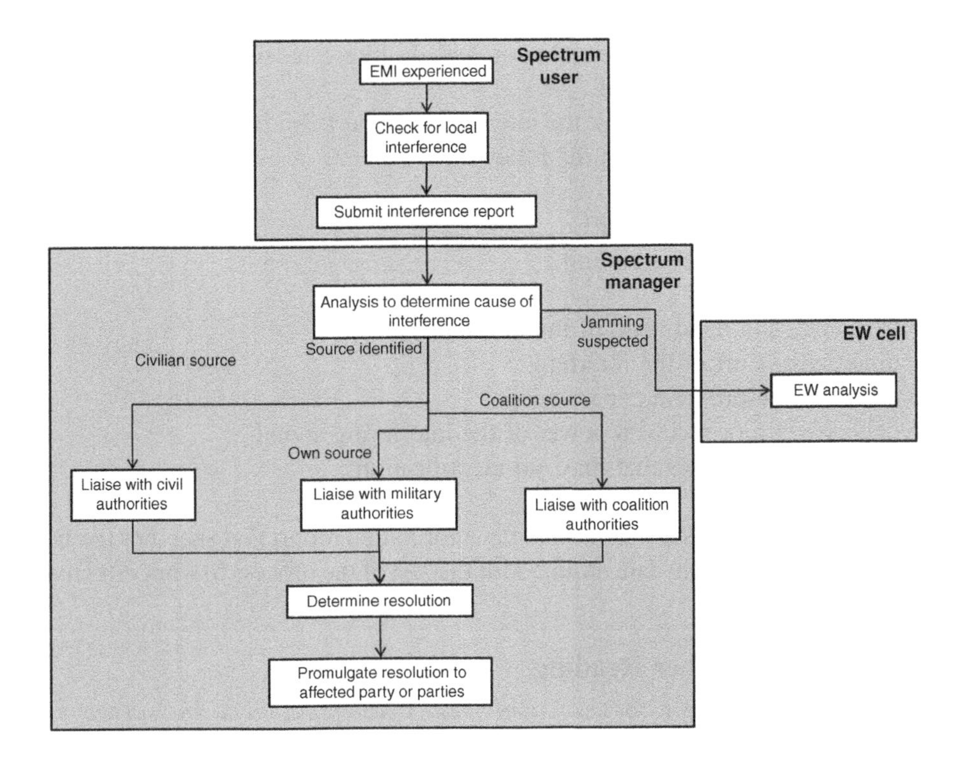

Figure 15.5 Simplified illustration of the interference method of the US JSIR process.

generated. Figure 15.5 shows a simplified version of the US JSIR (Joint Spectrum Interference Resolution) process.

Interference reports are used to report interference and more importantly to provide information that helps spectrum managers to identify the source. It is unlikely that the authorised spectrum user who reports interference will know where the interference originates, but spectrum managers should know the spectrum being used and the users assigned to specific frequencies.

The diagram in Figure 15.5 shows an approximate process to resolve interference problems. If Electro-Magnetic Interference (EMI) is experienced by a user, their first action is to determine if the interference is local and can be resolved locally. If no immediate resolution is found, the next user action is to submit an interference report. This is then submitted to the spectrum management process. The spectrum management team will try to determine the cause of the interference. If jamming is suspected, the report will be passed for action to the EW analysis cell. If the interference originates from civil sources, the military spectrum manager will liaise with civil authorities to seek a resolution. If it is caused by own military forces, it will be passed up the chain of command for resolution. If the military force is working with coalition partners and the source is traced to their systems, then again the problem will be discussed with those partners. Once a resolution has been determined, the interested parties will be informed of the mechanism to be adopted.

None of this can occur without the interference report. An interference report will normally contain as a minimum the following:

- the name of the interfered unit;
- the location of the interfered unit;
- the systems affected;
- the frequency and bandwidth of the interferer;
- the time on/time off of the interferer;
- the nature of the problem experienced;
- the field strength or received power of the interfering signal;
- any other characteristics that may aid identification.

Normally, such reports are submitted via a formatted report that specifies the fields to be used. This is the input data required for the rest of the process to work effectively.

References and Further Reading

Graham, A.W.; Kirkman, N.C.; Paul, P.M (2007), *Mobile Radio Networks Design in the VHF and UHF Bands: A Practical Approach*, John Wiley & Sons ISBN 0-470-02980-3.

Hess, G.C. (1993), *Land Mobile Radio System Engineering*, Artech House, ISBN 978-0890066805.

ITU (2008), *Radio Regulations*, http://www.itu.int/publ/R-REG-RR-2008/en.

US Joint Staff (2008), *Joint Spectrum Interference Resolution (JSIR) Procedures*, http://www.dtic.mil/cjcs_directives/cdata/unlimit/m332002.pdf.

16

Management of Interference at a Radio Site

16.1 Special Features of Radio Sites with Multiple Systems

16.1.1 Introduction

So far, we have considered interference from distant interferers. However, there is a special case where the interferers are close to the victim systems. This is where multiple transmit/receive systems are co-located within a short range of each other. This occurs in a number of real situations, including:

- military headquarters;
- military communications centres;
- airfields (Air Points of Departure [APOD]) and airports;
- ports;
- main Supply Route (MSR) communications centres;
- major radio sites;
- major relay sites;
- major surface warships;
- large aircraft.

In these cases, the number of co-located radio systems may be large, and the interference issues may be very difficult to resolve. In this case, the situation is normally described as an Electro-Magnetic Interference (EMI) or Electro-Magnetic Compatibility (EMC) problem.

Communications, Radar and Electronic Warfare Adrian Graham

16.1.2 EMC Issues

EMC is a major consideration for co-sited radio equipment. The problem stems from the difference between the power transmitted and the very small power received by radio systems. Despite attenuation provided by the transmit characteristics, filters and antenna responses, the power received even for signals far from the transmit frequency can swamp the signals transmitted from far away. It effectively causes an increase in the noise floor. The amount by which the noise floor is raised is not constant, nor is it amenable to simple calculation. The following sections describe the issues involved and how they can be accounted for in order to produce a working frequency scheme for a co-sited radio installation.

It is important to recognise that radio interference between co-sited locations does not always originate from the radio antennas alone. It can also be caused by sympathetic interference between cables, components, unshielded radio equipment, electrical wiring and unintentional features such as the classic 'rusty bolt'. Even where shielding is present, it will not provide total protection and will only provide a level of attenuation. Where the difference between transmit power and the wanted receive signal can be well over 150 dB, then attenuation of 100 dB may be insufficient to prevent interference occurring.

The next section considers the layout of a radio site and its implications.

16.1.3 Co-Site Radio Layout

Co-site locations may extend over many kilometres or may be present on a single radio mast. A radio mast may support dozens of radio transmitters and receivers, separated by (primarily) vertical distance and antenna directionality in the horizontal and vertical planes. The supporting transmit and receive cables need to run down the mast, and there is the possibility of currents inducing parasitic reactions in receiver cables. Even with a typical isolation over 80 dB provided by one cable, hence 160 dB between two cables a transmit power in one cable of 10 dBW would result in −80 dBW which is −110 dBm being present in the other cable. This could affect the noise floor of the adjacent radio system, albeit on the same channel. Any problems with cables such as partial breakages may make this problem far more acute.

Isolation between antennas also has to be managed as well, of course. ITU-SM.337-6 provides some simple formulae for assessing the isolation between dipole antennas. This can easily be modified for actual antenna gains. The three formulae provided are for vertical separation, horizontal and slant separation.

$$HI(dB) = 22 + 20log\left(\frac{x}{\lambda}\right)$$

$$VI(dB) = 28 + 40log\left(\frac{y}{\lambda}\right)$$

$$SI(dB) = (VI - HI) \cdot \frac{2\theta}{\pi} + HI$$

Where

HI = Horizontal isolation in dB.
VI = Vertical isolation in dB.
SI = Slant isolation in dB.
x = horizontal distance in metres.
y = Vertical distance in metres.
λ = wavelength in metres.
$\Theta = \tan^{-1}(y/x)$.

These values are suitable for conditions where x is greater than 10λ and y is greater than λ.

An example is shown for a frequency of 450 MHz for two vertically spaced dipole antennas.

This can be illustrated by an example. Consider two vertically spaced antennas operating with an EIRP of 25 W, which is 14 dBW (44 dBm). If the median faded signal power to be received is −110 dBm and the assumption is that the transmit signal is not faded, then in order to prevent de-sensitisation of the receiver, then the loss required from the transmitter antenna to the receiver is 154 dB. This equates to more than 950 metres – unreasonable for masts. The situation is different for antennas tuned to different frequencies. If the offset channel rejection were, say, 45 dB, then the minimum equivalent isolation would be 109 dB. The dipoles would only have to be spaced by at least 70 metres. The greater the isolation, the more closely the antennas can be spaced. This is illustrated in Figure 16.1.

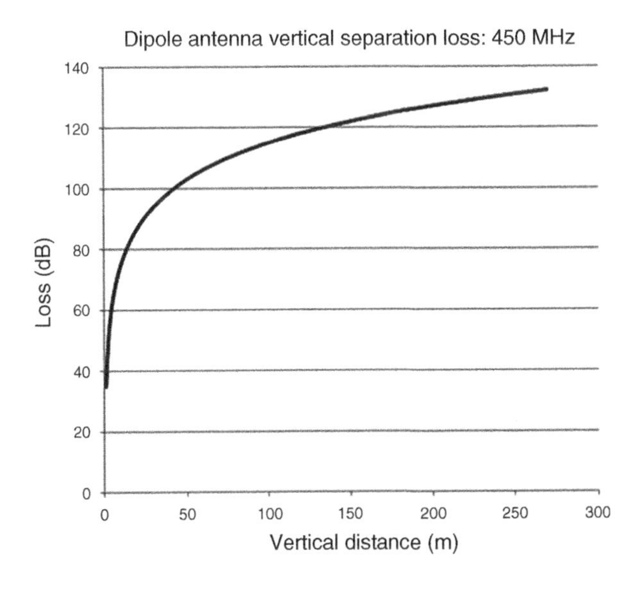

Figure 16.1 Vertical isolation for two typical dipole antennas.

For rejection of 70 dB, for example, the equivalent isolation would be 84 dB, leading to spacing of only 17 metres. These values are for standard dipoles. Stacked dipole antennas have greater gain in the horizontal direction and less gain in the vertical direction. This can make antenna separation more reasonable. For example, assuming the same figures as before with a 70 dB offset rejection and two stacked dipole antennas with an additional 25 dB vertical attenuation each, then the required equivalent isolation is 84–50 dB, which is 34 dB. This leads to a vertical separation of only one metre or so. This how masts can be loaded with many antennas in practice.

A more complex site may have multiple radio antennas as illustrated (not to scale) in Figure 16.2.

In this case, it necessary to consider horizontal and slant range as well as vertical spacing.

If we were to use the same parameters as before and for on-channel antennas we need isolation of 154 dB, then using the horizontal isolation formula, this equates to a horizontal distance of many hundreds of kilometres. In fact, this is not accurate and instead propagation models would be used to determine the actual horizontal distance. For the situation where there is 70 dB of offset rejection and the equivalent isolation is 84 dB, then the minimum horizontal distance is approximately 900 metres.

Notice the frequency dependence of the isolation calculations. In general, the lower the frequency, the lower the filtering possible and the further the isolation required. In particular, for HF systems, this may mean that systems have to be separated by several kilometres at least.

The simple analysis shown in this section provides two major insights into co-site radio installations. The first is the importance of as much filtering as possible for the radio systems and the second is the importance of planning the layout of such sites. Before considering these issues, it is important to identify the so-called spurious emissions and receiver vulnerabilities that affect the level of energy transmitted and the effect on the receiver. This is considered in the next section.

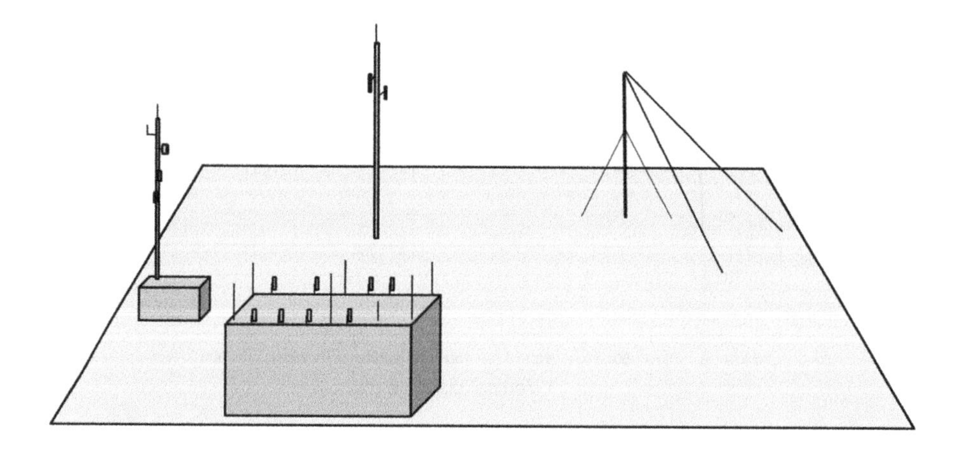

Figure 16.2 An illustration of a radio installation with multiple masts and antennas.

16.2 Sources of Interference at a Radio Site

When radios are used in close proximity, there are a number of separate issues to consider. These include:

- blocking;
- harmonics;
- intermediate frequencies;
- image frequencies;
- inter-modulation products;

These are considered in the next sub-sections.

16.2.1 Blocking

Blocking is the term used to describe the increase in the noise floor present off-channel due to transmitters. In effect, it is the post-filtered response that causes de-sensitisation in other receivers. This can extend over wide frequency ranges. For example, for co-sited HF and low-band VHF legacy voice systems, a rule of thumb of at least 2 MHz spacing has been used to account for blocking. For higher bandwidths, the spacing may be much higher, and for duplex systems, transmitters and receivers at based stations are more widely separated. TETRA for example uses 10 MHz spacing between base transmit and receive frequencies. The blocking scenario is shown in Figure 16.3.

If we use the 2 MHz minimum co-site frequency spacing described above and each channel in the frequency scheme is 50 kHz, then one transmission prevents reception in $20 \times 2 \times 2 = 80$ channels. This is based on 20 channels per MHz, 2 MHz blocked and the channels are blocked above and below the carrier frequency. This would mean for the VHF band 30–80 MHz, only 25 channels can be used at any one location. For HF, with frequencies of 3–30 MHz, this would give a nominal figure of 13.5 channels. However, given the constraints imposed by HF propagation, in practice the actual figure is much lower. This tends to mean that HF systems have to be widely spaced to allow practical use of many systems concurrently.

16.2.2 Harmonics

Harmonics are generated during generation of the carrier frequency. They are generated at multiples of the carrier frequency, i.e. 2, 3, 4, 5, 6, 7, 8... times the original frequency. Good transmitter design and filtering can reduce the level of harmonics, but even so they can cause co-site interference if a receiver is tuned to the harmonic of a powerful transmitter. In practice, the odd-numbered harmonics have stronger power than the even-numbered ones.

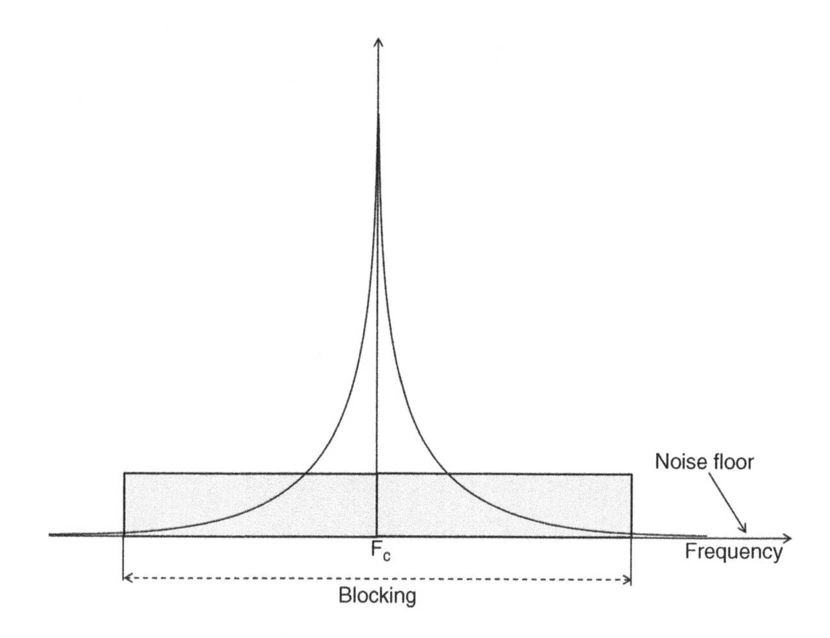

Figure 16.3 Blocking illustration. The transmitter is on frequency Fc, but even after filtering there is still energy present well beyond the transmit frequency. Although the level of energy is well below the on-channel power, it can still raise the noise floor over many kHz or even many MHz.

The bandwidth of the harmonic generated can be multiplied by the same coefficient as the harmonic, thus the 3rd harmonic can have a bandwidth three times the original bandwidth and so on. This is illustrated in Figure 16.4.

16.2.3 Intermediate Frequencies

Intermediate frequencies are used to simplify radio design. Fixed frequency local oscillators are used prior to the signal being converted to the actual transmission frequency. Without sufficient shielding, this energy can radiate into adjacent electrical equipment and can be passed into feeders and antennas. A VHF radio system might use a 10.7 MHz intermediate frequency. If an HF receiver is tuned to 10.7 MHz close to the VHF system then the receiver may be de-sensitised. Intermediate frequencies are a feature of radio system design and many other frequencies can be chosen according to the technology and design. Microwave systems may use 70 MHz (thereby potentially interfering with VHF reception) and radars can use 30 MHz for example.

16.2.4 Image Frequencies

Image frequencies affect receiver mixers that use a local oscillator to bring the carrier down to an intermediate frequency. The mixer works at an offset to the tuned

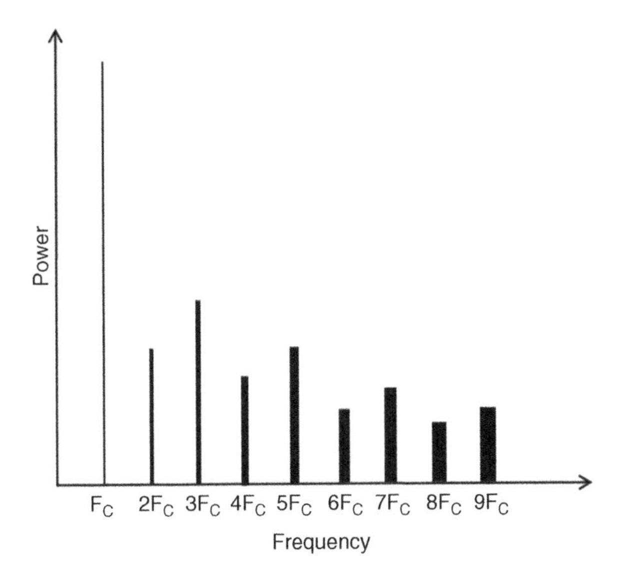

Figure 16.4 Schematic illustration of harmonics (not to scale).

frequency, but there is no way for the mixer to differentiate between a signal at an offset above or below the difference between the intermediate frequency and the tuned frequency. Consider a VHF system using a 10.7 MHz intermediate frequency. If the tuned frequency is 150 MHz, then an image at twice the difference between that and the intermediate frequency i.e. $150 - (10.7 \times 2)$ which is 128.6 MHz then a transmission on 128.6 MHz may cause interference to reception. This is illustrated in Figure 16.5.

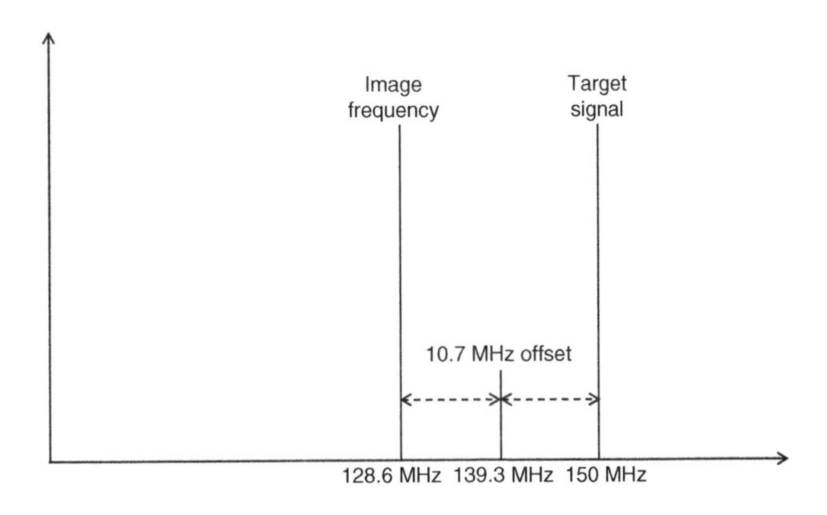

Figure 16.5 Illustration of an image frequency.

Filters can be used to reduce the power of the image frequency, but again for co-site operations, a higher power transmission may still cause problems at the receiver. It is best to avoid image frequencies for transmission by other systems at the same site.

16.2.5 Inter-Modulation Products

Inter-Modulation Products (IMP) can be a major nuisance at co-site radio installations. The following subsections describe the causes and effects of IMPs. The method of calculating their frequency characteristics is also described.

16.2.5.1 Causes of Inter-Modulation

IMPs are caused by non-linear interactions between elements at a radio site. They can be caused by coupling between the energy in transmit antennas, radios or feeders, overloading of receiver input stages or a phenomenon known as the 'rusty bolt' effect. The rusty bolt effect is caused between metallic joints energised by radio energy. This causes a rectifying effect similar to that of a diode, and this in turn causes radio energy to be transmitted.

IMPs are particularly difficult to resolve for a number of reasons. Firstly, they produce transmissions that are not at the frequencies of the systems that cause them. This makes the sources difficult to identify. Secondly, they are due to interactions between equipment that can be difficult to prevent. Thirdly, the generation mechanism can be difficult to identify; is it due to coupling between antennas, feeders or somewhere else in the system? Also, IMPs can start to occur later in the life of a radio installation. An obvious example can occur when bolts that were installed clean gradually become rusty. The last issue is that IMPs may or may not be present at a radio site, even if the calculations show that that may be.

The best way of avoiding this type of problem is to ensure that receivers are not tuned to channels that may suffer from IMPs. However, as we shall see this may not be possible in practice as the number of potential IMP blocked channels rises rapidly with the number of potential contributors.

16.2.5.2 Inter-Modulation Products

The best way of demonstrating the principles of IMPs is to examine the effects caused by just two transmitters on different frequencies. Second order products are shown in Figure 16.6. If transmission A is at 100 MHz and B is at 125 MHz, we can work out the potential IMPs.

There are only two first order IMPs. These are:

$$A + B = B + A = 100 + 125 = 225\,\text{MHz}$$

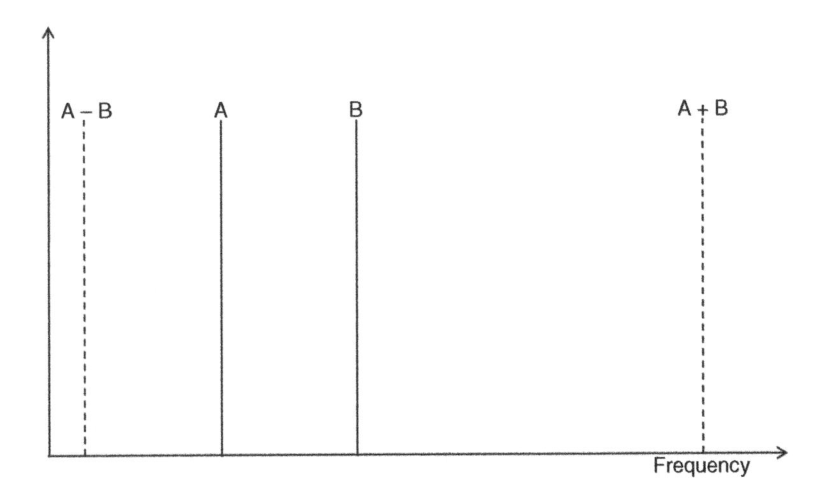

Figure 16.6 First order inter-modulation products.

$$A - B = B - A = 100 - 125 = 25\,\text{MHz (the resulting negative sign is unimportant)}$$

Note that even though the second calculation gives a negative result, only the difference is important; there is no such thing as a negative frequency.

The second order products are well away from the original bands of transmission. However, the high term is at the bottom of the UHF NATO air band and the lower term is within the HF band, so if sufficient filtering is not applied, problems can still occur. Of more importance in general are the odd-numbered products. The third order products are:

$$2A - B = 2 \times 100 - 125 = 75\,\text{MHz}$$
$$A - 2B = 100 - 2 \times 125 = 50\,\text{MHz}$$
$$2A + B = 2 \times 100 + 125 = 325\,\text{MHz}$$
$$A + 2B = 100 + 2 \times 125 = 350\,\text{MHz}$$

As can be seen, these are closer to the frequencies of the original transmissions. There are also four products rather than the two seen for the second order products. This gets worse when we consider the fourth, fifth, sixth, seventh and ninth order products. Fortunately, the power in these higher orders is typically lower the higher the order.

An additional complication is that IMPs are not restricted to interactions between two sources. There may be three or more sources although again the power of the resultant IMPs reduces with the number of sources. Typically, only interactions between two and three sources are considered in most IMP calculations.

A similar process can be conducted for three frequencies, with the third one being termed ‘C’. The second order products are:

$$A + B - C$$
$$A - B + C$$
$$A - B - C$$
$$A + B + C$$
$$B - A - C$$
$$C - A - B$$

The same process can be followed to extend the third and fifth order and further for higher orders. It can be seen that the number of potential two-source and three-source IMPs rapidly increase with the number of potential sources at a co-site location.

Between two sources there are two second order products, four third order products, seven fourth order products, eight fifth order products, 11 sixth order products, 12 seventh order products, 15 eighth order and 16 ninth order. This gives a total of 75 IMPs for two sources. If there are three potential sources, then all combinations of two sources must be considered and so on for more sources. This would give three times as many; four would give six times and so on.

There are computational methods of not only determining the inter-modulation products but also identifying IMP-free schemes for many channels of the same bandwidth and channel spacing.

16.3 Methods of Managing Interference at Radio Sites

16.3.1 Prevention

Prevention of problems is always better than curing them once they have arisen. The first prevention method is to design radio systems with sufficient integral filtering and suppression to avoid problems at the transmission and reception stages. Also, using feeders with high attenuation can help prevent the induction of cross-interference. A major additional advantage can be achieved by ensuring the best layout of radio installations both internally and by positioning antennas properly. One simple method is to ensure that as far as possible feeder cables are not laid next to one another. If they have to cross, then they should cross at right angles. Frequency assignment is used at co-site locations to further minimise noise and the other issues discussed. And, of course, rusty bolts should be avoided!

Where interference is found or suspected of being possible, there are methods to help solve the problem.

16.3.2 Curing

A useful guide to minimising interference is ETSI EG 200 053, which describes EMC and radio spectrum matters. This includes a number of methods of minimising interference above those described above.

Some methods include:

- antenna distribution networks, in which the receive feeders are separated widely from the transmit feeders;
- receiver distribution networks where filters are used to reject interference from transmitter circuits;
- ferrite circulators, which provides 20–40 dB isolation at the output part of the transmitter;
- cavity resonators, which reduce wideband noise from transmitters;
- spectrum dividing filters to limit the out of band responses of transmission lines.

References and Further Reading

Graham, A.W.; Kirkman, N.C.; Paul, P.M (2007), *Mobile Radio Networks Design in the VHF and UHF Bands: A Practical Approach*, John Wiley & Sons ISBN 0-470-02980-3.

Recommendation ITU-R M.739: Interference due to Inter-modulation Products in the Land Mobile Service between 25 and 1000 MHz.

Recommendation ITU-R SM.337: Frequency and Distance Separations.

17

Communications Electronic Warfare

17.1 Introduction

Communications Electronic Warfare (CEW) is designed to exploit or counter enemy communications and actions taken to counter the threat posed by enemy CEW assets. In practical terms, CEW systems include:

- radio detection systems used to detect the presence of target transmissions;
- radio intercept systems used to demodulate and decode target transmissions;
- radio direction finding systems used to localise target transmissions to a given degree 2 of precision;
- radio jamming systems designed to prevent enemy units communicating effectively with one another.

Systems designed to radiate in order to disrupt enemy communications are termed Electronic Attack (EA) systems and passive systems designed to exploit enemy communications are known as Electronic Support (ES) systems.

In this chapter, we will look at these systems in some detail.

Historically, most CEW systems were designed specifically to combat military tactical communications systems operating at HF and above. However, these days nearly any type of radio communications system is a potential CEW target. Modern civilian communications systems can be as or more complex than military systems and adversaries can use any system that suits their needs. This is exemplified by insurgents using simple PMR systems for communications. This has become known generically as 'Icom' systems, albeit that Icom is simply the name of a successful PMR radio manufacturer. In the discussion that follows in this chapter, we will consider any type

Communications, Radar and Electronic Warfare Adrian Graham

of radio communications system as a legitimate target. We will do this in order to examine the appropriate physics rather than considering any political or reasons of legitimacy.

In this section, as with others in the book, I am using loose interpretations of some terms deliberately. The reason for this is that different sets of people use different terms for the same thing and this can make it seem as though we are talking of different things. Thus, whereas a communications expert may use the term 'transmitter', an EW specialist may always use the term 'emitter'. They are the same thing; the only difference is the terminology. If all can use terms that mean the same thing interchangeably, this can only lead to greater cross-pollination of ideas across this spurious divide. I make no apologies for this, even though it may cause some to grind their teeth!

17.2 Detection and Intercept Networks

17.2.1 Introduction

Detection systems are designed to detect the presence of targets operating in the radio spectrum. They may also be used to characterise transmissions but do not attempt to decode the original method; this is what intercept systems do. Neither detection nor intercept networks are designed to localise target transmissions although modern systems may provide all three aspects. In this section, we will only look at signal detection and interception. Unless otherwise specified, we will not normally be considering satellite systems at this stage.

As with all radio systems, we are seeking to differentiate between a signal and noise. Figure 17.1 shows a simplified diagram of noise measured in 1-Hz bands across the target search frequency range. The diagram is simplified in that the noise values would be variable both in the very short term and over longer periods. However, it is worth making the simplification in order to make the basic principles clearer. In the top diagram of the figure, noise is shown over a wide range. As noise tends to decrease with frequency, the noise response falls off as the frequency increases as shown. By applying inverse gain, matching the response but in the opposite manner, we can further simplify the diagram to show normalised noise as a flat response. This is useful to reduce the required dynamic range of processing systems downstream of the gain correction process. It also means we can examine the relative strength of noise versus target signal without considering variations in noise.

In this example, we have used a 1-Hz band as an illustration. In practice, we can sum noise over any sub-band, with the noise increasing as the bandwidth increases as has been described several times previously. The type of display shown is typical of a graphical output from a modern detection system. If a detectable target appears, we will see it poking above the noise floor as shown in Figure 17.2. In this case, the target is obvious, although we would want to be able to detect targets as small as possible.

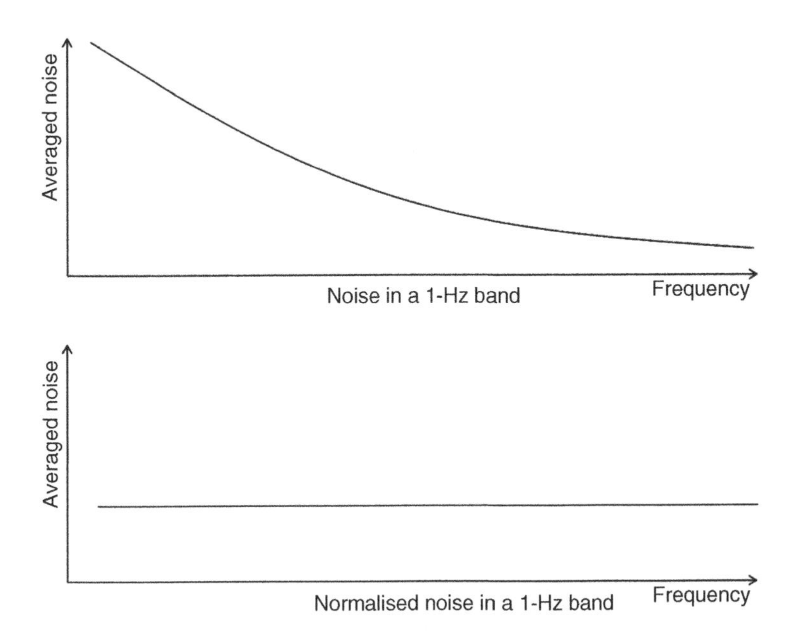

Figure 17.1 Normalisation of noise by applying inverse gain to produce a flat response.

From this detection, all we can tell is that there is something, somewhere, radiating RF energy (as long as we are not seeing some artefact of internal system noise or adjacent electrical system that adds as RF component into the system transmission lines!). All we can tell at first glance is the target central frequency, some information about the bandwidth – the amount depends on how much of the energy exceeds the detection threshold, but we should be able to see the 3-dB bandwidth at least, given that the signal is fairly conspicuous. However, given the frequency and bandwidth we may

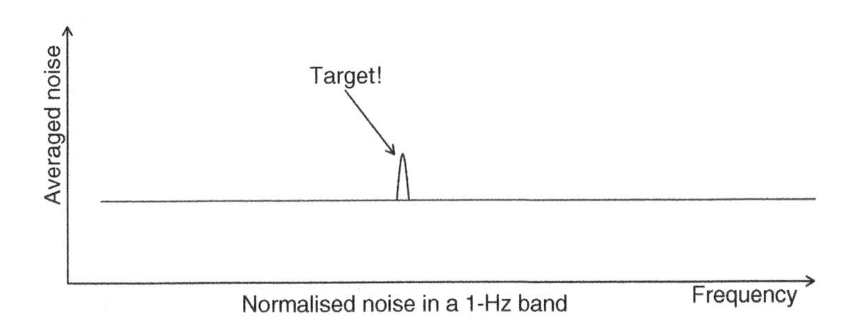

Figure 17.2 A target transmission shown on a normalised noise display.

be able to analyse the signal to determine at least something about the transmission. For example, let's say that the signal has the following characteristics:

- The central frequency is 383.025000 MHz.
- The 3 dB bandwidth is in the region of 15 kHz.
- The envelope of signal strength is constant averaged only a very short timescale.
- The central frequency does not vary to the degree of precision of the detector.

Are there any conclusions we can draw? Well, we can relate the signal characteristics to systems of which we know the standard parameters.

The analysis may conclude the following:

- In the country where the signal was detected, we know that the band 380–400 MHz is set aside for the TETRA system.
- The 3-dB bandwidth is about right for TETRA voice calls.
- We know TETRA base stations transmit continuously.
- We know that the band 380–385 MHz is used by TETRA downlinks (from the base station to the mobile).
- Because the signal shows no Doppler variation (changes due to radial movement of transmissions), we can make the assumption that the transmitter is fixed.

The conclusion is that the detected transmission is a TETRA base station, although as yet we do not know where. We may be able to estimate range based on the knowledge of typical TETRA transmit powers and losses in the type of environment we are operating in. Additionally, we may have access to a database of TETRA base stations and therefore may be able to identify exactly which TETRA base station we are seeing. Even if we don't have such a database, we may be able to tie in other information in order to identify a likely location. For example, if we know that the TETRA system is used by the local police and that there is a police headquarters within 10 km, then it is possible the signal originates from there. However, we would have to follow up such an assertion with further enquiries. Also, we may need to consider such factors as 'spoofing', where a sophisticated emitter is used to pretend to be another system entirely to fool the enemy. We would have to consider the likelihood of this in our analysis.

All of this analysis depends on knowledge of the TETRA system. It highlights one fundamental precept of both CEW and EW; databases of known system parameters are essential.

The type of display shown in Figure 17.2 is acceptable for steady transmissions, but consider the images shown in Figure 17.3, which show the same display for different snapshots in time. Image (a) shows the same display we saw before. Image (b) shows a new, smaller signal 10 MHz above the original signal. In image (c) , the transmission has gone. The signal has only been detected for a short length of time. If the operator

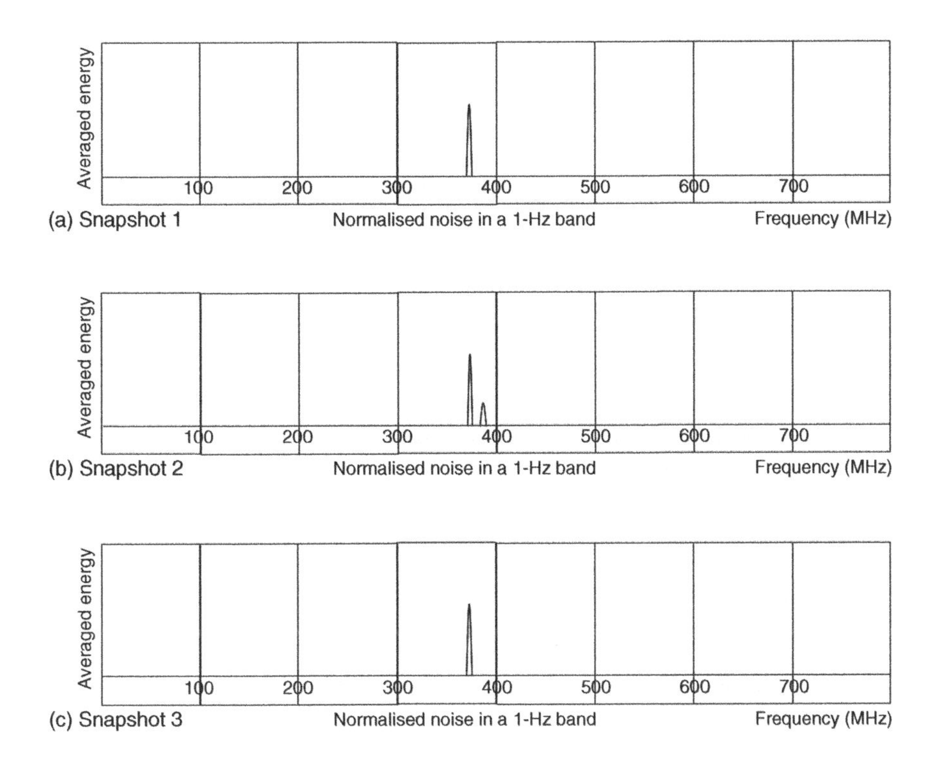

Figure 17.3 Short duration transmission signals.

were not paying attention, the signal may be missed completely; in any case, short-lived signals may not be detectable long enough for their parameters to be determined. Also note that the 'improved' display shows a frequency scale for easy reading.

A more useful display, and one which is common in ES equipment, is known as the 'waterfall' display as illustrated in Figure 17.4. In this case, the vertical scale is not signal strength but time. The signal strength is shown by the boldness of the signal line, and the bandwidth by the width of the line. The display is continually refreshed, usually from the bottom, so that older information is continuously discarded. However, the captured information is normally recorded so it can be retrieved. The continuous signal is a constant line at the same frequency and the transient signals are more readily seen and analysed.

This can be seen in Figure 17.5 where the operator has drawn a box around the transient signal to be analysed to study it in more depth. The more detailed analysis shows small variations in frequency. This could be due to system instability but are more likely to be due to radial movement of the transmitting antenna reference the detection system antenna. This most likely indicates that the transmitter is moving, and this movement is causing Doppler shift. It could well be that the transient signal is a

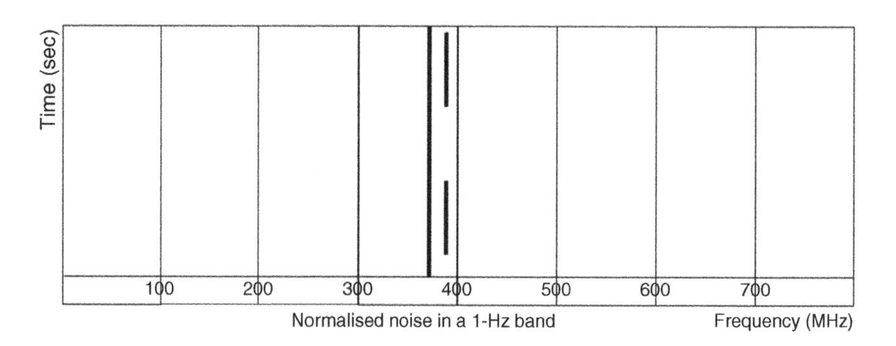

Figure 17.4 A waterfall display.

mobile member of the network. Analysis of the Doppler shift will allow the radial speed to be determined and from that the type of platform can be estimated. In this case, it looks like a car talking to the base station and transmitting only when the operator is speaking.

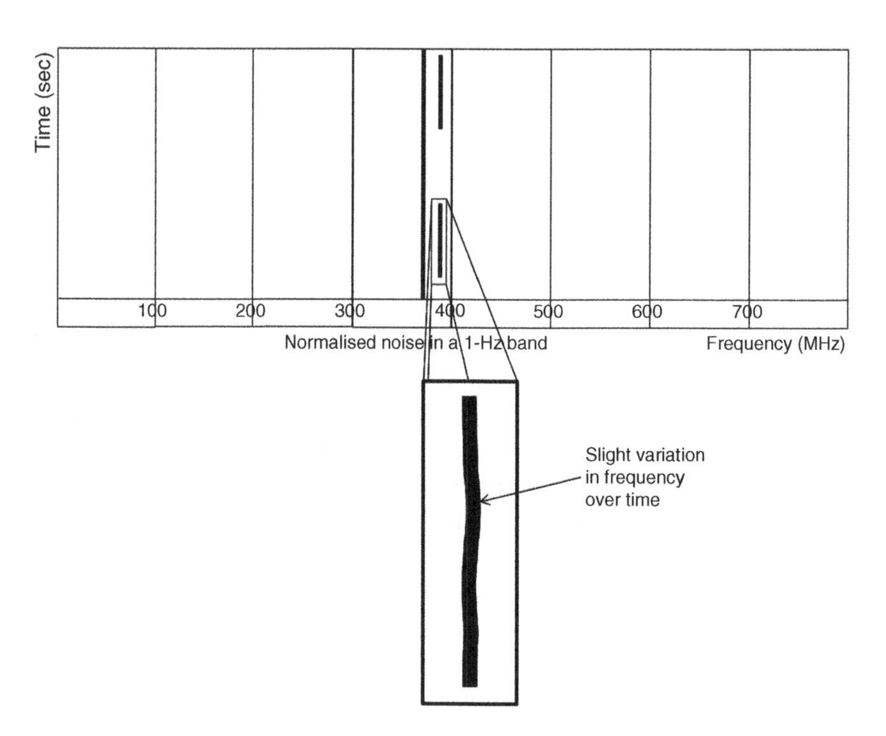

Figure 17.5 A waterfall display with a zoom function to examine signals of interest.

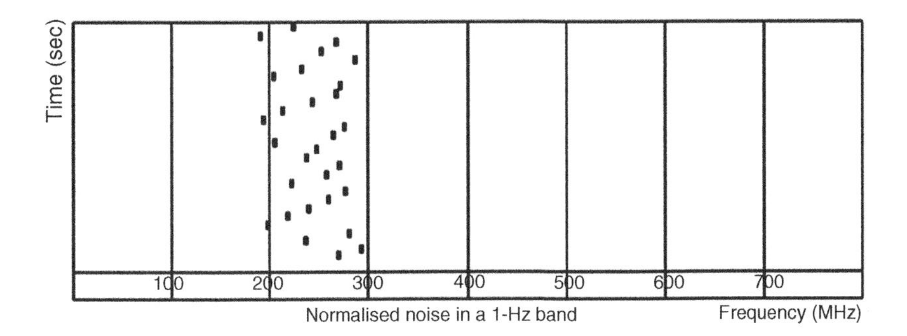

Figure 17.6 Frequency hopping signals in the NATO UHF air band.

Consider the display shown in Figure 17.6. It appears to show random transmissions occurring over 180–300 MHz. Are they related or not? There are distinct characteristics that may help us to decide. Firstly, each transmission is of the same duration. Also, their strength is roughly the same (normalised). No two overlap in time. They are also in the NATO UHF air band, which supports frequency hopping systems. All of this suggests that the transmissions are related and that they belong to the same frequency hopping system.

Now that we have looked at how a signal can be displayed, it is worth looking at a typical modern system's performance.

17.2.2 Detection Capability

Specially designed ES detection receivers have different design considerations to those of communications receivers. These include:

- the ability to cover wide frequency ranges;
- the ability to rapidly scan for signals;
- signal analysis capability;
- user-defined configuration of search characteristics, including the ability to search for pre-selected channels and signal types;
- ability to analyse a wide range of signals covering different signal bandwidths;
- ability to detect many multiple signals simultaneously;
- signal and noise averaging capabilities.

Such receivers can usually record and playback detected signals and often they can be combined with other systems to provide inputs to intercept systems and via DF antennas and processing equipment, to assist in localisation.

Despite these differences between communications receivers and ES receivers, much of the planning and performance prediction is carried out in exactly the same ways as for any other RF system.

For example, we can look at sensitivity figures in much the same way as we have done before. As you may recall, we can look at the minimum sensitivity of the system by building up the receiver noise floor from:

$$P_n = F_r - 204 + B$$

Where
Pn is the receiver noise floor in dBW.
B is 10 log (bandwidth in Hz).

A specific ES receiver has the noise figure values shown in Table 17.1.

We can also use this to examine the noise floor in different bandwidths as shown in Figure 17.7(a) and (b). The values shown here are in dBm, not dBW as above).

This can be graphed as shown in Figure 17.7(b).

Of course, this is assuming that there is no additional external noise so to achieve these figures the receiver antenna must be in a very quiet location. Also of course these noise figures are nominal; actual values will change and the noise must be averaged over time to achieve these figures.

The level above which a signal will be detected will depend on the system design and the configuration selected. However, it should be possible to detect that a transmission is present if it is a dB or more above the noise floor and is constant. However, such a low signal could not be decoded and it may be difficult to establish the parameters of the transmission.

We will now look at the requirements for intercepting a signal and decoding the baseband transmission.

17.2.3 Intercept Systems

Intercept systems are more complex than detection systems. They need to be able to identify a particular modulation scheme and then demodulate it in order to retrieve the baseband transmission. If the signal is not encrypted or coded in such a way as to prevent easy reconstruction (such as requiring the right spreading code) then the

Table 17.1 Typical noise figures in dBW

Band	Typical noise figure (dB)	Noise in 1-Hz band
9 kHz–32 MHz	12	−192
32 MHz–2000 MHz	10	−194
2000–3000 MHz	12	−192
3000–3600 MHz	15	189

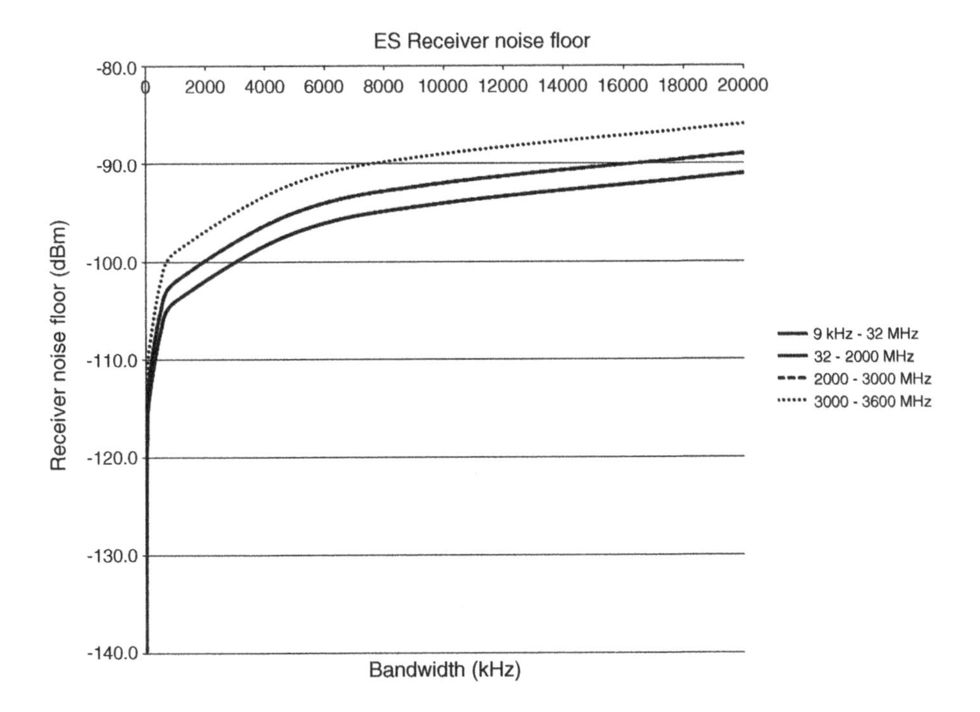

Figure 17.7 (a) Noise floor for different bandwidths and frequencies. (b) Receiver noise floor figures.

original message will not be recoverable without substantial effort. In such cases, recordings of the signal must be sent to a central facility such as the NSA in the US or GCHQ in the UK. However, first the signal has to be captured and this is done in the same way as for a communications receiver in each case.

A typical ES detector would normally be able to demodulate:

- AM;
- FM;
- PM;
- Pulsed;
- I/Q (e.g QAM);
- TV;
- USB;
- LSB;
- CW.

In all of these cases, the signal must be strong enough to be demodulated. We can determine the minimum threshold by adding the required margin for the bandwidth and modulation scheme.

For example, FM generally requires a signal of at least 17 dB above noise and interference to be decoded successfully. For a bandwidth of 15 kHz, we can calculate the threshold as:

Noise floor for 15 kHz = −122 dBm.
Add detection threshold of 17 dB gives a value of −122 + 17 = −105 dB.

The same process can be performed for any other modulation scheme. This means we can build a standard link budget and perform system predictions as for any other type of radio system. As always, we need to ensure we include margins for fading as appropriate. We will see more of this in the next section on planning.

17.2.4 Planning Detection and Interception Networks

At first sight, planning detection and intercept systems seems to be very different from planning other types of radio systems. Usually, we may have a fixed base station in a position over which we have control. This station transmits to mobile network subscribers and receives transmissions back from those subscribers. Although we do not know the location of individual subscribers, we know they have to be in the service area in order to be able to successfully communicate with them. How can we compare this to planning for detection networks?

Let's look at the scenario. We have a passive system that does not transmit. We can position and configure this system within operational and logistics constraints. We have no control over the mobile (enemy) elements, but we should have some idea of the targets we want to detect.

Let us assume that our targets have the following characteristics:

- operating in the VHF band 30–88 MHz;
- using voice only, with a channel spacing of 25 kHz and an occupied bandwidth of 20 kHz;
- manpacks or vehicles only; no air targets;
- transmit power of vehicles = 15 W out of the antenna;
- transmit power of manpack = 5 W out of the antenna;
- targets are using omni-directional antennas.

We would know this type of information from other sources. These need not be based on intelligence effort but rather open source information based on industry news ('*...the Army of the Republic of Nowhere has ordered 100 sets of VHF tactical radios from TactiRadio Corporation...*') or other sources such as local news, pictures from news reports and so on. This may be augmented by classified Electronic Order of BATtle (EORBAT) database information gathered prior to the operation.

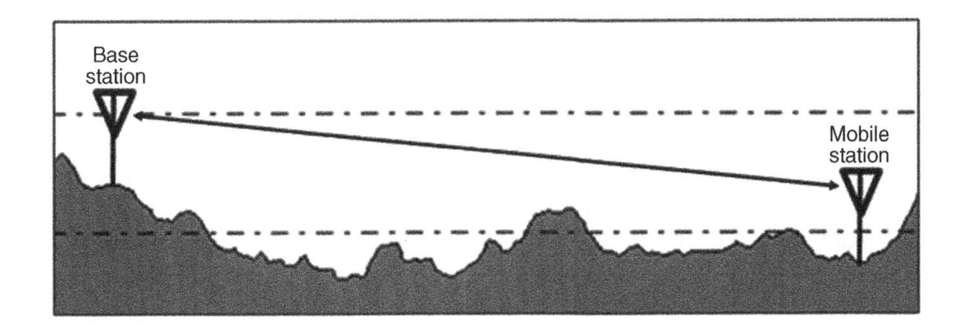

Figure 17.8 A basic communications link.

Now that we know the type of equipment we are trying to find, we can start looking at the operational environment in order to start the planning process. But before we can do that, we need to address the issue of placing a passive sensor to detect active transmissions from unknown locations.

Figure 17.8 shows the standard communications scenario. Often we will look at both links, starting with the downlink from the base station to the mobile system.

Figure 17.9 shows the situation for a detection or intercept station trying to intercept an unknown target, which is very similar in nature to a normal mobile station that we do not know the explicit location of.

So, in planning terms, what is the difference? The answer is in fact very little. This is due to the theory of reciprocity. This states that the loss between two antennas is the same in BOTH DIRECTIONS. This means that, in principle, a transmission from a base station to a mobile experiences the same loss that occurs in the other direction. So, if the detection station is not transmitting, we can still model what in the communications example would be the uplink direction. This means we can use exactly the same planning methods that we would use in communications. We do have to be cautious, since just because the loss between the two antennas is identical, we cannot say that both sides receive the same workable signal. Noise at the location of the

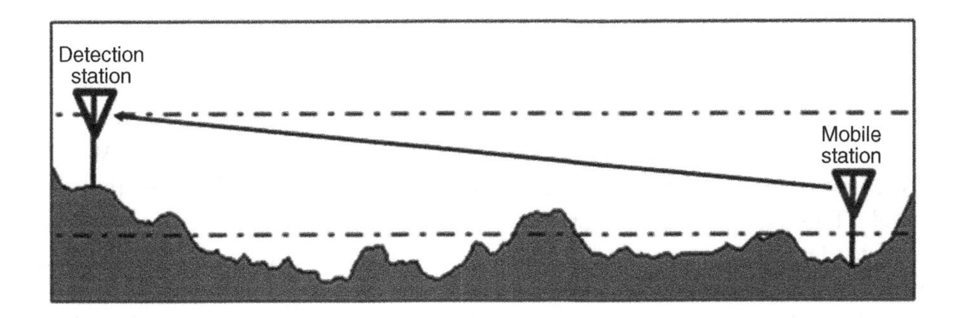

Figure 17.9 The EW detection system link is very similar to the basic communication links.

detection antenna may prevent reception of the mobile signal and clutter in the vicinity of the mobile may reduce the energy radiated long distances. However, with this caveat in mind, we can proceed with confidence. And do not forget that the positioning of the detection/intercept antenna is under our control and, in general, we have the ability to place the expensive ES asset in a better location than most mobiles operating in a tactical environment will benefit from.

We can also play another trick, should we need to turn to a non-EW radio planning system to perform EW predictions (it happens). The losses from one antenna to another depend on the transmission path. The signal received at the receiver location depends on that plus the effective radiated power. So if we were to use the antenna heights of the EW asset and the target as they are (it is important not to mix these up) but put the transmit power of the target at the location of the EW asset *as though it were transmitting, even though it is not*, we can do a straightforward prediction just as though it were a transmitter. This is a straightforward coverage plot. If the environment model includes clutter and noise, then the problems associated with targets in different environments (e.g. within trees or town centres) then we overcome the caveats imposed by different noise levels. We must however retain the minimum noise level at the EW asset. Now we will apply this to our example.

Let us consider a defensive position, where we are operating from an airfield in a potentially dangerous environment. We want to be able to detect any enemy transmissions within the area of the airfield. For safety, the detection/intercept system is within the airfield perimeter. We will assume that the area around the airfield is relatively flat, but the discussion that follows will work for variable terrain. We will also assume that the local ground is moderate in terms of conductivity and permittivity. This is illustrated in Figure 17.10.

Let us assume our own equipment has the following parameters:

- frequency range = 20 – 2000 MHz;
- antenna height = 15 metres above local ground;
- antenna gain = 5 dBi;
- receiver noise figure = 10 dB;
- no losses from antenna to receiver.

First, we want to determine the maximum effective range of the detection/intercept system. We do this as normal by creating a link budget, bearing in mind that we are looking at a passive system attempting to receive enemy transmissions. We have identified two types of target. First we will look at the more difficult target, that of the manpack. This has a lower power than the vehicle fit and will therefore be more difficult to detect. The power of 5W out of the antenna equates to approximately 37 dBm. This is shown in Table 17.2.

From the table, we can see the maximum loss the system can sustain without losing the signal or being able to demodulate it. However, we have not considered fade

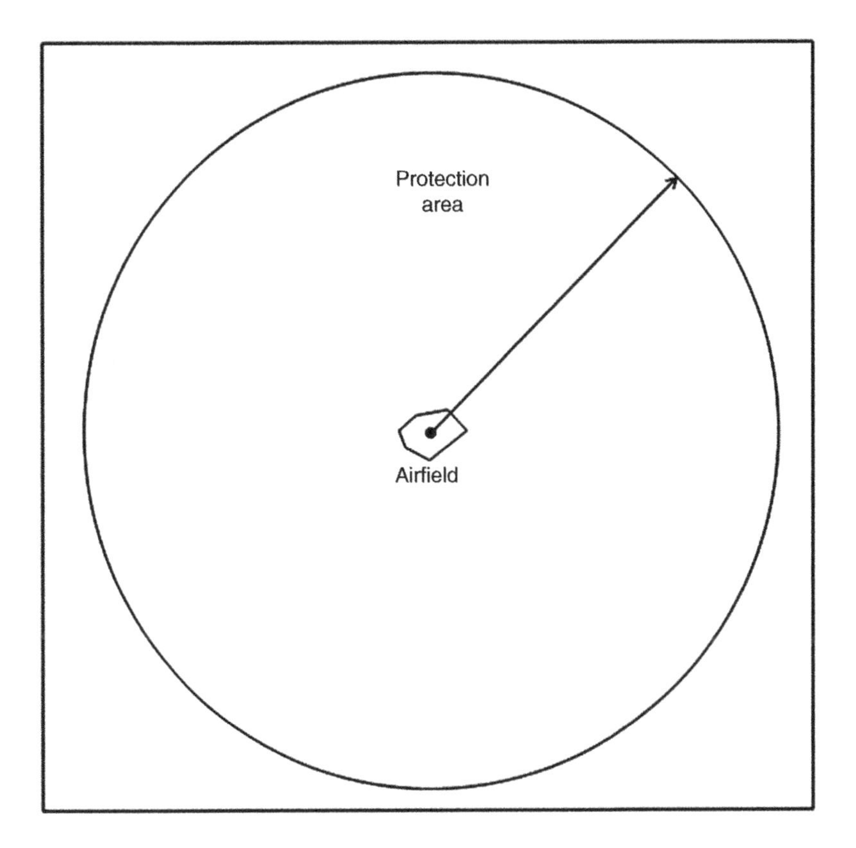

Figure 17.10 Protection area around an airfield.

Table 17.2 Link budget for detection and intercept

Parameter	Value	Units
Radiated power	37	dBm
Receiver noise floor (20 kHz BW)	−130	dBm
Receiver noise figure (VHF)	10	dB
Detection threshold	1	dB
Noise floor at antenna location[a]	−135	dBm
Antenna gain	5	dBi
Detection threshold (unfaded)	−124	dBm
Maximum detection loss	**161**	**dB**
FM sensitivity (for 10 dB SINAD)	19	dB
Intercept threshold	−105	dBm
Maximum intercept loss	**142**	**dB**

[a] We can assume this is a measured value. It is below the receiver noise floor, so it can be ignored in this case.

Table 17.3 Fading corrections for different probabilities of detection or intercept

Probability of detection/intercept[a]	Approx. rayleigh fading margin (dB)
50%	0
90%	10
99%	20
99.9%	30

[a] We add this margin to the detection threshold or intercept threshold as appropriate.

margins as yet. In this book, we are using either Rayleigh or Ricean fading. The worst case is Rayleigh fading, so it would be prudent to use this and any extra performance we get above it is a bonus. Also, since we have not specified the type of environment (perhaps because we do not yet know what it is like yet), Rayleigh fading would be the safest course. As a rough rule we can use the figures shown in Table 17.3.

Now we can determine the system loss we can tolerate for a given probability of detection or intercept. This is shown in Table 17.4.

Now we have figures from which we can estimate the maximum expected range. If we have a terrain and clutter model, and a suitable propagation model, we can determine expected range explicitly around the airfield, but even if we do not we can make some assumptions to give us an estimate. This may often be the case as in the following situations:

- We as yet have no clear idea of the operational environment.
- We are trying to produce metrics for nominal range in a specific type of environment without looking at so-called site-specific range.
- We are comparing the performance of systems and want to use a common method of determining loss in a typical environment so we can compare them methodically.

The aim of the exercise is illustrated in Table 17.5. We want to be able to fill in the grey boxes giving us nominal ranges for the low, mid- and top-part of the frequency band, for different levels of POD/POI.

But if we have no terrain and clutter data, how can we estimate loss? We need to determine as accurate a model as possible but within the limitations we have, and we have to understand the risks associated with the methods we choose. There are a

Table 17.4 Maximum system loss calculations

Probability of detection	Max. system loss (dB)	Probability of intercept	Max. system loss (dB)
50%	161	0	142
90%	151	10	132
99%	141	20	122
99.9%	131	30	112

Table 17.5 Nominal range calculation sheet

POD	Max. Loss (dB)	Range (km)		POI	Max. Loss (dB)	Range (km)	
50%	161	30 MHz		0	142	30 MHz	
		60 MHz				60 MHz	
		88 MHz				88 MHz	
90%	151	30 MHz		10	132	30 MHz	
		60 MHz				60 MHz	
		88 MHz				88 MHz	
99%	141	30 MHz		20	122	30 MHz	
		60 MHz				60 MHz	
		88 MHz				88 MHz	
99.9%	131	30 MHz		30	112	30 MHz	
		60 MHz				60 MHz	
		88 MHz				88 MHz	

number of considerations we need to take account of in order to select an appropriate model. These include:

- frequency of operation;
- estimated path length;
- type of environment, including terrain variability and clutter;
- antenna heights/altitude above ground.

In our example, we can consider the issues shown in Table 17.6.

There may be a variety of models available to us, but in this case, we will use a bald Earth diffraction model (generated from the ITU program GRWAVE). The path loss for 30, 60 and 88 MHz is shown in Figure 17.11. This does not include the effects of terrain or clutter.

Can we use the values from the graph to determine maximum range without any further correction? We could, but at the risk of being too optimistic. Since we have

Table 17.6 Issues to consider in estimating nominal range

Metric	Comments
Frequency	30–88 MHz. This is a relatively low frequency and we must avoid models that are not appropriate due to frequency, e.g. Okumura Hata (minimum frequency 150 MHz)
Path length	At low band VHF, we can expect transmissions to extend beyond line of sight into the diffraction regime. Therefore we need a model that includes diffraction
Environment	We have stated that the environment is relatively flat, and the soil is moderate, giving permittivity of 4 and conductivity of 0.003
Antenna heights	The antenna heights are low (15 metres for the ES asset; 1.5 metres for man/vehicle height – estimated), so clutter may well be an issue

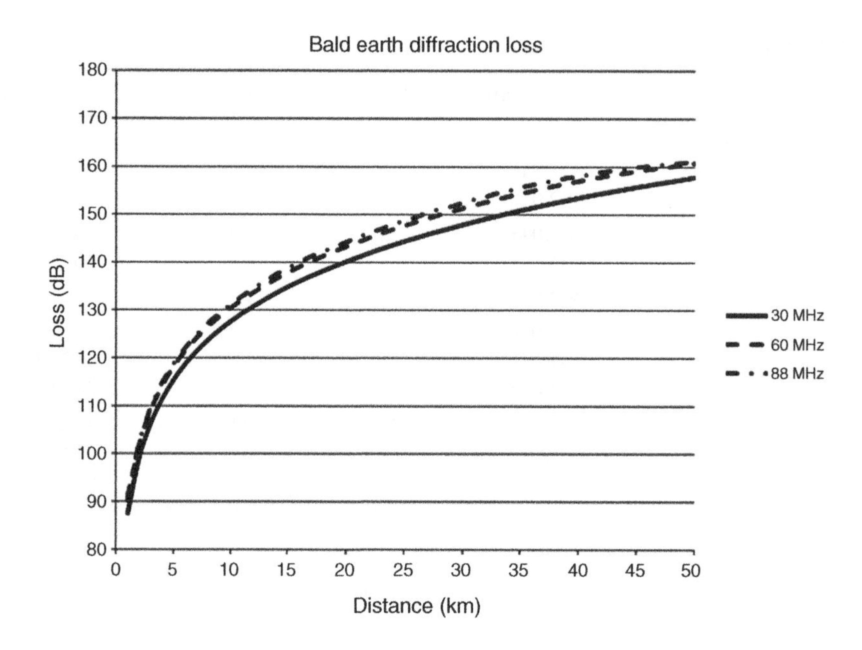

Figure 17.11 Path loss values for 30, 60 and 88 MHz using the ITU GRWAVE model.

determined that the antennas are low to the ground, terrain fluctuations will be important. We can account for this by adding a margin for terrain shadowing. Typical values for shadowing are in the region of 8–10 dB. We will use 8 dB since we know the terrain is roughly flat. This is an assumption and as with all assumptions, it carries with it risk. We will look at the risks later. We have already included fading loss, so it is important that we do not add it again. We can include our margin of 8 dB to the maximum permissible loss by subtracting it from the figures we have already determined, as illustrated in Table 17.4. We can also determine the maximum estimated range for the system based on the path loss figures from Figure 17.11. This is shown in Table 17.7.

In terms of our objectives, we have now determined the radius of the protection area offered by the ES asset for different types and levels of protection.

Notice the large differences in range between 50% POD/POI and 99.9% POD/POI. Also, there are differences between the bottom and top of the frequency band, but these are less pronounced. We did not include clutter, so we are assuming that the target and the receiver are in relatively open areas.

We have made a number of assumptions in our analysis and we need to bear this in mind when using the figures for real planning. What we have not done is identify exact detection ranges; instead, these are no more than figures that can be used for initial

Table 17.7 Completed nominal range calculation sheet

POD	Max. loss (dB) Inc. shadowing	Range (km)		POI	Max. loss (dB) Inc. shadowing	Range (km)	
50%	153	30 MHz	39	0	134	30 MHz	15
		60 MHz	33			60 MHz	13
		88 MHz	31			88 MHz	11
90%	143	30 MHz	24	10	124	30 MHz	9
		60 MHz	20			60 MHz	7
		88 MHz	19			88 MHz	7
99%	133	30 MHz	14	20	114	30 MHz	5
		60 MHz	12			60 MHz	4
		88 MHz	11			88 MHz	4
99.9%	123	30 MHz	8	30	104	30 MHz	2.5
		60 MHz	7			60 MHz	2.2
		88 MHz	7			88 MHz	2.2

guidance until better ones are available. However, it is useful to have an idea of expected performance because this can help with:

- determining the viability of the system to warn of enemy attack, and in which phase of the enemy operation this is likely to occur;
- determining whether other methods need to be used to provide protection if the ES asset cannot do so effectively. This might be in the form of patrols, checkpoints and other physical barriers;
- estimating how many assets will be required to meet operational objectives.

Even with these nominal range figures, we can use them to put into place a first-cut deployment plan. Let's say we have decided that 7 km range is insufficient to provide adequate protection and so we will need to place more sensors to improve it. We will start by working out the planning metrics to use. We do not want to use them all, because it adds much extra nugatory work. We will plan on the following:

- We will use the figures for the top end of the band since they are the worst.
- We will use the figures for 90% POI. This gives us a nominal range of 7 km. Clearly the POD will extend beyond this range.
- We will look at trying to place asset on higher ground to improve performance. As a rough guide, if we were able to place the antenna 35 metres higher with respect to ground, the range would increase to 12 km; for 85 metres higher, it would go out to 18 km.
- We will specify a target area and then plan to cover as much of it as possible, and to tie it into other defensive measures. This will be a refinement of the original protection area.

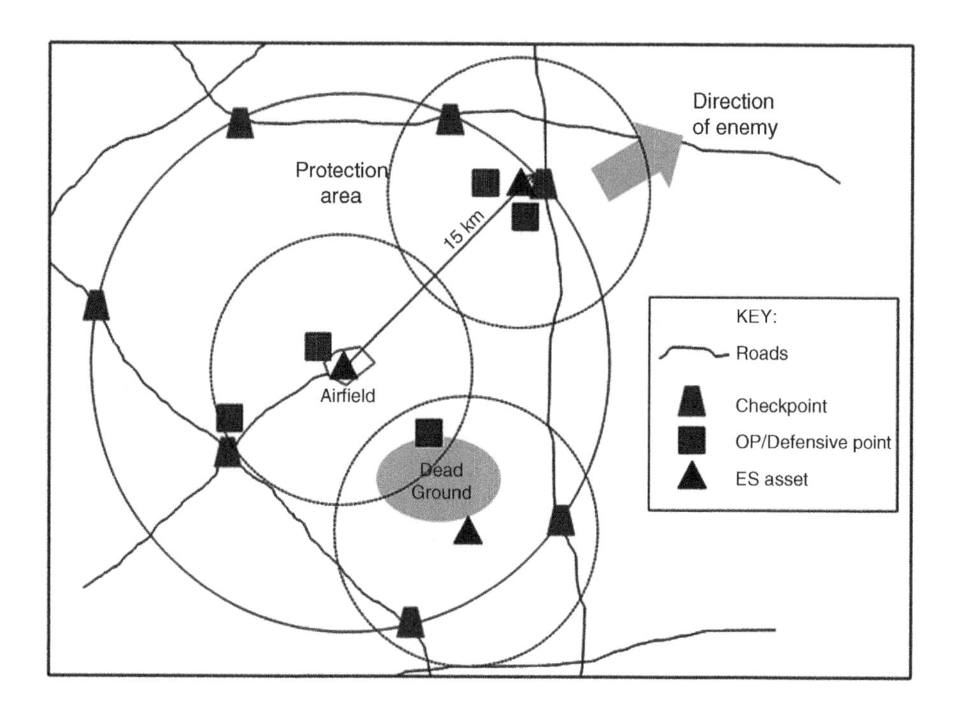

Figure 17.12 Enhanced view of the protection area.

The result is shown in Figure 17.12. This figure shows an enhanced view of the original area to be protected. The enemy is assumed to be primarily to the Northeast; however, we must also protect against attacks from different directions. First, physical checkpoints have been set up at important road junctions and where they penetrate the protection area. Then, we have added combined Observation Points (OPs) and Defensive Positions (DPs) to cover significant areas. These are the airfield itself, the dead ground to the Southeast, near important checkpoints and in the direction of the enemy. These locations can also be used to mount patrols to cover areas without fixed defences, such as the area to the Northwest of the airfield, which can be carried out by the DP close to the airfield and at the checkpoint to the Southwest.

Three ES assets have been used; one at the airfield itself, one providing additional coverage of the dead ground and one in the direction of the enemy. We have in this case been unable to mount the ES antennas any higher above local ground. The range for 90% POI is shown as dotted circles. Note that the intercept and particularly the detection network will provide additional coverage beyond that shown. To provide total coverage of the protection area, we would need at least 5–6 ES assets but this is not particularly efficient and in many cases the demand for assets will outstrip supply and hence fewer assets have to provide the best coverage, with shortfalls picked up by other methods, such as the physical security used. The crucial point is that in order to make the best use of scarce ES assets, it is not necessary merely to think about the RF issues but the operational conditions and the other assets available.

Some additional design considerations for ES assets are listed below.

- The site of the system must be as quiet in noise terms as possible. This may mean siting away from other assets that may radiate noise or interference.
- The antenna must be clear of obstructions and as far above the terrain and clutter as possible.
- Directional antennas (rather than the omni-directional antennas used in this example) will provide good coverage over more important sectors, so long as they are oriented correctly.
- High ground can extend the range of the system considerably. However, in some circumstances coverage nearby will be reduced. This is illustrated in Figure 17.13. The grey area shows the vertical coverage of the antenna, taking account of the vertical polar pattern. The clear area shows a distinct coverage gap close to the antenna due to the siting of the antenna on the slope of a high, steep slope.

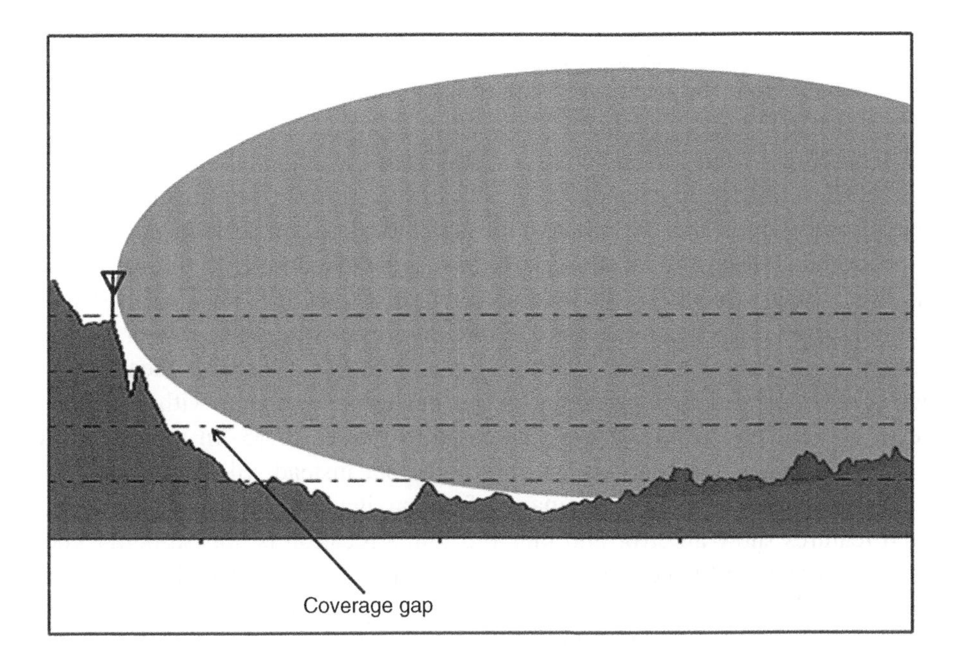

Figure 17.13 Coverage gap due to high siting of an antenna where there is a major terrain dropoff.

In terms of designing an intercept/DF network, normally the following can be used to optimise the design.

- Try to minimise system coverage overlap unless it is necessary for other reasons. That way the least number of assets can provide the widest possible coverage.
- Determining the operational requirements in particular identifying high priority targets and areas will help the design. For example, if targets are primarily urban, then rural coverage may be wasteful and should be avoided.
- In all cases, it is very likely that each of the ES sensors will have to be able to communicate with each other. These links must be planned along with optimisation of sensor locations. There is no point in having the best possible coverage of enemy territory for tactical detection and intercept if the information gathered cannot be disseminated. However, passive strategic systems or systems designed for later analysis may not require Battlespace communications. Instead, data may be recorded and returned physically at certain intervals.

We now go on to look at direction finding systems.

17.3 Direction Finding Networks

17.3.1 Introduction

Direction Finding (DF) is a method of identifying the bearing of a transmission via detection means. If this is achieved by a number of correctly-positioned DF sensors simultaneously, then the location of the transmission can be determined. An ideal case is shown in Figure 17.14. In this illustration, there are three DFs in a line – called the 'DF baseline' – and each receives a signal from the target emitter. Using the DF techniques described in this chapter, they each get the bearing of the transmission. When the data from all three DFs are collated, the target location can be deduced. Normally, this means that at least one location has to have a data feed from all three sensors.

As ever, things are rarely this straightforward in practice. Physical and operational influences often mean that the DF baseline is not straight as shown, but more importantly, the direction of the transmission will be less clear. This means that the target is not localised to a particular point, but rather a region within which the target is most likely to be. This is illustrated in Figure 17.15. In this – far more common – scenario, the bearings will not agree but instead will lead to a 'cocked hat'. This is an old naval term originating from navigation errors in which bearings of land features show an error and thus the ship's location is not explicitly known but there is a region of uncertainty which is known as a cocked hat. The most likely location is within or near the ellipse shown in the figure. In many cases, this is still operationally useful information because it is not always necessary to have target location to a high degree of precision. Also, further analysis can be performed

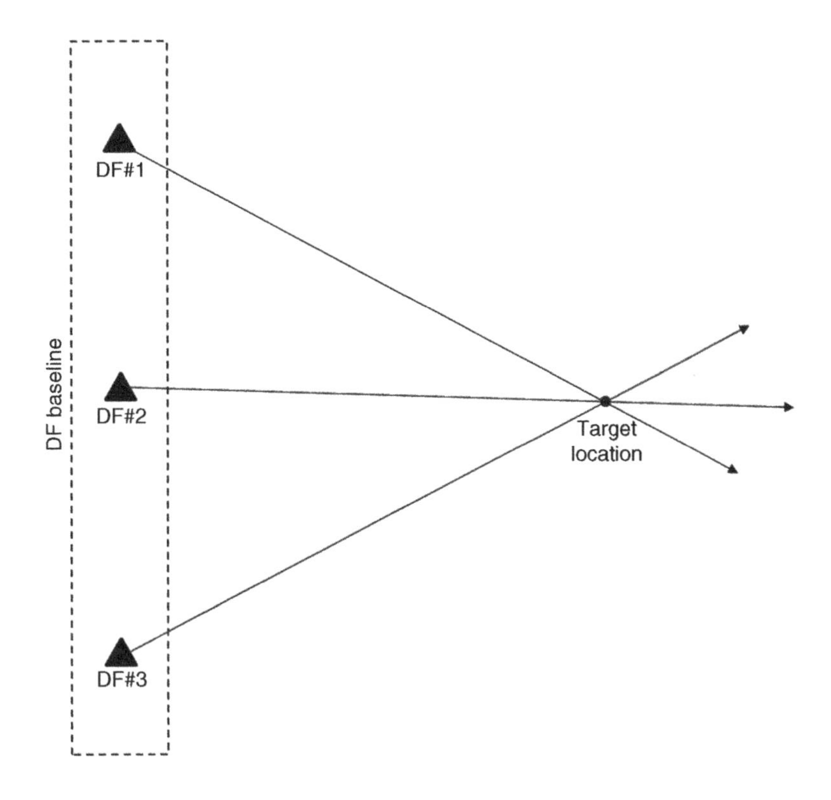

Figure 17.14 The ideal DF baseline.

to localise the emitter more accurately or an aircraft can be sent to visually identify the target.

The location uncertainty is due to a number of factors all simultaneously having an effect. These include:

- DF system precision;
- DF system positioning and configuration;
- signal reflections from terrain or clutter such as large buildings;
- atmospheric variations;
- interfering transmissions.

In this section, we will look at the history of direction finders, the types of DF available and some practical examples.

17.3.2 History of Direction Finding

The principles of Direction Finding (DF) have been known since the early days of the scientific study of electromagnetic waves. Heinrich Hertz found out about the

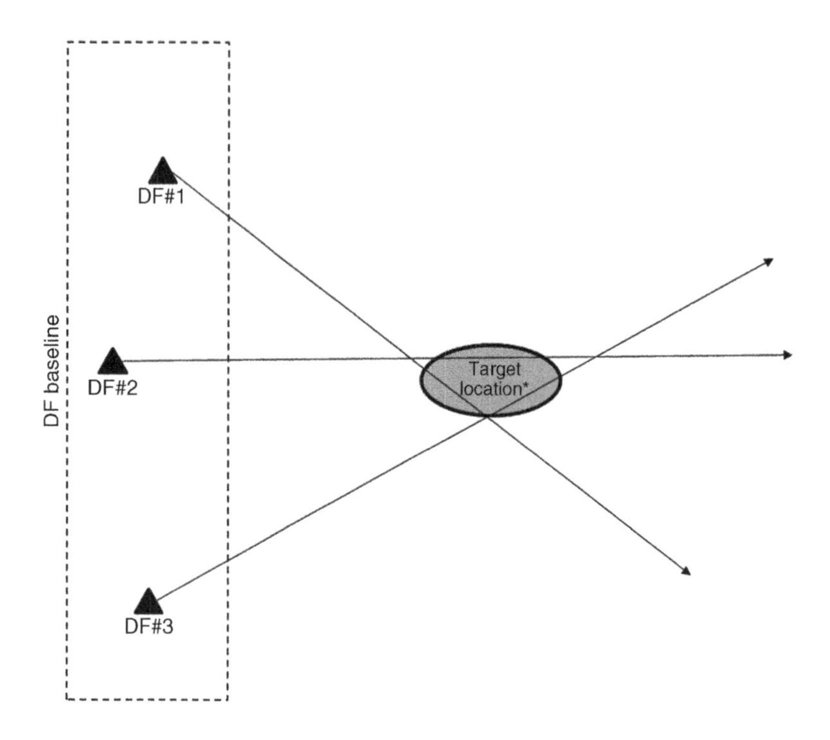

Figure 17.15 The more likely scenario; the 'cocked hat'.

directivity of antennas as early as 1888. A patent on a homing DF was proposed in 1906 by Scheller. Developments of DF continued such as the use of polarisation DFs, of which the rotating-loop DF is one of the best known types. These were the most frequently used type of DF during the First World War. In the inter-war period, Sir Watson-Watt developed non-mechanical DF systems using crossed loop antennas. These and and three-channel Watson-Watson direction finders were fitted on British naval vessels from 1943, known colloquially as 'huff-duff'. Meanwhile, DF systems based on Adcock antennas first appeared in 1931 and these were used by both Britain and Germany.

In practical terms, camouflaged direction finders were available from about 1931. They were used in the localisation of radio transmissions of spies. DFs based on the Doppler principle first appeared in 1941. Remote wide-aperture circular array DFs first appeared in about 1943.

Since the Second World War, DF has been continually developed for civil and military applications. One major application is in aviation and since the 1950s airports have been equipped with VHF/HF Doppler DF systems for air traffic control. DF systems have generally migrated from analog to digital starting from the early 1970s. Digital signal processing advances in the early 1980s have allowed interferometry DFs to be developed. Meanwhile, of course, the signals that the DFs have been designed to detect have also continued to change, to include frequency hopping and spread spectrum systems. These developments are discussed further in this chapter.

17.3.3 Operational Uses of DF Systems

Direction finders are used in a wide variety of applications, including:

- radio monitoring;
- passive EW against enemy transmissions;
- air traffic control;
- maritime traffic control;
- radar Warning Receivers for platform self-protection;
- localisation of specific platforms on the battlefield, particularly systems such as air defence and surface search radar, also enemy command and control, special forces, etc;
- localisation of network subscribers in a mobile network. This can be achieved by determining which base station is serving the mobile, but direction finding provides a far better estimate of location.

We will now look at the basic principles of DF.

17.3.4 Basic Principles of DF

Direction finding is based on the theory that at a distance from a radiating antenna, the advancing wavefront is effectively perpendicular to the direction of travel at any reasonable distance, as illustrated in Figure 17.16. The waves radiate out in all

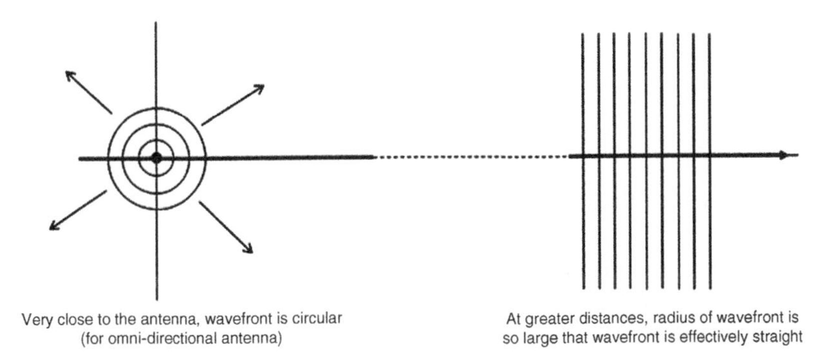

Figure 17.16 At long ranges, the radius of emission becomes essentially a straight line. This can however be modified by diffraction around terrain and building features.

directions, modified by the antenna directivity. Very close to the transmitter, this forms volumes of spheres. However, very quickly the radii of these spheres become so large that any small portion of the edge is effectively a straight line. In practice, this ideal situation will be compromised by physical effects such as terrain and clutter features; however, these factors do not prevent successful operation of DF systems.

Any kind of DF system falls into one of two method categories:

- measurement of the direction of the electric or magnetic field vectors (polarisation DF);
- measurement of the orientation of surfaces of equal phase (phase-based DF).

Polarisation DFs use loop or dipole antennas and this category includes the classic rotating-loop DF. Modern systems using polarisation and Watson-Watt discrimination are used in circumstances where antenna size is an issue. This is most often found for aircraft, ships and vehicles when the HF band is the target.

Most modern systems are however based on the phase-based DF method. This produces a summation of the advancing wavefront at different points based on individual antennas. If these signals are in phase, then the summed output is a maximum and the angle of incidence is at a minimum. A more complex version is based on samples taken at various points in the field, to which complex mathematical operations are applied. This method includes interferometry, Doppler and rotating-field DFs. Systems based on digital signal processing are capable of processing many signals in parallel. These can be used for complex beam-forming.

The main design aspects of DF systems are that they provide an accurate estimation of the angle of incoming signal while being capable of detecting small signals and signals that are only present for a short space of time. Additionally, they must have a wide dynamic range to deal with large variations in target signal strengths. For scanning DF systems, high scanning rate is also a major design consideration.

It is also worth considering the operational role of DF systems when considering how accurate they must be. This varies considerably between applications. Table 17.8 shows some applications and key design considerations.

17.3.5 Types of Direction Finders

DF systems consist of the following components:

- An antenna system with antennas that can help in the determination in the direction of arrival of incoming signals. The size of the system depends on the frequencies under consideration. There may also be a separate antenna to determine signal parameters.
- A DF conversion system. This converts the signals arriving at the antenna into directional information.

Table 17.8 Key design considerations for DF systems

Application	DF Accuracy Needed	Sensitivity Needed	Comments
Radar Warning Receiver (RWR)	Low – quadrant sufficient	Low – to avoid false alerts	System must detect threat signals very quickly and must discriminate between threats to the platform and other engagement transmissions
Single bearing HF-DF	High – both azimuth and elevation	High to detect low level signals	System must have capability to model current HF prediction conditions to calculate estimated transmission locus
Radar localisation system	High for horizontal accuracy (to detect land or sea based radar), reasonable vertical accuracy for aircraft system	High	Radar localisation accuracy determined by purpose of system. Medium for system identification, high for targeting purposes
Communications localisation system	High for horizontal accuracy (to detect land or sea based radar), reasonable vertical accuracy for aircraft system	High	As for radar systems, medium for identification and for force disposition determination, high for targeting

- A discrimination system used to determine the angles of arriving signals. This is used both to determine the angle and to reject unimportant signals.
- A display system used to present the information to the operator. This can display the time and angle of arrival of the signal, plus parametric information about the signals detected.

The selection of these components differs according to the type of direction finder. In this section, we will look at the following DF systems:

- Watson-Watt;
- Doppler;
- Interferometer.

17.3.5.1 Watson-Watt DF System

Watson-Watt DF systems use Adcock or loop antennas, with Adcock being the most common. The system comprises four spatially displaced monopole or vertical dipole antennas, with the differences in amplitude of the received waveform antenna being used to determine the angle of the incoming signal. Typically, an additional centre antenna is used to resolve bearing ambiguities. The system is shown in Figure 17.17, as it would appear on the roof of a car or other platform.

The way that direction is determined is by comparing the amplitudes of the signals received by each of the antenna. Figure 17.18 shows the antenna array orientated so that there are antennas on the north, east, south and west directions. The difference in amplitude between the north and south antennas gives the Y-axis angle and the difference between the east and west antennas provides the X-axis angle. The amplitude differences are obtained as voltages out from the antenna. There can

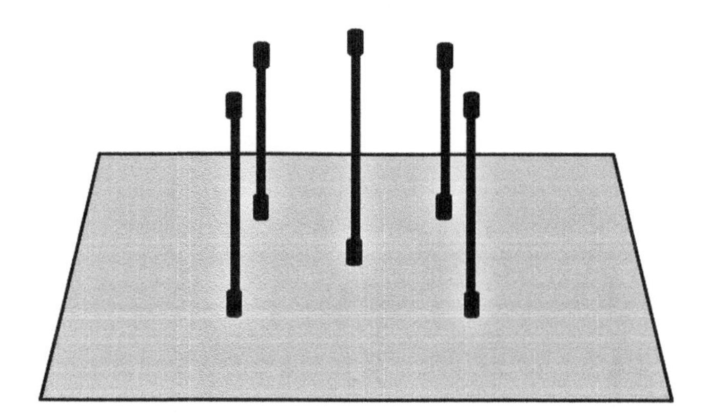

Figure 17.17 Watson-Watt antenna arrangement.

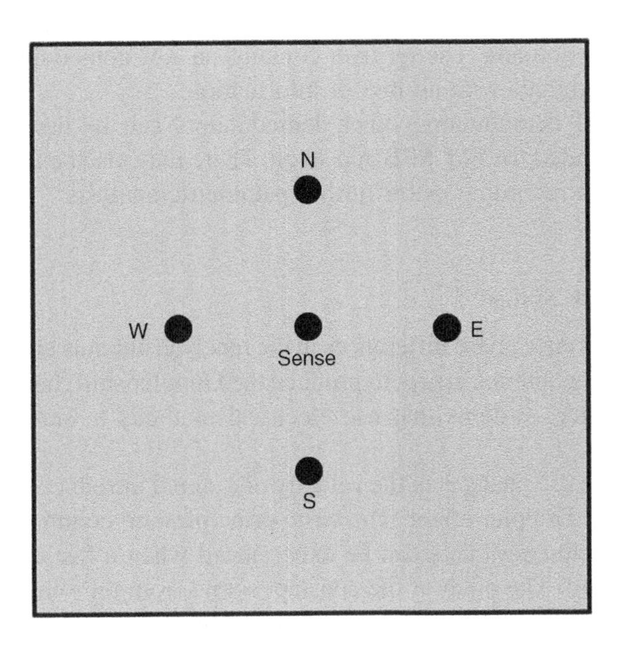

Figure 17.18 Watson-Watt DF array orientated to the cardinal points.

be 180-degree angle ambiguity, which is resolved by the central sense antenna. Some systems use vectorial summing of all the outputs to perform this function virtually.

The antenna arrangement is capable of operating over a wide frequency range, with the limitation being imposed by the spacing of the array. Closely spaced antennas produce a better angle response than wider ones, but lower sensitivity. A spacing of 1/3 of the wavelength of the highest frequency to be covered is about the maximum, and spacing of about 1/10 of the lowest frequency is about the minimum. Sensitivity is also affected by the antenna responses, which will vary over the frequency range. Typically, a frequency ratio of about 3– or 4–1 is about the maximum to achieve good response over the whole band. One way of extending this is to use extra antennas; the use of eight antennas in a uniform circular array allows a full spacing of one wavelength rather than the 1/3 limit for four antenna systems. This is of course more expensive and also introduces additional design constraints. An alternative method is to co-locate two four element arrays on the same base plate. Four of the antennas can be cut to have a good response at a different frequency than the other four, thereby extending the frequency range. Potential interactions between the two sets of antennas can complicate design in practice.

The DF receiver can work on a three-channel system or, more often, a single channel method. This requires the output of the antennas to be AM modulated to allow extraction of the X-axis, Y-axis and sense signals. The X-axis and Y-axis outputs are modulated at different frequencies and then combined into the single channel with the output of the sense antenna. The receiver contains an AM demodulator that recovers the three original signals with no loss of information.

Beyond the AM demodulator, other demodulators can be used to allow audio outputs to be generated for FM, SSB and so on. There may also be further circuitry to integrate and allow recording or for further parametric analysis.

17.3.5.2 Doppler DF System

Doppler DF systems use phase differences in the received antenna array. True Doppler systems use rotating antenna arrays to produce the Doppler shift, but modern systems are 'pseudo-Doppler' systems that use electrical methods to simulate mechanical rotation.

Doppler systems use changes in the velocity of a signal introduced by the 'rotating' array to induce the Doppler effect. The basic principles are commonly known by its effects at audio frequency. This can be experienced when a fast car, train or plane passes at high speed. The pitch of the constant sound from the vehicle appears to be higher as it approaches and lower as it speeds away. In principle, the way a DF antenna arrays work could be simulated with a constant noise source by rapidly rotating the head around. However, this is not recommended and it looks very stupid to bystanders.

Pseudo-Doppler DF uses typically four equally spaced antennas positioned on the circumference on a circle. To produce the same effect as mechanically rotating the antennas, each one is switched on and off in sequence very rapidly. This process is easier to achieve, more controllable and can produce far greater frequency variations than mechanical systems. Figure 17.19 shows a pseudo-Doppler DF system in plan view. An incoming signal is shown arriving from the north. The effect of switching on and off the radial antennas in a clockwise direction means the antenna at the east position is effectively rotating away from the source. The Doppler effect means that the frequency of the signal is reduced. For both north and south antennas there is no Doppler effect, and for the west antenna the rotation is in the direction of the signal and therefore the received frequency. Analysis of this situation proves that the antenna is coming from the north.

The output of the DF antenna is fed through an FM demodulator and then the phase difference measured to determine the angle of the original signal.

The example shown uses four antennas, but many more can be used to improve system performance. These are often used for air traffic control applications.

17.3.5.3 Interferometry DF Systems

Interferometry also uses phase differences to determine the angle of the transmitted signal. Interferometry systems provide good angle determination but are more expensive than the other systems described. The main difference between Doppler

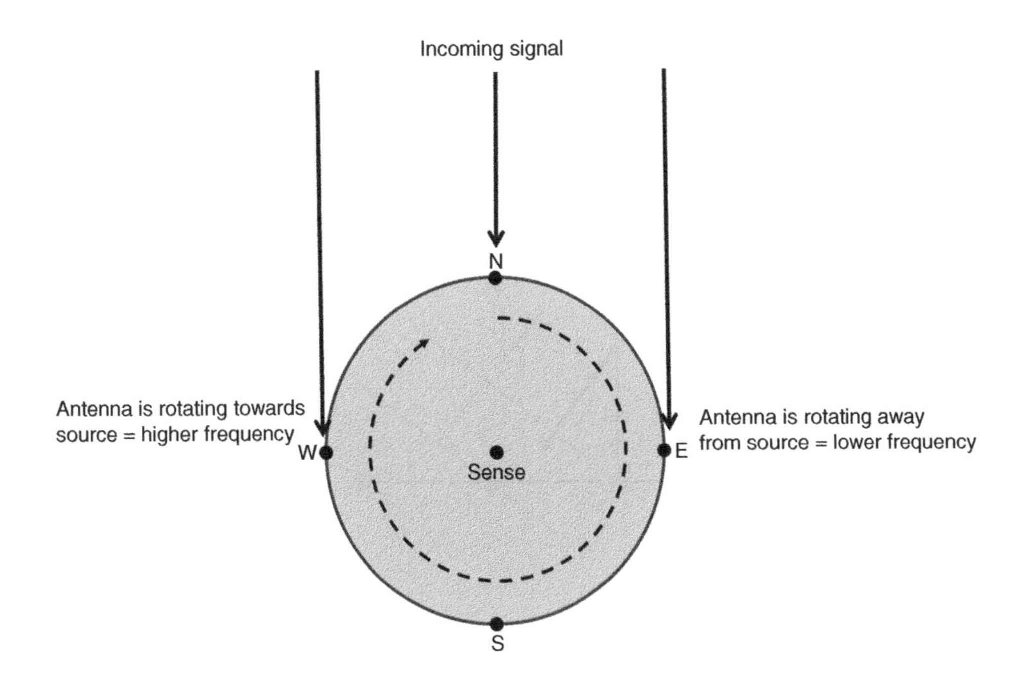

Figure 17.19 Pseudo-Doppler DF antenna system (plan view). Each of the antennas is turned on and off in a very fast sequence. In the system shown, the sequence would be N-E-S-W, repeating.

and interferometer DF systems is that the latter have a separate receiver circuit for each antenna. Each of the outputs are converted to the same intermediate frequency for further processing. To prevent systematic errors, each path from antenna to the DF process must be exactly the same length. This can be difficult to achieve in practice over a wide range of temperatures and signal strengths.

The easiest way to demonstrate how interferometry works is to consider the effects of two or more antennas equally spaced in a straight line, as is found in a simple phased array system. This situation is illustrated in Figure 17.20. The antennas are the black dots, each of which is spaced equally in a flat line array. The direction of the arriving signal is shown, with the advancing wavefront at right angles to this direction. If the signal were a very short duration pulse, then the time of arrival could be used to determine the angle of arrival using simple geometry. However, the signal is most likely to be a continuing signal and so the wavefront lines can be considered as points of equal phase. Since we know the frequency of transmission, we can work out the phase differences very accurately for each element in the array. Once we know the phase differences and the distance between each element in the array, the angle can be calculated using simple geometry.

If the antenna array is circular or conformal, geometry can be used to correct for the non-linearity of the array elements using more complex geometry.

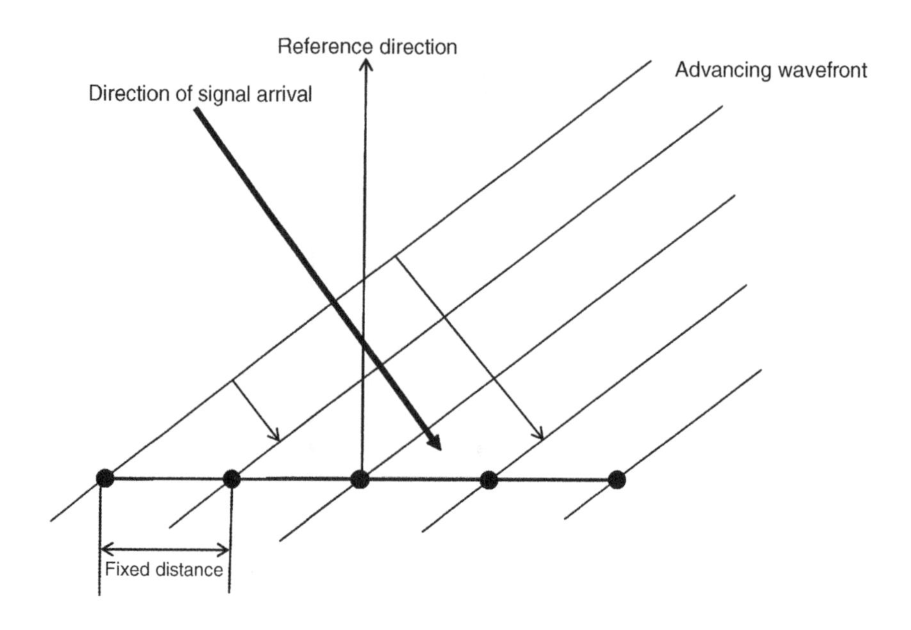

Figure 17.20 Illustration of a flat array using interferometry.

17.3.6 The DF Baseline

Individual DF systems can provide a bearing for a detected signal. For some systems, known as single bearing DF, this can be enough to provide a degree of localisation of the enemy transmission. Historically, this has principally been used for HF DF detection particularly for naval applications. When this data is entered into a system capable of modelling the current HF propagation conditions, this gives an estimated location for the target. This is illustrated in Figure 17.21. The HF-DF sensor (aka 'Huff-Duff') measures both the azimuth and elevation of the incoming transmission. Given knowledge of the conditions of the ionosphere, the emitter location can be estimated in both range and direction. Although the localisation will often be far from exact, it is sufficient in some cases to allow passive Over The Horizon Targeting (OTHT) which would allow missiles to be fired from a ship towards the target area, where the missile's own seeker would home in on the target.

This is shown in Figure 17.22. The grey ellipsoid shows the estimated location of the enemy ships. Note that in addition to considering the localisation shown by the DF system, enemy course and speed will also need to be accounted for; a missile travelling at 300 m/s out to a range of 60 km will take 200 seconds to arrive. If the ships are travelling at 20 knots, they will travel about 2 km during the missile flight time. At point (1) a multi-ship salvo of three ships firing 4 missiles each is launched. After initial pop-up, they fly at a low enough altitude to avoid detection by enemy radar (although, in many cases the enemy will not be emitting radar until a threat is perceived). All of the

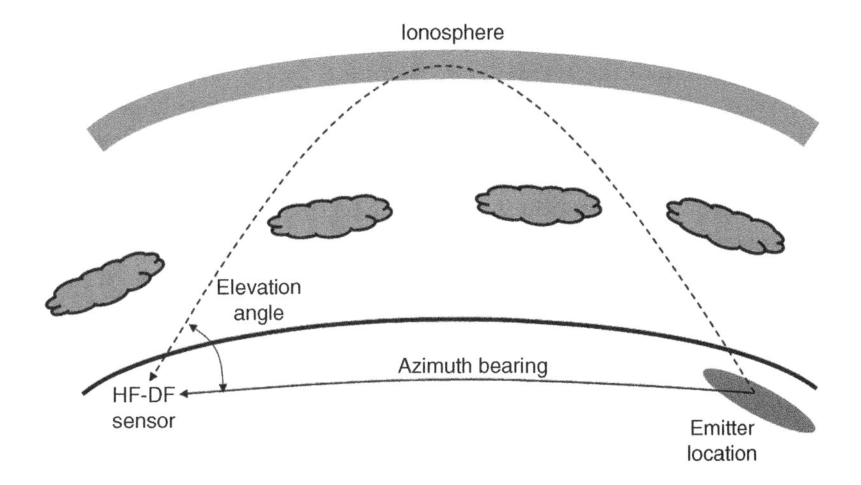

Figure 17.21 Illustration of a single bearing HF DF system.

missiles are programmed with a dogleg for two purposes; firstly, to ensure the missiles do not arrive from the direction of the firing ships thus giving an idea of their location, and secondly to ensure that the salvo arrives simultaneously at the target. Once the missiles enter the localisation area, their own radar or IR sensors are activated (3). The missiles continue to fly either in a straight line or a search pattern until the ships are detected (5, shown for only one missile) or the missile fails.

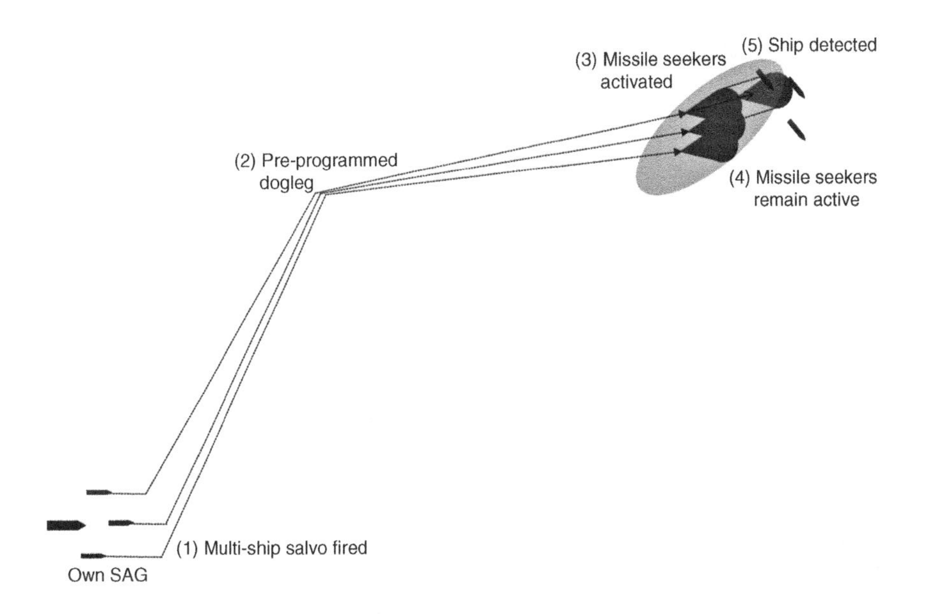

Figure 17.22 Using information from a long range HF DF system for missile targeting.

Since this is entirely passive from the attacking force until the missile sensors are activated, the attack does not reveal their location.

In most cases, however, target localisation depends on simultaneous detection by two or more DF sensors as previously described. This is illustrated again in Figure 17.23, which shows the classic DF localisation scenario.

As the figure shows, localisation is unlikely to be precise in the real world. Effects such as multipath will mean that the transmission location can only be approximated. This is not inherently a problem; the precision has to be related to the operational requirements.

17.3.7 Optimisation of the DF Baseline

Just as for detection networks, optimising the design of the DF baseline will make the best uses of generally scarce resources. However, the design considerations of DF baselines are different from those of detection networks, primarily because we need to provide overlapping coverage from a number of sensors for the system to work properly. This is illustrated in Figure 17.24, which shows six enemy transmissions, a 3-DF baseline with the coverage of each overlaid so that the darker areas show

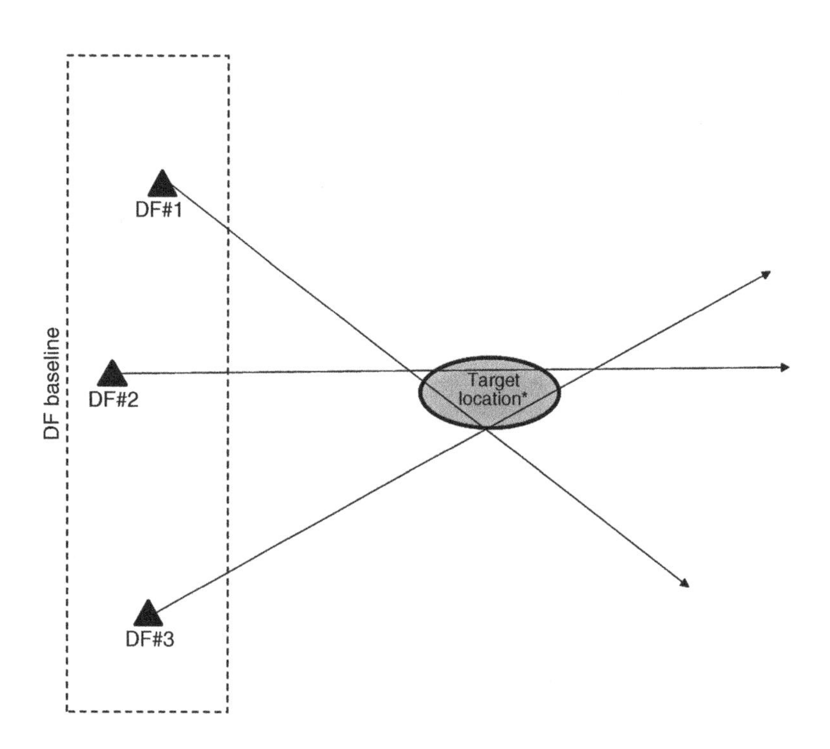

Figure 17.23 The classic DF baseline, with target being localised to an area rather than a specific point.

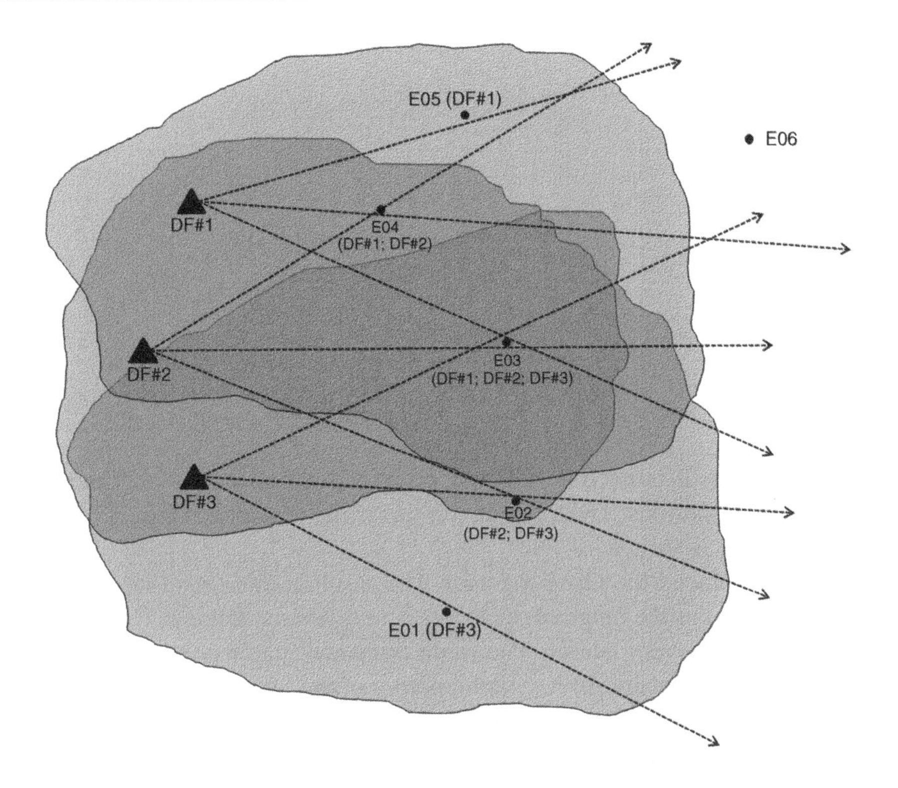

Figure 17.24 Overlapping DF coverage illustration.

where more than one DF provides coverage. The mid grey means that two (any two) of the DFs provide coverage and the darkest area is covered by all three. As usual, the design is performed in the same way as described in Section 12.14, which discusses radio optimisation, with the link budget being calculated for the target transmission all the way into the DF receiver.

The enemy transmissions are shown numbered as E01–E06. Also shown are which DFs pick up the transmissions. Of these, E06 is not detected by any of the DFs since it outside the coverage area. Both E01 and E05 are picked up by only one DF each. This prevents localisation because there is only one bearing to work with. E02 and E04 are each picked up by two DFs. This does allow localisation but with significantly less accuracy than if three DFs provided the localisation data. Only E03 is picked up by all three DF sensors and this provides the best localisation opportunity.

This illustrates one of the main DF baseline considerations; we want to provide coverage by at least three DFs over the widest possible region of the search area. This is distinct from the aspirations in detection network design where we want to provide the maximum coverage without overlap.

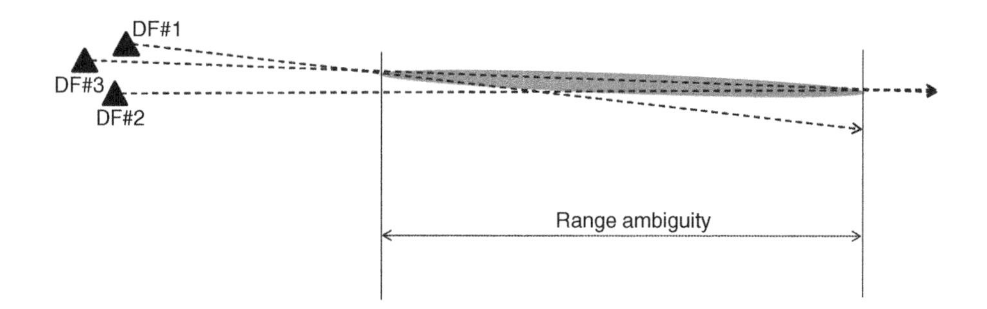

Figure 17.25 A poor DF baseline.

Additionally, we also need a suitable baseline that provides sufficient angle separation between stations. The rationale for this is illustrated in Figure 17.25, where a poor DF baseline is shown.

These DFs are placed too close together. The result is that target bearing will be detected as normal, but the range of the target is very poorly defined. This means that the localisation is not very useful. The same thing will happen even in a good DF baseline if the target is nearly in line with another of the DFs of a three-DF baseline. The ideal case is shown in Figure 17.26. In this case, the grey circle is the search area. The DFs are well spaced out and provide good DF capability against targets in the search area (assuming that the DFs have sufficient range).

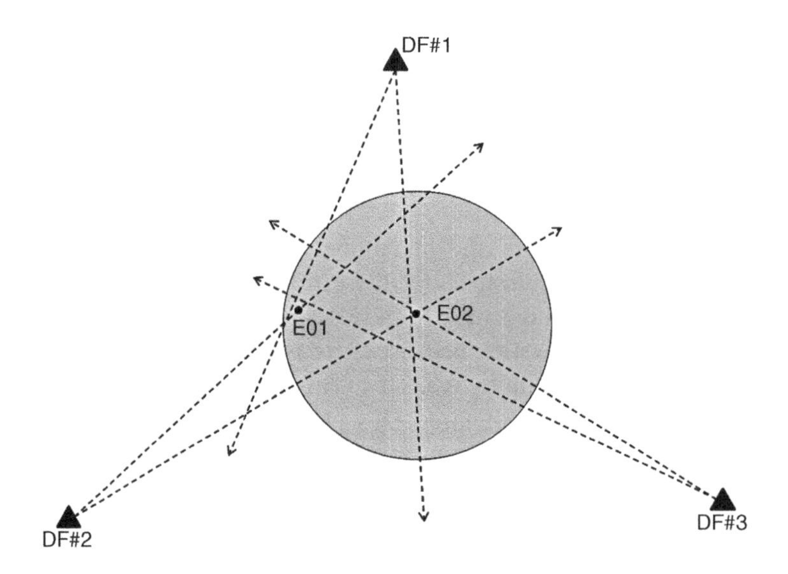

Figure 17.26 Ideal DF baseline where the target is in a circular area and it is possible to place DF sensors around the search area.

Apart from the design considerations to optimise the DF network coverage, it is also important that the DF system is sited in the best possible manner:

- The system should be placed in as quiet a radio noise environment as possible, subject to the antenna having a clear view all around.
- The antenna must be clear of obstructions and as far as possible above the terrain and clutter as possible.
- The antenna must be well away from power lines and large bodies of water (stand fast, maritime systems).
- Each asset in the DF system either must be able to communicate their data to each other or to a remote control centre (often called EWCC – Electronic Warfare Coordination Cell). This is required to collate the data to localise targets.

17.3.8 Airborne DF

Airborne DF systems have a couple of natural advantages over ground based systems. The first is that by flying they can be well away from terrain and clutter obstructions, giving them a tremendous range advantage. The second is that the aircraft can form its own baseline within a short timescale, allowing a single asset to localise enemy transmissions as shown in Figure 17.27. In order for this information to be timely,

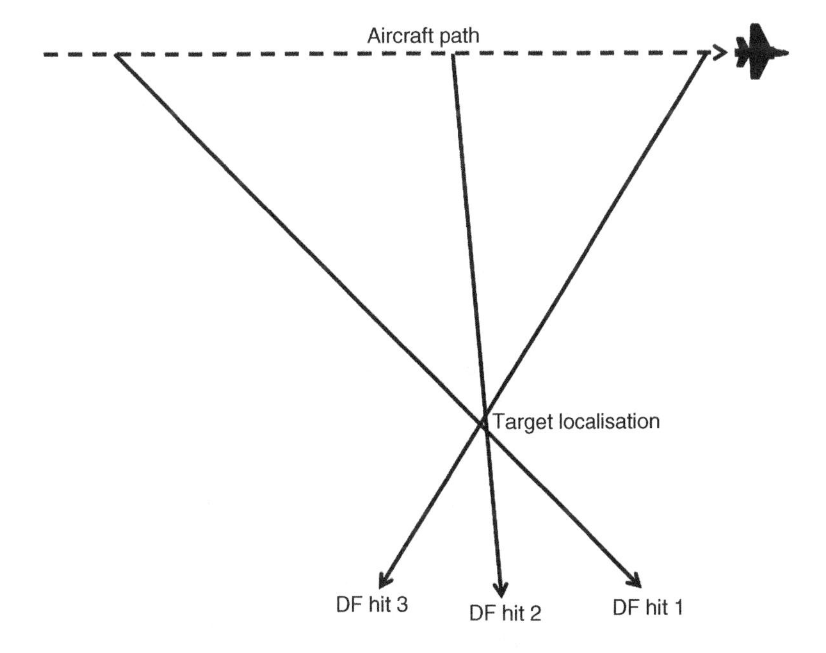

Figure 17.27 Illustration of a single aircraft forming its own DF baseline.

it has to be communicated to other assets very quickly either during the flight or shortly thereafter.

Aircraft are very good DF platforms, but they do have a number of potential drawbacks. They are vulnerable to being engaged by the enemy, and even UAVs are in short supply. Large UAVs are very expensive and require good communications in flight. However, UAVs do have the obvious advantage that there is no pilot to lose. Recovery of downed pilots is both demanding and risky, involving large numbers of aircraft and personnel.

Aircraft systems require a high level of type approval before they can be fitted to their platform. They also have to be as small as possible and, especially in the case of UAVs, to have low power consumption.

The operation of aircraft is expensive and again requires substantial resources. They also have limited time on target and if a continuous watch is required, at least four available aircraft would be typically required even for tasks of only a few days' duration. Longer tasks would require more aircraft still to allow for aircraft and crew down-time. This is why maritime and land-based systems still have a large role in surveillance and target localisation.

17.3.9 DF Assets and Communications

It is important to recognise that DF assets are of limited benefit unless they can communicate their findings, ideally in real-time. This means that DF assets need good communications during their on-task time. The communications channels must be sufficiently robust as to allow the transfer of all relevant data and may require redundancy to reduce outage time. This means that the consideration of establishing and maintaining good communications is as least as important as the design of the DF system itself.

17.4 Communications Jammers

17.4.1 Introduction

Communications jammers are designed to prevent enemy transmissions from being received successfully. One popular misconception is that jamming affects a transmitter; in the vast majority of cases, it is the receiver that is jammed because that is where the signal is at its most vulnerable and the least jamming power is required. Another misconception is that the jamming signal has to be higher than the enemy signal power. Many jammers work by raising the noise floor to an extent that the target signal cannot achieve the required signal to noise ratio. Others target particular vulnerabilities in specific systems, but again the jammer power may not have to be higher than that of the target system. This section looks at some important aspects of communications jamming.

17.4.2 Legality of Jamming

It is worth considering the legality of radio jamming. Much of this is subject to the interpretation of lawyers, and so the following discussion only addresses the key points, and can only be taken as a general guide. Any errors or omissions are those of the author attempting to understand the issues involved.

Attacks on the infrastructure of another country can be construed as an act of war, and this includes offensive jamming of radio infrastructure. Article 51 of the United Nations charter allows any member nation to take whatever actions necessary to protect their sovereignty, so this proscription does not cover offensive jamming to defend against attackers within their own borders or threatening their integrity. The UN is also mostly unable to act against actions taken by a government within its own borders against their own people, without a separate UN special resolution. However, Article 41 allows UN operations to take any non-military action against countries in breach of the charter. This specifically mentions actions against telecommunications and other radio services.

It must be noted that radio signals do not respect international borders and thus jamming directed against a hostile country may affect radio systems in adjacent friendly or neutral countries. Additionally, for military operations in host nations, there is also the possibility of disrupting local essential services. Both cases may cause international tension. This may undermine host nation resolve to continue in the operation and may cause adjacent countries to become more belligerent.

Another crucial issue in the modern era is the disruption that may be caused by friendly jamming disruption. This is where intentional jamming of enemy targets also results in disruption of own force communications. Consider the following contemporary possible examples.

As part of a major exercise in international waters, jamming is used as part of the exercise. Despite careful adherence to the avoidance of JRFL frequencies, a jammer is malfunctioning and has been pushed into non-linearity. As a result, some air and maritime navigation, communications and emergency channels are blocked even beyond the closure area. A civil accident results. What is the responsibility of the military forces conducting the jamming?

A number of patrols are in the same area, perhaps patrolling the same route but separated by a number of kilometres. The lead force has counter-IED radio jammers operating and these are also causing disruption to the other patrol's radio systems. Suppose a second force comes under fire from insurgents and because their radios are jammed, they cannot call for the reinforcements they would normally be able to rely on. Heavy casualties result. How does this affect the duty of care that their army is supposed to be able to provide?

An RCIED team are using jammers to prevent attacks by insurgents. The jammers cause a bomb being placed by a roadside to detonate prematurely, while heavy civilian traffic is passing. Heavy civilian casualties result. What is the legal position?

Countries such as the US protect their soldiers from foreign legal positions; however, what if one of these events happened in the CONUS? Other countries' armed forces do not necessarily have the same protection, nor do those operating in police and other services that may come to make use of jamming systems.

The following sections describe the use of different communications jammers. In all cases in this section, we have ignored antenna directionality and assumed omni-directional antennas. Of course, if necessary corrections can be made to account for antenna directionality.

17.4.3 Spot Jamming

Spot jamming is used to jam a pre-selected frequency that has been determined as a target of interest. The process may be along the following lines.

An intercept/DF network picks up a transmission that is determined not to be from own forces or from neutral operators. Analysis identifies specific parameters for the situation such as:

- transmission central frequency;
- 3-dB and 10-dB bandwidth;
- modulation scheme in use;
- signal strength at the detecting receiver(s);
- direction or localisation position if known;
- times of transmissions;
- how frequently the channel is used;
- duration of transmissions;
- association with other systems (activity analysis). This is where the behaviour of an enemy network is studied to determine its structure;
- specific identifying features such as frequency instability, operator Morse behaviour, the same voice on successive intercepts or transmission of formatted messages.

These factors can be analysed with the benefit of knowledge already known, such as association with known weapon systems and technical parameters of known enemy systems. The detected transmission may for example be associated with a specific air defence command system or be of a type known to be used for command and control. Transmission behaviour may also reveal the relative importance of the channel. If is frequently used, then it may be an important command and control channel. If it is noticed before artillery shells arrive, then it may be the artillery command network. If troop movements are correlated with the transmissions, then again it may be a command net. If other transmissions on other frequencies follow on quickly after the initial transmission, then again it might be part of the command network and so on.

Once the transmission is determined to be a high priority target, an EW tasking mission may be issued to the spot jammer. This will include the technical parameters and any other pertinent information, such as the duration of the jamming task.

The actions of the jammer detachment will be to tune the jammer to the required frequency and bandwidth if this is adjustable. The antenna will be pointed in the direction of the enemy transmission. Before jamming, the channel can be listened on to determine whether it is still transmitting (there is no point in jamming an unoccupied channel). If it is still in use, then jamming can commence. The jamming can consist of un-modulated or modulated noise. Un-modulated noise will raise the noise floor of the enemy receiver, preventing them being able to communicate. Modulated noise does the same thing but also disrupts audio reception of the transmission signal making it impossible for the receiving operator to hear the message.

Periodically, the jamming signal will be turned off so that the jammer operators can listen to determine whether the enemy is still using the channel or have changed to a different one.

How can success of the jamming task be measured? The enemy are unlikely to tell us. One method for unencrypted voice is to use the look-through phase to determine what the enemy are saying once the jamming has stopped temporarily. If they are complaining about the jamming, then that would be an obvious sign. If the frequency is no longer being used, this could indicate that the jamming has been successful but that the enemy have moved to a different channel.

If this type of information is not available, then the effective jamming performance can be estimated in the following manner. Figure 17.28 shows an example of jamming a 60 MHz system that has a nominal range of approximately 17 km without jamming in the environment in which it is being used. If a 25 W spot jammer is placed 20 km away from the target receiver and the jammer to signal power required to disrupt the signal is -7 dB down on the received level, then the estimation shows that the nominal range is reduced to just over 4.5 km. Note this is the distance to the receiver; the distance to the jammer is not important, nor is the topology relating the transmitter to receiver. In other words, it does not matter if the transmitter is closer to the jammer or further away. Note that this analysis considers only one direction of the link. To determine whether the link is jammed in both directions, we would need to apply this to both ends of the target link in turn. What we can say is that there is less likely to be effective if both terminals are less than 4.5 km apart. At that range, the enemy system has the ability to burn through the jamming.

This analysis can be completed for different ranges of the jammer from the receiver as shown in Figure 17.29. The further the jammer is away from the receiver, the longer the burn through range that the radio link can achieve.

This analysis only applies to the input parameters of the target system, the jammer and the type of environment. It would need to be re-done for different systems in different environments.

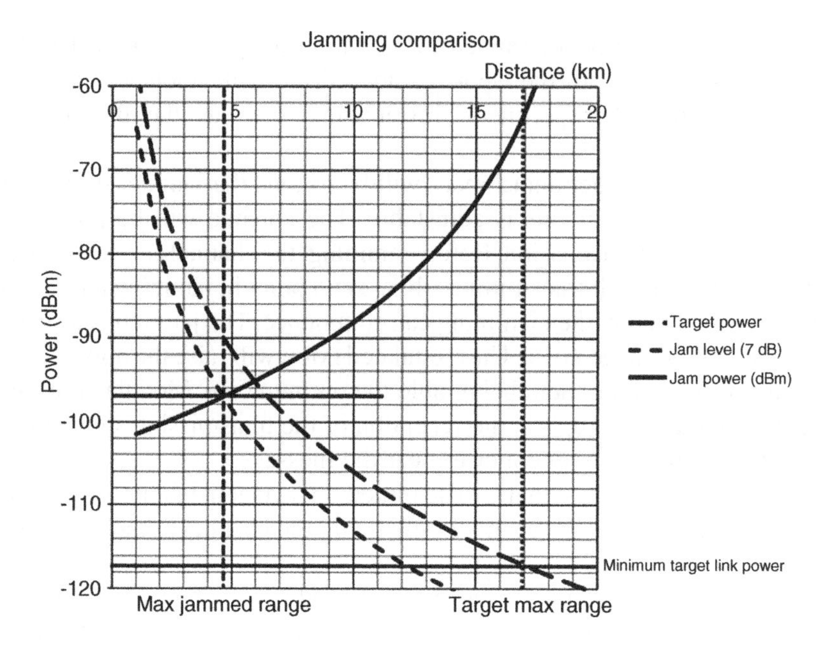

Figure 17.28 Effect of a jammer at a nominal range of 20 km from a receiver within an enemy radio link. The jammer reduces the nominal effective range from 17 km to about 4.5 km.

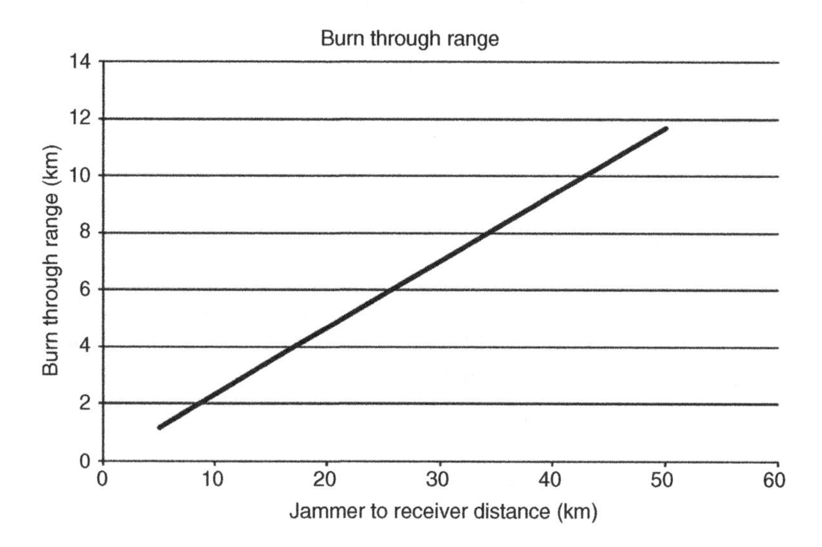

Figure 17.29 Estimation of burn through range for the same radio link and jammer as shown in the previous figure with different distances from the jammer to receiver.

17.4.4 Barrage Jamming

Barrage jamming is used to deny the enemy of the use of a portion of spectrum. This can be because enemy forces are frequently changing channels or that they are using full frequency hopping systems. Compared to spot jammers, barrage jammers need to supply jamming power into a number of channels rather than just one. On the rather simplistic assumption that the barrage jammers deliver exactly the same jamming power into each jammed channel, and that the effective jamming power has the same effect as the spot jammer, it is easy to calculate the power reduction for the number of channels jammed. This is shown in Figure 17.30.

The same type of analysis as described in the previous section can be carried out for barrage jammers with two corrections. The first is to apply the corrections for the number of channels being concurrently jammed. It may also be necessary to apply corrections for the antenna response over the frequency range. The second is to apply corrections for the frequencies being investigated as the propagation conditions may change over the range of the jammer and the frequencies being used by the enemy target network. If the jamming range is a relatively low percentage of the transmission

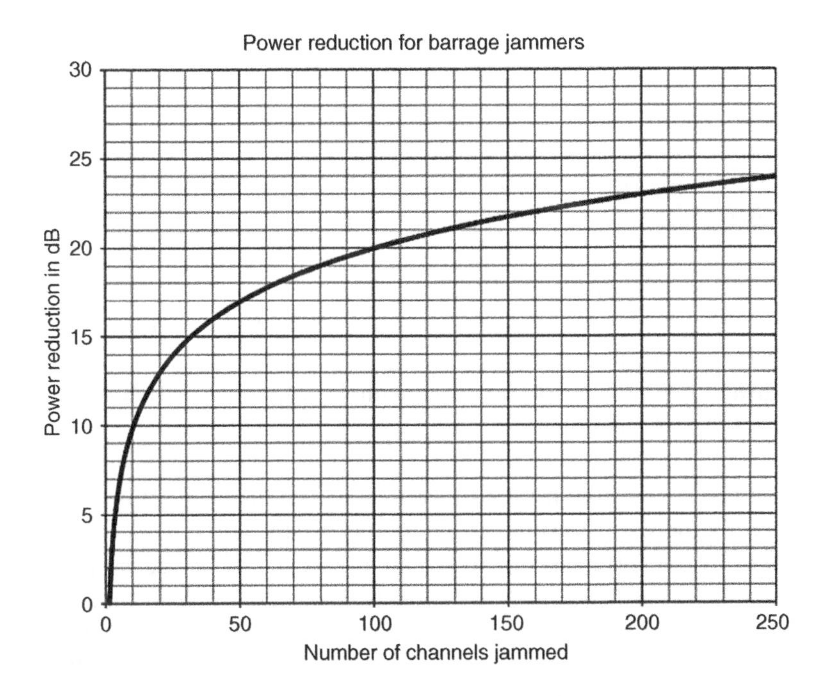

Figure 17.30 Power reduction when increasing the number of channels jammed.

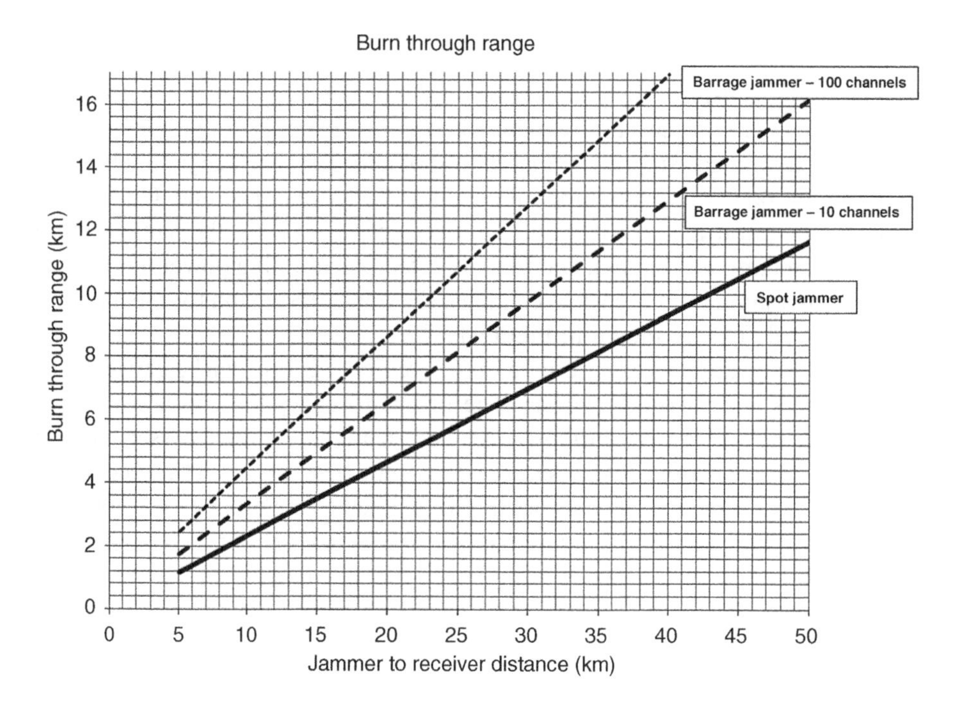

Figure 17.31 Comparison of the burn through range of an enemy network against a spot jammer and a barrage jammer of the same total power, tuned to jam 10 channels and 100 channels.

frequency, for example 50–55 MHz or 800–900 MHz then the propagation changes are relatively small. Figure 17.31 shows the burn through range of;

- the 25 W spot jammer of Figure 17.28;
- a 25 W barrage jammer set to jam 10 channels, which has 10 dB less power per channel than the spot jammer;
- a 25 W barrage jammer set to jam 100 channels, which has 20 dB less power per channel than the spot jammer.

Notice that the burn through range increases as the barrage jammer tries to jam more channels for the same power. In order to improve the performance of the barrage jammer to meet the performance of the spot jammer, it would be necessary to increase the total power of the barrage jammer by 10 dB to ensure the same power is available in each channel. This would mean increasing the total power radiated from 25 W to 250 W. For the 100 channel barrage jammer, the correction would be 20 dB, meaning the power would have to be increased from 25 W to 2500 W. Another way of achieving the same improvement could be obtained by increasing antenna gain above the spot jammer by 10 dB and 20 dB thereby improving power radiated in the direction of the network to be jammed. This would be at the expense of the area that could be effectively jammed.

It can be seen that barrage jammers need to have more power to jam larger number of channels simultaneously. For this reason, the trade off between the range of the band to be jammed and the jamming power needed is an important design consideration. It is for this reason that partial band jamming is often considered. This is where only part of radio range used by the enemy is used. For example, if a jammer can jam 25% of the channels used by a frequency hopper, then the jamming should be effective. Some frequency hopping systems may be able to identify the jammed bands and therefore avoid them. If this happens, the barrage jammer can be regularly re-tuned to skip around the band to make jammed channel avoidance more difficult.

17.4.5 Responsive Jamming

Responsive jammers have an RF detection capability that allows them to scan for threats and jam those of interest. To illustrate how a responsive jammer may work, we will consider a simple example using generic parameters to make the calculations easier. These parameters are not based on any particular system, but have been chosen to make it easier for the reader to understand the calculations.

Assume there is a frequency hopping system with the following parameters:

- frequency range of 40–60 MHz;
- system uses a raster channel structure of 25 kHz spaced channels;
- hop rate of 100 channels per second.

This means that there are 1000/25 = 40 channels per MHz and 800 channels available (based on the 20 MHz band). We will assume that the rise and fall times for the hopping system are negligible. This means that the dwell time for each hop is 10 milliseconds. This is too fast for most systems to tune to, so it is worth considering the design issues involved for this type of system.

The process to be carried out is:

- scan and identify the presence of the signal;
- identify the signal as a threat to be jammed;
- tune the jammer to the required frequency;
- fire up the jammer on that frequency.

A basic rule of thumb is that at least 25% of the dwell time must be jammed for the jammer to be effective. Using this figure, the jammer system needs to perform all of these tasks in 7.5 milliseconds including the time for the signal to reach the jammer from the target and for the jamming signal to reach the target.

The scan time for a modern system should be many 10's of GHz per second, so to scan 20 MHz should be achieved in a time period of 2–3 milliseconds. The tuning time for a modern jammer should be as little as 50 microseconds, so this should be less important. The rise time for the jammer to achieve full power also needs to be considered, as is the both-way propagation time delay. If we assume a time of 3 milliseconds for scanning and tuning, then the jamming signal must reach the target in less than 4.3 milliseconds after it reached the jammer receiver.

For two-way propagation out to 100 km, the round trip propagation time is approximately 0.66 milliseconds, leaving 3.64 milliseconds as the maximum tuning and rise time in order to reach a target at that range.

17.4.6 Adaptive Jamming

Adaptive jamming is an extension of responsive jamming but with the potential to jam several targets at the same time. It provides an improved method to achieve the same effects as barrage jamming but in a far more focussed manner. For example, if a barrage jammer is attempting to jam 100 potential jammers in order to effectively jam only four targets, then the radiated power in each channel is 20 dB down on the total power transmitted. The adaptive channel, splitting its power equally into the four channels is only 6 dB down. A 6 dB gain antenna can still cover a wide or even omni-directional horizontal area far more easily than a 20 dB antenna, and can still have a relatively wide vertical gain.

17.4.7 Smart Jamming

Smart jamming is the term used to describe jamming aimed at network vulnerabilities rather than simply raising the noise floor or causing unacceptable audio or data performance.

Methods of smart jamming are aimed at particular types of network such as GSM, UMTS, paging systems and many others. A smart jammer designed to attack GSM will have less or even no effect against other networks that may be present so it is important to be able to identify the exact type of network in order for it to work. Some particular network vulnerabilities include:

- pilot channels;
- synchronisation channels, time slots or data;
- paging channels or time slots;
- error correction checksums;
- acknowledgement or Not Acknowledged messages.

The purpose of smart jamming is to prevent normal performance of a network. This may be by denying subscribers the ability to log on to the network by causing base station overload, disrupting signals telling subscribers that they have a call, preventing successful call initiation or disrupting communications once a link is established by causing the system to successively re-send packets of data due to Not Acknowledged or error checksum faults.

This type of jamming is relatively new compared to the other methods and much work is being done in the literature at present.

17.5 The Role of Unmanned Airborne Vehicles

17.5.1 UAV Roles

Unmanned Airborne Vehicles (UAVs) have become an important part of modern warfare. They can provide a range of functions, including:

- remote sensing, including optical and radio surveillance;
- direction finding;
- over the horizon targeting;
- communications relay;
- stand-on jamming;
- interdiction, using Unmanned Combat Airborne Vehicles (UCAVs);
- local surveillance, using very small UAVs.

Hundreds of types of UAV are in service or development, covering everything up to strategic roles. Design and operational considerations vary considerably according to size, role, performance and endurance. Most UAVs require a command and control link and also a telemetry link to transmit the data they are collecting. In many cases, there will also be a vehicle telemetry system to transfer flight data back from the aircraft to the control centre. The potential options are shown in Figure 17.32.

Communications to and from the UAV may pass via satellites for strategic systems, via airborne communications relays or straight to and from ground control stations. Airborne communications relays may be aircraft or other UAVs. The routes shown in the figure may have more than one set of links; there may be others for redundancy.

17.5.2 UAV Advantages

UAVs have clear advantages over other systems in certain cases. Compared to ground systems, they have the advantage of altitude and unobstructed paths to and from potential target systems. As they are unmanned, they can be sent forward to areas that are more dangerous to perform stand-in jamming, surveillance or radio localisation tasks. They are cheaper to operate than manned aircraft. The scalability of UAVs allows them to be used for a variety of tasks. Very small systems can be used at the tactical level by units as small as squads to provide intelligence collection over the next

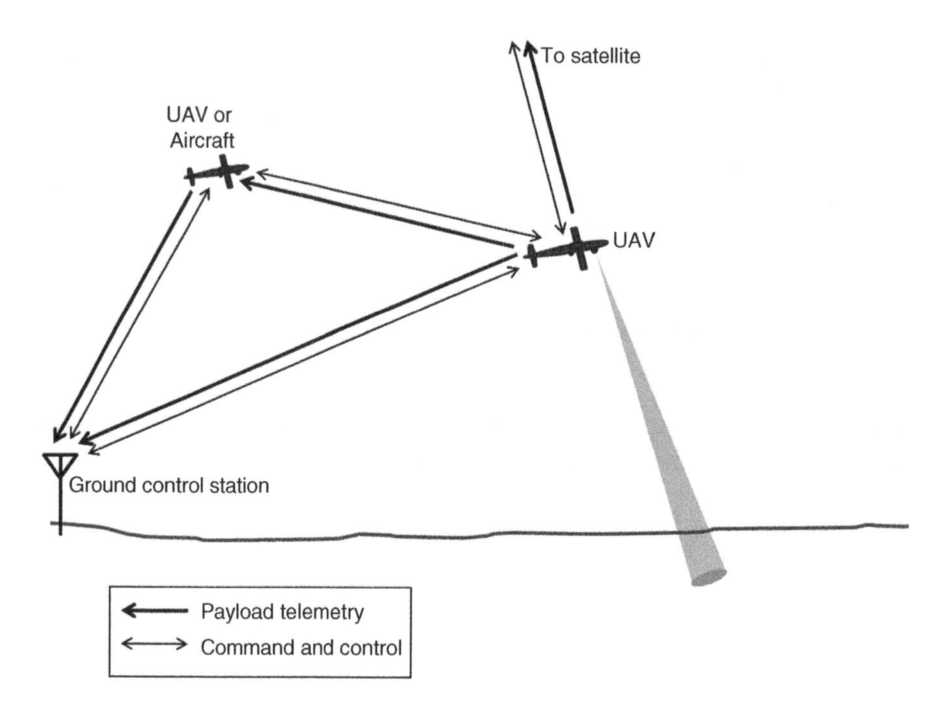

Figure 17.32 Communications routes to and from UAVs.

hill or around the next street corner. Low-cost, simple to use systems can be widely issued to units for tactical tasks of a few kilometres. Larger systems can be used to provide surveillance tens of kilometres into enemy territory at the expense of larger cost, logistics and operational issues.

UAVs are also finding roles in other areas such as policing, border monitoring, surveys of power lines and areas of interest. It is not difficult to imagine the role they could play in providing real-time images for emergencies and for natural disasters.

17.5.3 UAV Disadvantages

If UAVs have such obvious advantages, do they have any disadvantages? As with any other system they do. Large UAVs are more vulnerable to ground fire and they would stand no chance against airborne interception aircraft. If control of them is lost due to damage or system failure, large UAVs can cause significant damage on the ground and recovered UAVs are intelligence gold for technical specialists looking to find out how the most modern systems work.

UAVs are not totally autonomous and do rely on communications to function correctly. Interrupts to command control can be tolerated to a greater or lesser extent, but it communications are lost the mission will be terminated at some point.

The most sophisticated systems may be programmed to return to base or specific points to allow communications to be re-established if possible, but others will lose control and crash.

UAVS, in common with other aircraft, have limited time on task. Depending on their design, they can loiter for a certain length of time in the target area but then they have to return to base for re-fuelling and maintenance. Also, the aircraft are flown remotely and their pilots also require rest. This means that just like manned aircraft, many UAVs are required to maintain extended coverage.

Another key issue is the power available required onboard to supply the UAV systems and payloads. This places severe power management issues on the design of all vehicle sub-systems. Jammers in particular have limited power and must be managed to ensure that their power is not exhausted too early. This constraint adds to design cost and operational mission management.

UAV systems also pose potential airspace management problems with other airspace users. Since they are unmanned, they cannot react to other traffic in the same way that a manned aircraft can. It may be necessary to block off large parts of airspace and flight levels to ensure that other aircraft remain clear. In dynamic military environments, this is inefficient.

17.5.4 UAV Communications and Spectrum Management

Communications for UAVs are more important than for manned aircraft. The vehicle is remote both from the pilots that fly them and the customers of the services they provide. Without communications, the vehicle cannot be controlled or real-time data received. The importance of robust communications is clear. The design methodology and mission management for UAVs is the same as for other ground to air systems. The system is subject to the normal fading characteristics due to ground reflections, but because the UAV is continuously moving, this can be countered by fly-wheel operation that counters short time duration fades. An additional method employed in the most sophisticated systems to counter ground reflection fading is the use of horizontally and vertically directional antennas that block ground reflection and also offer high directionality to relay UAVs or ground control stations. These systems are relatively complex requiring motorised antennas to ensure that the moving UAV remains pointing at the other terminal. This approach does however have the advantage of reducing the potential for interference or jamming and extends operational range. This type of system normally requires line of sight operation to the other terminal, including satellite systems. The line of sight must be maintained as the UAV pitches, banks and rolls in normal operation. If this approach is used, it is possible to use free space propagation predictions to estimate system performance. Other approaches to maintain communications involve link or antenna diversity to increase the probability that one of more of the antennas in the system can communicate with the UAV and communicate to other nodes in the network. It goes without saying that air the air vehicle mounted communications, antennas and associated systems must be lightweight and low power.

UAVs pose particular challenges for spectrum management. Their communications channels must be well protected from friendly, enemy and other sources. There may be a requirement for multiple channels to be assigned to allow for backup channels if the primary ones are blocked. In addition, the UAV controllers may need a number of other channels to allow them to talk to other air users and for the different phases of the operation, including launch and recovery. They may also need to have communications with ground forces if the operation is in support of ground operations. This means that a single UAV mission may tie up lots of spectrum and as one of the advantages of UAVs is their flexibility, this may be required at short notice. The spectrum may be required at short notice and also released quickly once the mission is completed. This requires dynamic spectrum management – something that is emerging rather than being commonplace at present.

17.6 Countering Enemy Communications Electronic Warfare

17.6.1 Poacher-Gamekeeper Approach

Part of a gamekeeper's job is to protect the animals on their land from poachers intent on stealing those animals. The very best gamekeepers are those who really understand how poachers work; often this is because they have been poachers themselves. This is something that is also true of communications electronics warfare. The better an understanding of how radio systems work and how operators use those systems an EW operator has, the better they will be able to exploit technical and operational vulnerabilities. The better communications operators understand EW systems and how EW operators work, the better they will be able to protect against enemy EW action. To promote this, EW operator readers are recommended to read widely about radio communications, and radio operator readers are likewise recommended to read widely about EW.

17.6.2 EPM

One of the most important methods of countering enemy EW action is to prevent them ever knowing that they know your systems are there in the first place. Power management comes in. If you are not transmitting, it is impossible for the enemy to detect or intercept you. Obviously, this is not always possible, but minimising transmissions is important. Naval ships and aircraft often travel in electronic silence or use other methods described in the next section. They only start to transmit when they absolutely need to do so, to perform operations or to activate defence systems. In all systems, the use of minimum power for communications is also normally employed.

17.6.3 Tactical Methods of Countering Enemy CEW

The methods for countering interference and jamming described in previous sections of the book can all be used to counter enemy communications electronic warfare. These are summarised below:

- power minimisation;
- minimising transmissions;
- Using antennas as low as possible;
- use of directional antennas;
- orientating directional antennas away from the enemy;
- using low probability of intercept systems such as spread spectrum;
- using other traffic to mask own transmissions, such as using unused channels in GSM networks for example;
- making use of terrain and clutter shielding;
- using 'spoofing' (described below).

Spoofing is a method of using a radio or radar system to mimic the parameters of another system. Using this method, a warship can pretend to be a non-combatant or other vessel. The aim is to fool enemy forces into misidentifying the warship and ignoring it, until it is too late. This is a modern day version of a traditional method of ruse de guerre in which warships used to fly false colours to fool other warships.

References and Further Reading

Ademy, D. (2001), EW101: a First Course in Electronic Warfare, Artech House, MA, USA, ISBN 1-58053-169-5

Ademy, D. (2004), EW102: A Second Course in Electronic Warfare, Horizon House Publications, USA, ISBN 1-58053-686-7

Frater, M. R.; Ryan, M. (2001), *Electronic Warfare for the Digitised Battlefield*, Artech House, MA, USA, ISBN 1-58053-272-3

Graham, A. W.; Kirkman, N. C.; Paul, P. M. (2007), *Mobile Radio Networks Design in the VHF and UHF Bands: A Practical Approach*, Wiley & Sons ISBN 0-470-02980-3

Poisel, R. A. (2008), *Introduction to Communications Electronic Warfare 2nd Edition*, Artech House, MA, USA, ISBN 978-1-596934528

18

Non-Communications Electronic Warfare

18.1 Non-Communications EW

18.1.1 Broadcast

Jamming is not just carried out against enemy communications systems, but also against other types of radio including broadcast systems. Broadcast systems differ from mobile systems in that the transmission is one way only. The purpose of broadcast jamming is to prevent reception of the victim broadcast transmitter by receivers located within its service area.

Apart from the conceptual difference that the receiver does not transmit, in fact the jamming task is very similar to communications jamming for the mobile case. The process is to determine the received power at each point within the service area and apply suitable jamming power to prevent reception, or to render the output audio or pictures intolerable to subscribers. This is illustrated in Figure 18.1, which shows a directional jammer jamming part of the coverage area of a broadcast transmitter. Ideally, the complete broadcast area should be jammed. This would occur if the limit of jamming is somewhere past the broadcast transmitter as the signal power from the broadcast transmitter in this direction would rapidly fall off with distance, making jamming easier.

18.1.2 Navigation Systems

Navigation jamming, also referred to as navigation warfare, is the process of disrupting reception of navigation signals at certain areas. The purpose can be to deny acceptable reception or to disrupt accurate navigation by providing apparently real signals that actually result in the receiving platform believing that it is in one location whereas it is actually in another.

Communications, Radar and Electronic Warfare Adrian Graham

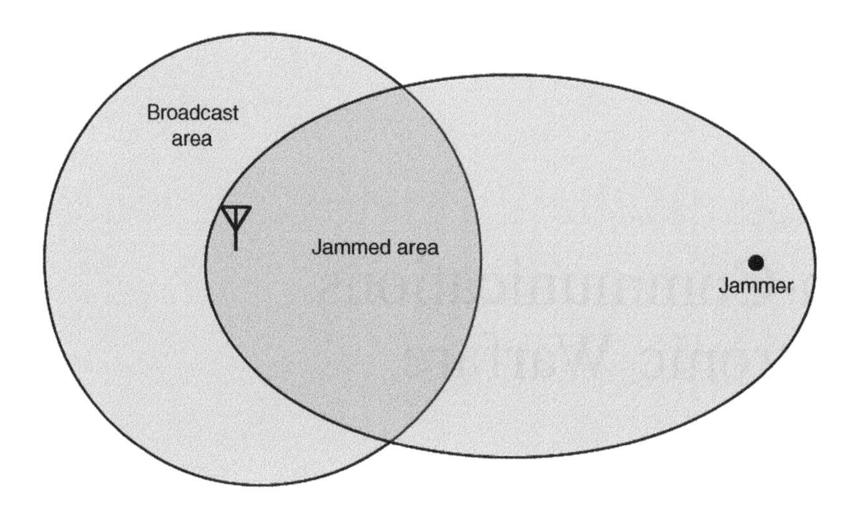

Figure 18.1 A directional jammer operating against a broadcast station. The idea is to jam the receivers – the transmitter receives no signal and therefore does not need to be jammed.

These days, GPS is the main focus of navigation warfare. GPS relies on highly precise timing to allow the signals from the overhead satellites to provide an accurate fix. GPS receivers also work on relatively small signals, making them vulnerable to noise, interference and intentional jamming. Trials have shown that GPS is vulnerable to white noise, CW, AM, FM and swept jamming. If the jamming signal achieves the necessary J/S ratio, it prevents the victim receivers obtaining location data.

Another more insidious method is to transmit GPS signals with inaccurate data. In this case, the victim receivers believe they are still receiving valid signals but in fact, they are being sent off course. This can be achieved by using a valid GPS transmission method but by playing with the timing synchronisation of other parts of the signal.

Much work is being done into both GPS jamming and anti-jamming techniques. There is much literature available on the subject, but a detailed analysis is beyond the scope of this book.

18.1.3 Secondary Radar Systems

Before going on to look at primary radar jamming techniques, we can look at jamming of secondary radar systems. Secondary radars are described in Section 7.3.11. The basic principle is that the ground secondary radar sends out a signal that is received by an aircraft. The aircraft then 'squawks' its identification and flight level details. Jamming on the ground to air frequency of 1030 MHz prevents aircraft receiving the interrogation signal. Jamming on the air to ground frequency of 1090 MHz would prevent the ground radar receiving the returned information from the aircraft. In this sense, jamming of secondary radar is similar to jamming any duplex radio system.

Jamming secondary radar would prevent the ground station from receiving all flight altitude data in the area.

Another method of exploiting secondary radar is to return a false return signal, allowing combatant aircraft to pretend to be normal civilian traffic. This is another version of spoofing.

18.2 Radar Jamming Techniques

18.2.1 Cover Jamming

Cover (noise) jamming works in the same way as for communications jamming. It desensitises the receiver and makes it more difficult for it to receive valid returns. The increase in noise will be picked up by the radar receiver Automatic Gain Control (AGC). This will cause the AGC to compensate for the increased input and in doing so it will cause low level targets to be rejected. AGCs are essential in radar design to allow the system to work over a wide dynamic range but retain sensitivity from moment to moment. Since the lowest signal returns are obtained towards the edge of the radar's coverage, these are the returns that will be affected first. This means that the range of the radar will be reduced. If the jamming noise at the radar receiver is sufficiently high, the whole system may cease functioning due to overload.

Figure 18.2 shows two phases of a jamming scenario. In the first, the radar is pointing directly at the jammer. The jammer to radar path is shown as it passes into the main beam of the radar. The radar transmit and target reflection paths are shown below the radar to differentiate them from the jammer path but of course they would really be overlapping on the radar main beam in practice. In this scenario, the jammer has the advantage because the reflection from the jammer aircraft skin would be very small. The jammer can transmit a relatively small signal that is larger than the reflection power and still degrade the radar receiver response. However, in the second phase, the radar is no longer pointing at the jammer and the jamming signal is in the sidelobes of the radar antenna response. In this case, the jammer advantage is gone due to the attenuation of the sidelobes compared to the main beam.

Jammer noise is difficult to reject, especially when it enters the main beam. There are potential methods to counter it, but all require trade-offs. The principle problem is due to the AGC in the receiver. The jamming in energy can be reduced by the use of pulse compression, which counters narrowband energy but in this case, the jammer can use barrage jamming to counter the counter-measure, or if the jammer is using high enough narrowband noise, it can still saturate the receiver.

One other point of interest in this scenario is burn-through. If the jammer power remains constant as the jamming aircraft approaches the target, its power increases in a square law function if we simply assume free space loss. The radar signal has to reach the aircraft and return, doubling the distance. Thus, burn-through will not be achieved if the jammer is heading directly towards the radar.

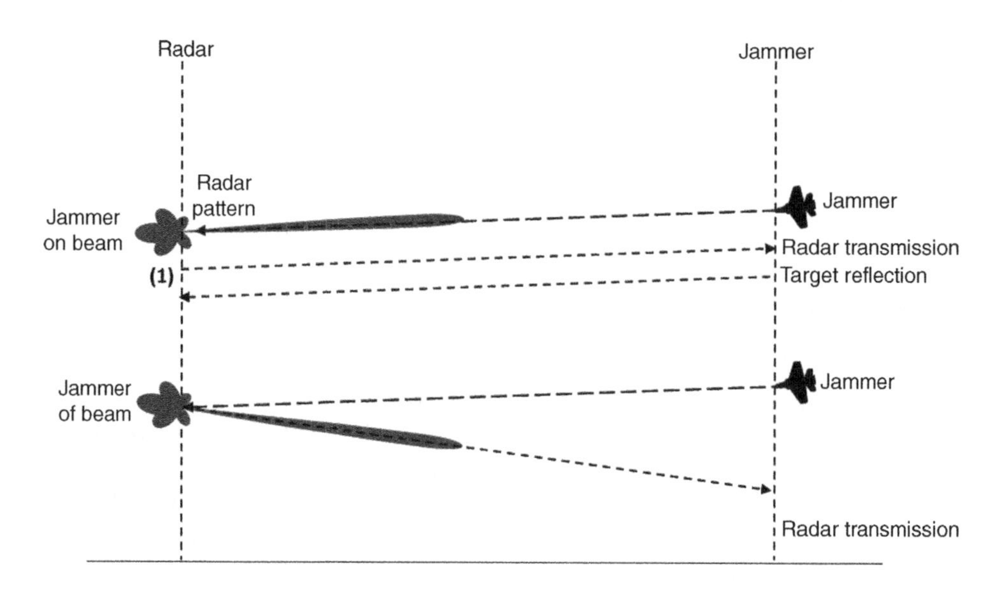

Figure 18.2 Two different stages in the sweep of a surveillance radar. In the first, the radar is pointing directly at the jammer and all of its energy enters the receiver. In this case, the jammer can emit a relatively small amount of power and still induce a lot of power into the radar receiver circuitry. In the second case, the radar main beam is no longer pointed at the jammer. The jammer power enters into the radar sidelobes and is more easily rejected.

It can be seen from Figure 18.2 that noise jamming has most effect when it is the main beam. This means that the effect is centred on the jammer and the angle width of the most effective jamming is dependent on the angle of the radar main beam. Since higher frequencies typically result in narrower beamwidths because of the relationship between the frequency and the antenna aperture, the higher the frequency, the narrower the beam effectively jammed. One interpretation of this is that a stand-in or stand-off jammer aircraft may or may not be able to shield an attacking force on the same inbound vector. Also, the direction of the jammer is revealed to a greater precision. This may be used to vector intercept aircraft onto the direction of the attack or cue missile defences. Tracking radars typically use higher frequencies than surveillance radars and a single bearing target may be engaged even if its range is uncertain by some missile systems.

It should be reasonably obvious to the reader that if the jammer aircraft were to modulate its jamming power to counteract the gain of the rotating radar antenna then a better jamming result would occur. This is in fact true and is covered in the inverse-gain jamming section.

18.2.2 Range Gate Pull Off and Pull In

Cover jamming is relatively 'dumb' compared to deception methods. It does not require advanced knowledge of the target radar parameters. All of the other methods in this section, including range gate pull off and pull in, do require knowledge of the specific radar type parameters in depth in order to deceive the radar; hence, they are known as deception EW methods.

Range gate pull off relies on the target being illuminated to create a copy of the incoming radar transmission and replicate it. The principle is that the target identifies the threat radar signal as it hits it. It then programs the jammer to the required frequency and character of that pulse. At first, it returns the signal along with its reflection. It increases the power of the jammer return to match and exceed the power returned by the platform it is fitted to. This de-sensitises the radar AGC so that the original target is lost in the noise. Then, it successively delays the return to make the target appear further away than it actually is. This 'walks' the return further back, making the target appear to be travelling towards the radar slower than it actually is.

Figure 18.3 illustrates the process. In Diagram (a) one of the radar pulse trains is shown. It is assumed that the radar continues to illuminate the target in the subsequent parts of the diagram, but this is not shown. The radar pulse bounces off the target and becomes an attenuated echo returning to the radar with similar pulse characteristics to

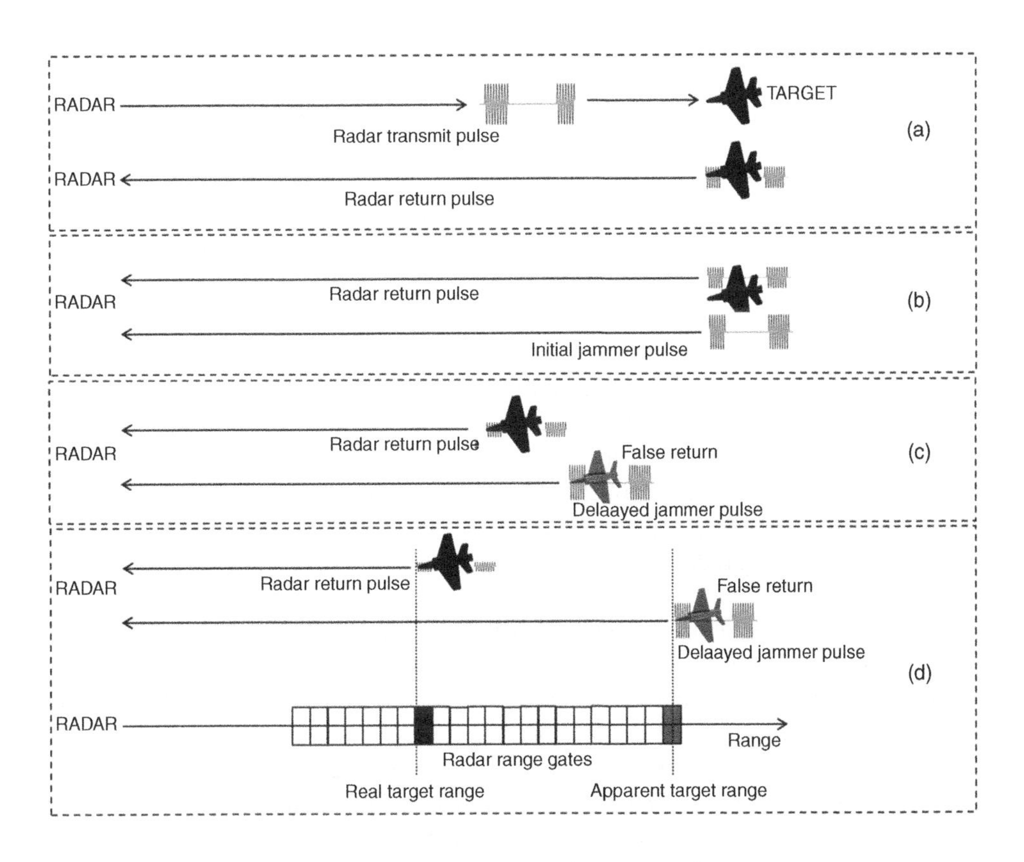

Figure 18.3 Range gate pull off illustration.

the original. Diagram (b) shows the jammer replicating the return but at higher signal power. This causes the radar AGC to compensate, making the original return appear smaller to the extent that it is rejected by the radar receiver. Diagram (c) shows the jammer delaying the radar return and in Diagram (d) some of the radar range gates are shown. Each range gate is the smallest range resolution the radar can detect. The real return is masked because of the changes to the AGC and is ignored. The real range is shown by the black bin, but the radar believes the target is in the grey box further away, giving the false target shown.

Range gate pull in works the other way, making the target appear to be travelling faster towards the target. In both cases, the range to the actual target is wrong. This may delay or advance the launch of countermeasures such as missiles. If missiles are launched too late at a target that is closer than it appears, then the missile detection system may not be switched on in time, in which case they will likely miss. If the missiles are launched too early, they will fall short of the target. This may not be an intuitive result. The popular view of missiles is that they launch and head directly and inevitably towards the target. In fact, most missiles have a short boost phase during which the rocket motors are running and then they glide, albeit at supersonic speed. This means that they are in some senses ballistic – initially rising then falling from their maximum altitude. If they are not launched in the correct attitude and direction, they have limited ability to correct their flight. Active missiles also have a short time to acquire their targets; if they do not do so, they will self-destruct or miss in any case. The effects of range gate pull in and pull off deceiving a SAM system are shown in Figure 18.4. Note that one or either is likely to occur, not both at the same time as shown in this diagram. The search radars of the guided missiles are not shown, but in either case they would form a cone shape in the front of the missiles in the direction of travel and in neither case would they detect the real target.

Beam-riding missiles may have a flatter trajectory, but once they are launched, their onboard seekers will attempt to lock on to the target (this is called front lock) while retaining control from their control radars (rear lock). If they cannot maintain both front lock and rear lock, they will be deemed to have lost control and will be destroyed. If the target is further way than expected, they cannot achieve front lock. If they have to turn too rapidly to maintain front lock because the target is closer than expected because the controlling radar has been pulled off, they will lose rear lock and likewise self-destruct.

Range gate stealing requires substantial knowledge about the transmission characteristics of the radar train transmitted. The illustration shows a very simple transmit signal but in reality, many characteristics can be changed on a pulse-to-pulse or train-to-train basis by varying pulse duration, frequency, jitter and a variety of other parameters. The jammer has to be able to adapt to these changes in real time, making the process very challenging. The difficulty is reduced if the jammer has a library of the modes of the target radar, and the ways the signal has been observed to be operated in the past. This is discussed further in Section 18.4.

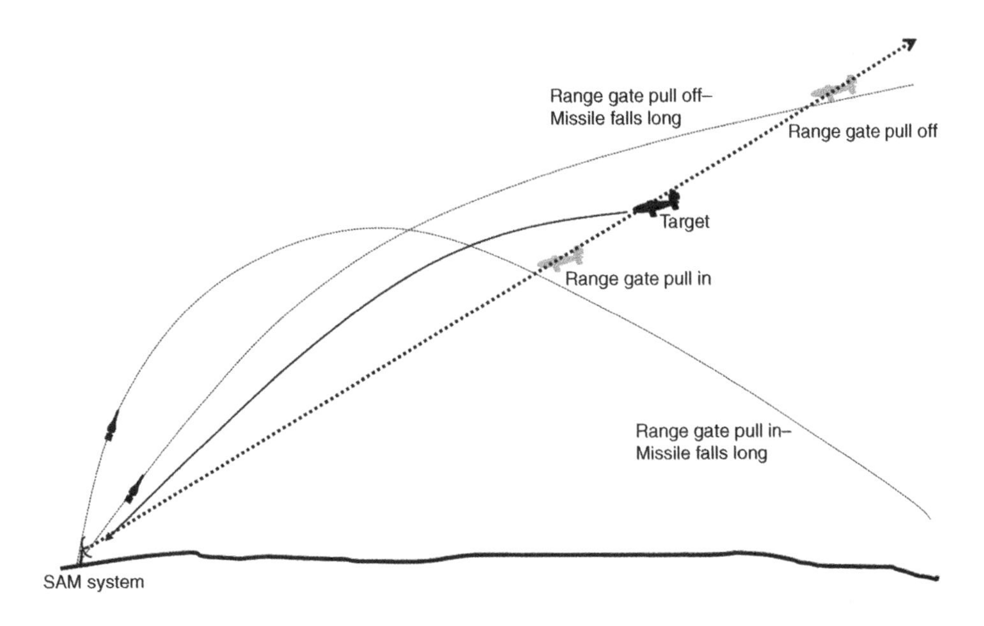

Figure 18.4 Effects of range gate pull off and range gate pull in on a missile engagement. The grey shadows are the false returns caused by either range gate pull in (closer to SAM system) or range gate pull off (farthest away). In either case, the missiles are launched on the wrong trajectory, and cannot recover to hit the real target (black).

18.2.3 Velocity Gate Pull Off

As well as target range, radar systems may also attempt to determine target direction and speed. This helps in predicting where the target is heading. This is vital for missiles or guns to intercept the target. We have already seen the use of pseudo-Doppler radar in determining the velocity of targets. Just like range gates, radars have digital velocity gates for determining which bin a target's velocity fits in. Velocity gate pull off works in a very similar way to range gate pull off, but the parameter that needs to be changed in this case is the Doppler shift of the target return signal. One method of achieving this is for the jammer on the target to provide a sweeping frequency return. This causes the radar receiver to believe the target is moving at a higher radial velocity than it actually is. If is successfully achieved, the radar believes the target is in a different velocity gate than it actually is. In some radars, the system will attempt to 'focus' on the gate that the target is in, in order to maximise the amount of information returned. If the jammer successfully steals the velocity gate, it can cause the radar to focus on the false return. When the jammer is switched off, the radar needs to go back through the whole process of re-acquiring the target. If this is repeated many times, the radar will be unable to accurately determine the true target speed. This process is illustrated in Figure 18.5. The left hand diagram shows the non-jammed situation, including the velocity gate for the target determined by the receiver. The right hand diagram shows a swept jammer increasing the

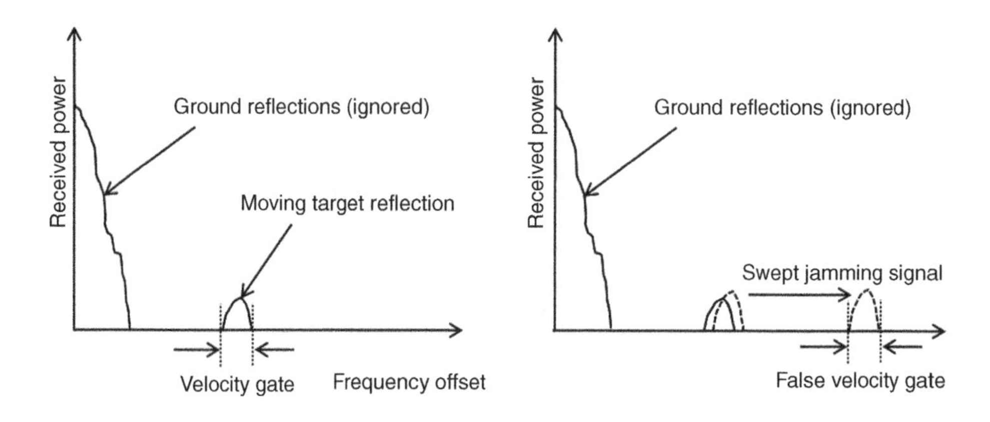

Figure 18.5 Velocity gate pull off illustration.

frequency of the return, causing the velocity gate to be pulled off from the real velocity of the target. When this jamming signal is switched off, the radar has to reacquire the target.

18.2.4 Angle Stealing

Angle stealing is the concept of tricking the radar into believing the target is on a different bearing than it actually is. It can be used against conical scan radars and is possible against other radar scan types. In order for angle stealing to work, the radar must be transmitting in a regular scan interval. This means that the target is picked up in a particular part of the cycle on each repetition of the scan. If a jammer provides a synchronised return signal, but is out of phase with the original transmission, the radar receiver perceives the target to be in a different part of the scan and hence at a different bearing. This is illustrated in Figure 18.6.

Diagram (a) shows part of a conical spiral scan, as seen from the seeker. The real target is shown along with a jammer return at a different phase in the scan cycle. The target appears to be horizontally and vertically offset due to the jamming signal. The jammer has to overcome the gain response of the missile seeker radar, but if it does so, it can cause the missile to steer towards the false target.

Using high power jammers it is also possible to add false targets to rotating radars. This is illustrated in Figure 18.7. The antenna pattern of the radar radiates energy in all directions, but it is attenuated by sidelobes in every direction except that of the main beam. It is important however to recognise that once energy gets into the receiver circuits, there is no way to determine which direction it has arrived from. If a noise jammer managed to inject enough energy into a radar receiver by overcoming the attenuation of the sidelobes, it would mask targets arriving from that direction. If it can inject the same level of power but replicating the expected return signal with a suitable delay, it can create a false target on the bearing of the main beam of the radar at the instant the signal arrives.

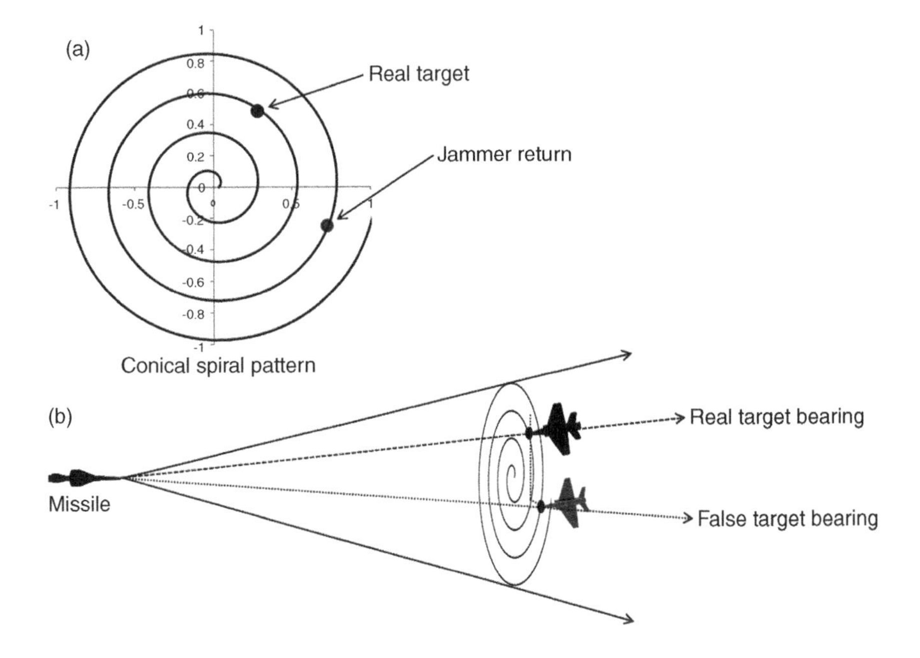

Figure 18.6 Conical spiral radar scan and a jammer returning the original signal but offset in the phase of the scan to produce a false target.

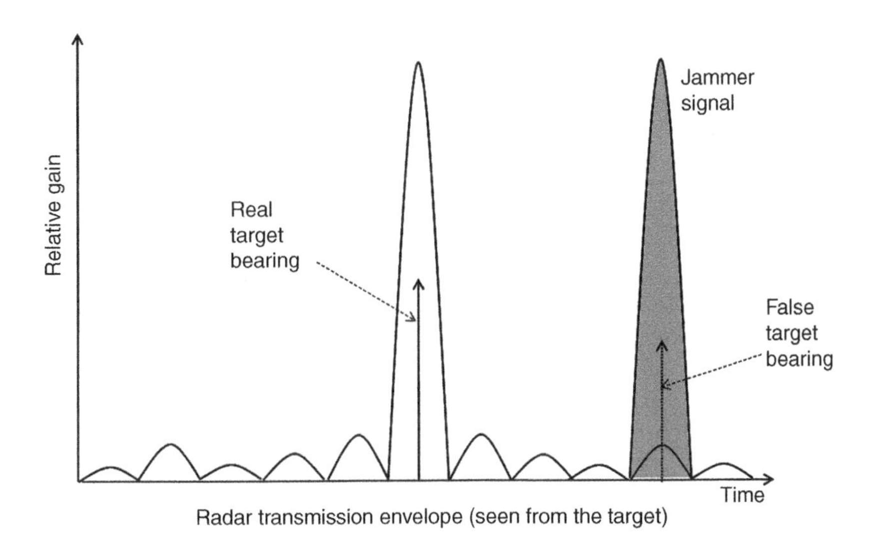

Figure 18.7 Angle stealing against simple rotating radars.

18.2.5 Inverse Gain Jamming

Inverse gain jamming is a method of denying the target radar angle information. In this case, the jammer does not produce a single alternate target but rather blocks all angles. Figure 18.8 illustrates again the scenario. Diagram (a) shows the signal from the radar as perceived from the target. Each of the pulse trains will be received not only through the main lobe but also via all of the sidelobes at a lower level. If the jammer transmits either noise or replicates the incoming signal with the opposite gain applied, it can confuse the enemy radar. This is shown in Diagram (b). Diagram (c) shows the non-jammed situation, where the target is clearly seen. If the jammer were to send additional replications of the original signal with the same delay from each transmission but during the radar sweep, the display on the radar would be as seen in Diagram (d). If the pulse trains are sent at other intervals, then targets would appear at other ranges and bearings. If noise is sent instead, then the radar would be de-sensitised and thus may not be able to detect any targets.

The main problem with inverse gain jamming is the power required to overcome the radar antenna directionality. This is likely to result in very low sidelobes and thus the jammer has to be very powerful. Additionally, the radar can be designed with

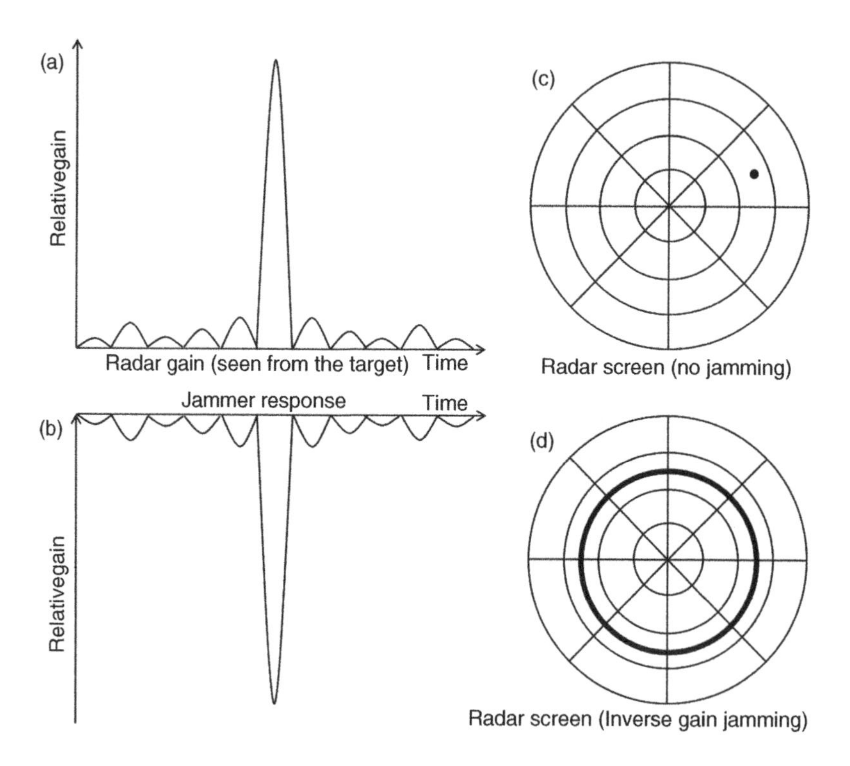

Figure 18.8 Inverse gain jamming.

sidelobe reduction. This makes it even more difficult for the radar to overcome the additional attenuation.

18.2.6 Other Jamming Methods

The jamming methods described are only a subset of those available. There are many others, including formation, blinking, terrain bounce, skirt, cross-polarisation and cross-eye jamming.

Formation jamming is used to screen the presence of more than one aircraft within a single radar resolution cell. Radars are unable to detect more than one target within one of their detection cells, and thus if two aircraft or more aircraft stay very close together, then only one target will appear to be present.

Blinking jamming again involves two targets but in this case, they perform active jamming against guidance radars. The two targets send returns to the radar alternately at a rate close to the servo bandwidth; typically a few Hertz. This can induce resonance in the target radar that can result in large overshoots in the radar's tracking. If this is carried out against missile radar guidance, the missile radar can be directed alternately to the two targets. The missile begins to yaw wildly and if the jamming is successful will miss both targets.

Terrain bounce can be used over flat, radar reflective surfaces such as water. The principle is illustrated in Figure 18.9. Diagram (a) shows the head on aspect of an aircraft target and the view from below. The view from underneath has a larger skin reflection area than the head on aspect. Diagram (b) shows the view from a missile approaching a target at low altitude. The guidance radar sees the return from the target but also a return reflected from the surface. The centroid of the total response is between the real target and the reflection. Diagram (c) shows that the missile may head towards the centroid, below the real target as shown by the dotted line. This can cause the missile to overshoot the real target, particularly as it gets closer.

Skirt jamming exploits the phase response of filters in the radar receiver by injecting a strong jamming signal into a region just above or below the filter frequency. This can cause non-linearity in the phase response across the wanted band, which can affect the radar's tracking circuitry.

Cross-polarisation jamming can be used against parabolic dish antennas. The jammer uses two antennas, which are 90 degrees out of polarisation; one can be vertical and the other horizontal for example. This causes the tracking antenna to respond erroneously to the response, causing tracking errors.

Cross-eye jamming uses phase shifts between two sets of physically spaced antennas. These can be mounted on the wing tips of aircraft. The two jamming signals replicate the incoming radar signal but 180 degrees out of phase. This causes the receiver to be given a returning wavefront that appears to be a null directly in the direction of the target as the two signals cancel out. The tracking signal is therefore disrupted.

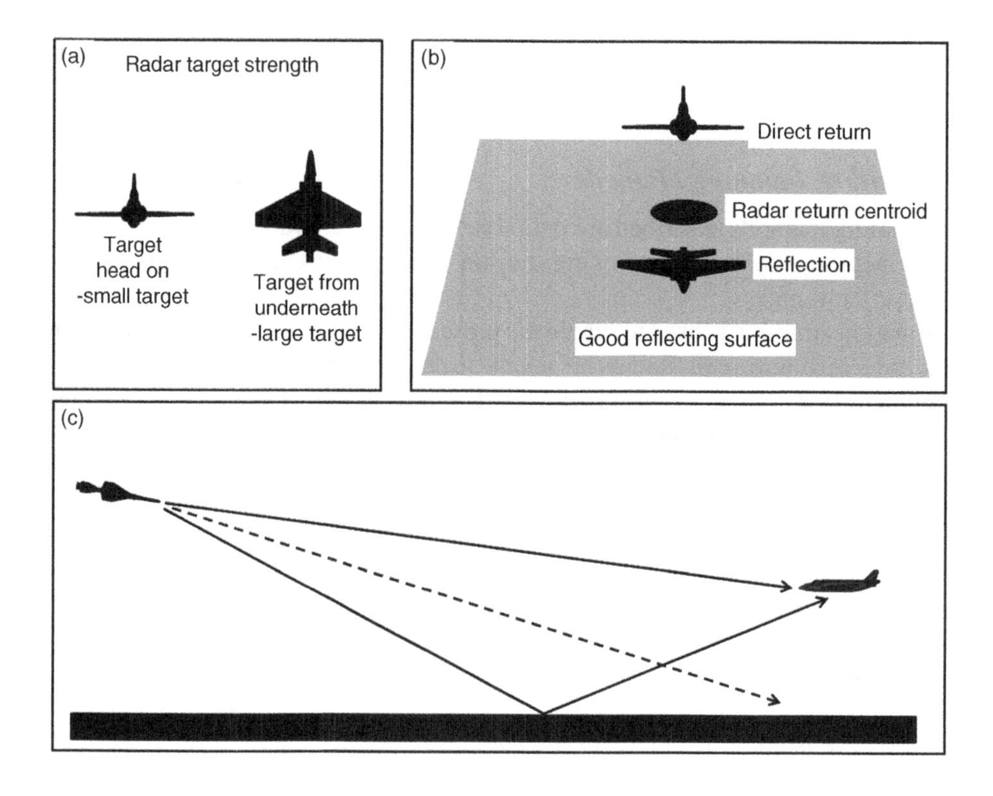

Figure 18.9 Terrain bounce principles.

18.3 Platform Self Protection Methods

18.3.1 Platform Types and Protection Required

One of the major uses of jamming is for platform self-protection. This is where an aircraft, ship or land vehicle uses jamming or other methods to counter enemy fire by missiles or radar-guided guns. All military platforms are vulnerable to enemy fire and there are a range of methods used to counter them. Taking the case of a surface action group of naval ships, self-protection can take a number of forms, including:

- stealth; avoiding radar or reducing the return from the platform;
- avoiding transmissions for the enemy to detect (EPM);
- combat air patrol to detect and warn off enemy aircraft or shoot down missiles at range;
- long range air defence missiles;
- short range air defence missiles;
- radar jammers to counter enemy missiles, using radar warning receivers to identify threats;

- chaff and other decoys;
- close In Weapons Systems (CIWS); typically high rate of fire radar controlled guns.

Land and air platforms will also use a subset of these options for self defence.

18.3.2 Decoys

18.3.2.1 Introduction

Decoys are devices used to seduce missiles away from the real target. There are various different types, including towed, expendable and independent. The purpose in each case is to provide an apparently more attractive target than the platform the decoy is intended to protect. Early decoys include radar chaff used as early as the Second World War. Since then the technology has developed enormously. The methods that decoys use can be based on seduction, saturation or providing independent targets. Decoys can be active or passive in nature.

Seduction decoys provide an apparently more attractive target to missiles than the real one. This can be because they provide a larger target return, appear to be a larger version of the target by active means, or are positioned in a position where they will form the first major target the missile sees.

Saturation targets overload the capability of missile or other systems by providing too many targets to engage in the time available. In this case, at least some of the real targets are likely to get through. This has been used in multiple re-entry vehicle ballistic missile systems for some years. These high value, very difficult to engage targets can incorporate a number of dummy warheads each of which is indistinguishable from the real warheads. As they travel towards the target at high supersonic speed, there is a very short time for any anti-ballistic missile systems to position themselves in an intercept position. When there are many false targets to track, the probability of all real warheads being successfully engaged is substantially reduced. With a very small engagement timescale, there is no chance to launch additional missiles to counter the remaining warheads.

Independent targets are launched from a platform and follow their own pre-programmed course in order to provide a new target for defences to have to deal with. These are designed to mimic the real target characteristics. These can be deployed in advance of a real attack or as decoys when the platforms themselves come under attack.

18.3.2.2 Airborne Decoys

Aircraft can deploy radar decoys such as chaff, which is used to create a temporary new target with a greater RCS than the real targets. They can also use active radar decoys that are either towed or expendable. In both cases, the aim is to prevent successful enemy engagement of the target. This occurs after the real target has been detected and engaged; deployment of decoys before this stage will only cause the enemy to become aware of a threat they may not have known existed.

Since the decoy is initially co-located with the target, it is likely to be within the same radar resolution cell. As an expended decoy moves away from the position of the target, it can cause the threat radar to ignore the real target and focus on the cell still containing the decoy. This allows the target to escape. If the decoy return is in the same cell, it still disrupts the centroid of radar return, preventing more detailed localisation of the real target. As an enemy missile radar approaches, the decoy return will spread over a range of cells, again preventing the real target from being tracked.

Towed targets can be active, seducing an approaching missile to attack the towed body rather than the aircraft itself. As the decoy is small compared to a real aircraft, it may well survive detonation of the missile warhead, if it is triggered. More likely, the fuse will not be activated because a viable target will not be detected. The missile will then pass both target and decoy, missing both.

18.3.2.3 Ship Decoys

Ship-based decoys are aimed at protecting against anti-ship missiles. There are a number of methods of doing so. The ship can fire a salvo of chaff decoys, effectively hiding the main target. This is illustrated in Figure 18.10. The ship is likely to be manoeuvring at high speed and changing direction constantly. As it moves, it fires chaff in many directions. These explode, generating billows of chaff. The chaff rapidly forms new clouds generating radar returns. As the missile approaches, it is likely to target the chaff, ignoring the ship and passing through the clouds of metal strips. The missile warhead is unlikely to fuse because there is no impact, nor detection of a large metal structure. If a salvo of missiles has been launched, the chaff works on each missile in the same way. Chaff is dispensed liberally in order to counter missiles potentially arriving from any direction.

Chaff is passive, meaning that the chaff dispensed may not allow targets far away to detect them, unless the ship is being targeted by active radar from range. Both aircraft and ships may not be targeting directly to avoid counter-detection and attack. This would mean that the ship is not exposing itself to further detection beyond that already achieved.

Chaff can also be used in a smarter way to protect the main platform. This depends on knowing the radar scanning method of the inbound missiles. Anti-ship missiles scan the horizon for their targets, based on the fact that ships very seldom fly (the author has seen 4000 ton warships leave the water, but they don't fly; they glide very, very briefly).

Figure 18.11 shows an illustration of a missile approaching a warship. The missile radar in this case is scanning in raster mode, scanning across the arc of its available search angle. The scan is not entirely horizontal, but instead scans above the horizon as well to account for variations caused by waves and by the missile's pitch.

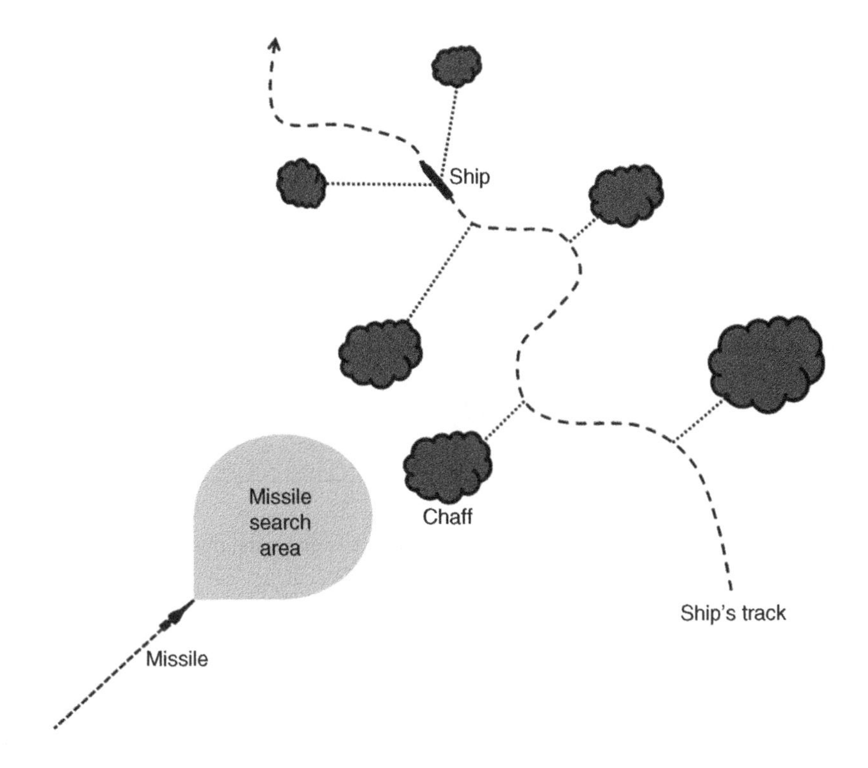

Figure 18.10 Use of chaff to counter enemy missiles by saturation and seduction.

If the target ship launches chaff as shown in Figure 18.12, the raster scan is likely to pick up one side of the chaff before the real target. With luck, the missile is seduced by the chaff clouds and the ship manoeuvres at high speed to clear the area.

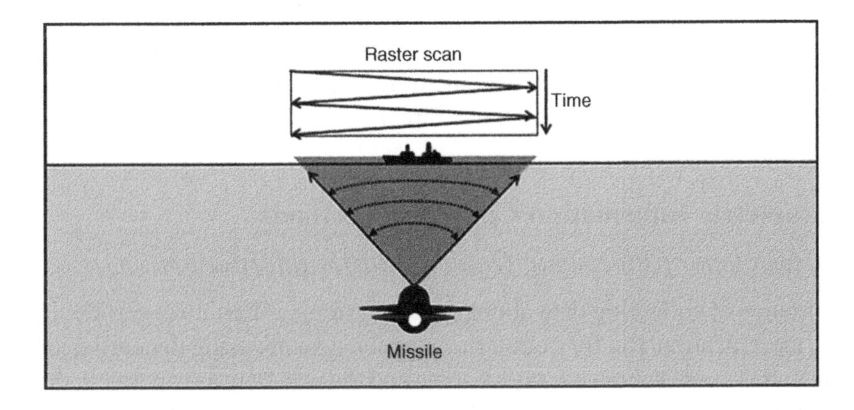

Figure 18.11 A missile scanning the horizon for a target.

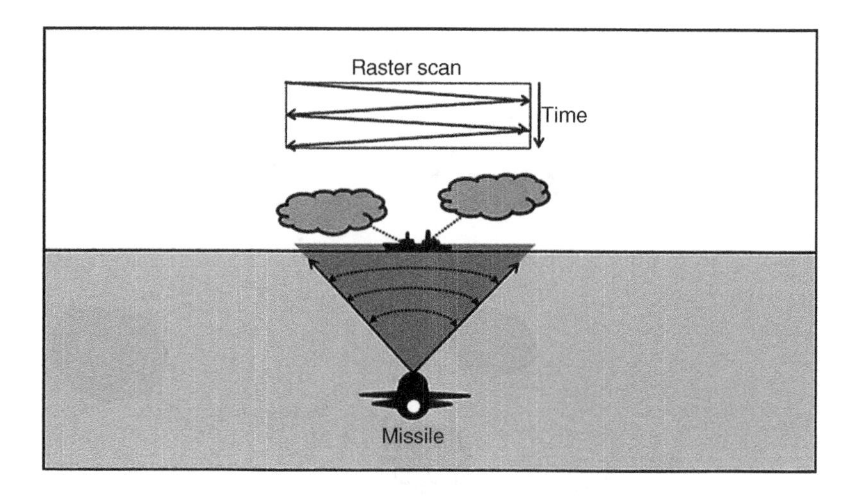

Figure 18.12 Target launching chaff, as seen by the missile.

As well as launching chaff to counter the seeking scan of the missile radar, the ship can also be using active EW to use RGPO or angle stealing to direct the missile towards the chaff cloud.

There are a wide range of other approaches that can be taken by ships under attack; the ones discussed are just an indication of the possibilities. Ships also need to be able to defend themselves against infra-red and optical systems; they also carry decoys for this eventuality.

18.3.2.4 Terrestrial Decoys

Terrestrial decoys can use the same basic methods as for ship defence; however, the smaller size and reduced capability to carry expendables and EW jammers restrict their use to some extent. Development is however being carried out to better protect land assets beyond the application of chaff expendables. Land targets are also more prone to be attacked by a wider range of threats including optical and infra-red, so additional effort has been put into countering these threats as well.

18.4 Parametric Information Collection Methods

18.4.1 Collection of Electronic Order of Battle Information

The importance of collecting data about enemy systems is hard to overstate. It is vital for threat identification and for countering enemy systems using deceptive jamming. The more that is known about enemy systems and the way they are operated, the better. This is important for communications EW but far more so for radar EW. The process of collecting data from potential enemy systems is not a short-term activity to be carried out only when the threat arises; it must be carried out over many years in order to glean the vital information that arises from active electronic collection (Measurement and Signature Intelligence; MASINT), Human Intelligence (HUMINT), COMINT, IMINT and open source material. When collected over time, this information can grow into a database of the Electronic Order of Battle (EORBAT) used by all friendly systems to identify enemy threats and to counter them.

MASINT is covered later in this section. HUMINT is a resource that cannot be overlooked. While except in the most important cases most human intelligence agencies will be focussed on different targets, principally those of strategic importance, there is still the potential for a wealth of data to be obtained. Information can be readily obtained from defectors and refugees who happened to work on system design or operation and in many cases the subject will be very keen to help. Information can also be obtained using legal means from captured enemy forces. It is not very uncommon for documentary evidence to be handed over in return for money or extraction from an unfriendly country. Data collected from HUMINT sources is likely to be fragmentary (except in the case of documents), but it all adds to the picture.

COMINT will help to identify a number of important informational factors. This includes the location and disposition of enemy command and control systems. This will also be true for radar-related systems such as air defence and offensive missile systems. Many fixed installations will have landlines rather than radio communications; these can be intercepted by using Special Forces tapping into the lines themselves, or it may be that part of the system uses microwave links at some point allowing interception. As an alternative, the original landlines can be disrupted. This forces the enemy to use radio command and control as a backup. Mobile systems may use landlines or radio communications. If radar systems have a dedicated communications net assigned to them, it provides a useful method of localising the radars, associated missile systems and their command and control nets. This is because the communications nets are likely to be tested and used far more regularly than the radars themselves.

IMINT is the process of determining enemy disposition by photographic methods. This was famously demonstrated during the Cuban missile crisis of the early 1960s, where over flights by US U2 aircraft showed the construction and preparation of Soviet nuclear missile silos in Cuba. IMINT is good at identifying platforms, constructions and large equipment. To be useful for technical analysis, a resolution of about 10 centimetres or better is required, however for force disposition analysis a resolution of a metre or so in fine. Many people believe that satellite systems can pinpoint any target at will; the truth is far different. In fact, it is more like finding a needle in a haystack. Although satellites can cover vast swathes of area, they cannot do it at the required resolution. In general, satellite and aircraft systems need to be cued by other means. This is relatively easy in a limited combat zone, but not when looking for systems more widely distributed. Other methods such as HUMINT or COMINT can identify areas for further study. After this, aircraft or satellites can be sent to look for further evidence and can successfully find and identify target systems.

These days, open source material can be the best method of identifying new systems. There is an aggressive market in modern military equipment, and much data can be obtained directly from the design company by acquiring brochures or other promotional material. More can be determined by talking to the people on the stand of an exhibition. They are often keen to talk about their system's capabilities, and will talk to anyone who appears to be interested. News cameras are often the first on the scene of events and their broadcasts can often help in the identification of systems in place as well as providing valuable information about what is happening on the ground. Of course, there is always the possibility of misdirection but when this information is added to what is already known, the information can be interesting in a number of ways.

18.4.2 Management of EORBAT Information

EORBAT information should be managed in the same way as other intelligence data. It should be centralised and the distribution controlled appropriately. Historically, intelligence has been hindered by the existence of a number of agencies that do not talk to each other. This leads to confusion and error. Instead, a central mechanism for managing information should be adopted (hence the 'Central' in CIA, for example). One organisation should take and maintain primacy in data management, processing and dissemination. In the UK, this is the Defence EW Centre (DEWC). This avoids the inexcusable scenario where vital information is known to one organisation but not communicated to those who need it urgently – or worse, that information users received contradicting information. Where disparate agencies exist such as foreign, home, defence and communications intelligence, this is not a problem as long as their roles are agreed and that they contribute to a single managing agency.

There are two aspects to the EORBAT; the first is the technical parameters of each type of radar and communications system and the second is the known locations of particular emitters.

18.4.3 Radar Threat Characteristics

A vast array of data can be collected against radar systems. Some of the most important are shown in Table 18.1. There are two types of data; those that define the boundaries of the system such as maximum and minimum values, and those that define the characteristics observed during a particular encounter. Unless data can be obtained from other sources, the minimum and maximum values can be deduced from multiple observed characteristics and from basic limitations imposed by the system design.

Table 18.1 Radar characteristics

Characteristic	Comments
Name/Type	Nomenclature of system or name or station
Location	Coordinates of station
Associated systems	Any other systems associated with emitter
Platform	Type of platform to which system is fitted
Mode	Observed mode
List of modes	List of available modes
Output power	Observed output power in dBW, dBm or W
Maximum power	Maximum output power in MHz or GHz
Minimum power	Minimum output power in MHz or GHz
Output device	Type of system used to generate the output power
Tx Frequency	Observed centre or reference frequency in MHz or GHz
Minimum frequency	Minimum supported frequency in MHz or GHz
Maximum frequency	Maximum supported frequency in MHz or GHz
Tunability	Type of tuning e.g. continuous, raster, etc
Tuning step	Tuning step in kHz or MHz
Frequency Tolerance	Variation around tuned frequency in Hz
Frequency hopset	Hopset for frequency hopping systems
Hop rate	Number of hops per second
Dwell rate	Duration of dwell in milliseconds
Occupied Bandwidth	The bandwidth can be expressed as a single figure or it can be expressed as part of the power spectral density
Antenna type	Category of antenna
Horizontal aperture	Horizontal aperture
Vertical aperture	Vertical aperture
Diameter	Diameter of a dish
Beam type	Type of beam emitted
Horizontal beamwidth	Observed horizontal beamwidth
Minimum horizontal beamwidth	Minimum system horizontal beamwidth
Maximum horizontal beamwidth	Maximum system horizontal beamwidth
Vertical beamwidth	Observed vertical beamwidth
Minimum vertical beamwidth	Minimum vertical beamwidth
Maximum vertical beamwidth	Maximum vertical beamwidth
Antenna gain	Observed gain of transmit antenna(s), normally reference either an isotropic (dBi) or dipole antenna (dBd)
Minimum gain	Minimum antenna gain
Maximum gain	Maximum antenna gain
Antenna polar pattern	Directional performance of antenna (see next section)
Antenna gain response	Frequency gain response of the antenna. Normally expressed in dB down on the highest gain response
Antenna height/altitude	Height of the centre of the radiation pattern above local ground or altitude above sea level. Usually expressed in metres or feet
Antenna azimuth	For directional antennas, the direction of the main beam in degrees or mills reference grid or magnetic north
Antenna tilt	The electrical or mechanical vertical tilt of the main beam of the antenna. Usually expressed in degrees or mills
Rotation speed	Observed rotation speed in rotations per second

(continued)

Table 18.1 (*Continued*)

Characteristic	Comments
Minimum rotation speed	Minimum rotation speed
Maximum rotation speed	Maximum rotation speed
Horizontal scan speed	Observed horizontal speed
Horizontal scan rate	Observed horizontal scan rate
Horizontal scan type	Scan type
Vertical scan speed	Observed scan speed
Vertical scan rate	Observed scan rate
Vertical scan type	Scan type
Phased array number of main beams	Number of beams in the phased array
Phased array number of elements	Number of elements in phased array in horizontal and vertical directions
Power spectral density	The spectral shape of the transmitted energy. Can be used for interference analysis between dissimilar systems. Normally expressed as dB down from the main power
Tx antenna Polarisation	The polarity of the transmitted signal
Modulation	Observed modulation schemes
Modulation types	System modulation types
Pulse burst rate	Observed number of bursts per second
Minimum burst rate	Minimum system burst rate
Maximum burst rate	Maximum system burst rate
Pulse burst duration	Observed duration in microseconds
Minimum burst duration	Minimum system burst duration
Maximum burst duration	Maximum system burst duration
Number of pulses per burst	Observed number of pulses per burst
Minimum number of pulses per burst	Minimum system number of pulses per burst
Maximum number of pulses per burst	Maximum system number of pulses per burst
Pulse burst off time	Observed time in microseconds between pulse bursts
Minimum pulse burst off time	Minimum system pulse burst off time
Maximum pulse burst off time	Maximum system pulse burst off time
Burst type	The type of pulse format
PRF	Observed Pulse Repetition Frequency
Minimum PRF	Minimum system PRF
Maximum PRF	Maximum system PRF
Duty cycle	Observed duty cycle
Minimum duty cycle	Minimum system duty cycle
Maximum duty cycle	Maximum system duty cycle
Power per pulse	Observed power per pulse
Minimum power per pulse	Minimum system power per pulse
Maximum power per pulse	Maximum system power per pulse
Pulse compression method	Type of pulse compression
Pulse compression ratio	Observed pulse compression ratio
Minimum pulse compression ratio	Minimum system pulse compression ratio
Maximum pulse compression ratio	Maximum system pulse compression ratio

Table 18.1 (*Continued*)

Characteristic	Comments
Rise time	Observed rise time
Minimum rise time	Minimum system rise time
Maximum rise time	Maximum system rise time
Fall time	Observed fall time
Minimum fall time	Minimum system fall time
Maximum fall time	Maximum system fall time

Most easily collected data relates to emitter data rather than the receive characteristics. However, analysis can identify likely values for the receive side.

Emitter characteristics can be grouped into specific areas such as:

- physical data such as location, size and so on;
- output power during various phases of transmission and the relationship between transmit and non-transmit periods;
- antenna characteristics, including gain, rotation, scan rates and types;
- frequency parameters, including tuning and frequency drift;
- spectral characteristics such as occupied bandwidth and transmitter power spectral density;
- time-based data such as pulse characteristics and variations between separate pulses, PRF, PRF variations and signal rise and fall times;
- modulation characteristics;
- associated radio/radar systems.

Data collected should be collated and analysed in a central processing facility to allow all information to be included in the threat assessment. These days, the data would normally be stored in a database structure, allowing retrieval of system parameters for authorised users. Dissemination of the data to tactical operators can be achieved by creating written EORBATs or by secure remote access to the core database, provided that technology and operational procedures allow.

18.4.4 Communications Parameters

As well as collecting radar parameters, it is important to collect communications and other non-radar systems, especially where they are targets of interest or are associated with them. For example, tactical radar systems will require coordination. This means that communications between radar systems and a central control point will be required. If the existence and ideally content of such links can be detected, then this

Table 18.2 Communications characteristics

Characteristic	Comments
Name/Type	Nomenclature of system or name or station
Location	Coordinates of station
Associated systems	Any other systems associated with emitter
Platform	Type of platform to which system is fitted
Mode	Observed mode
List of modes	List of available modes
Output power	Observed output power in W, dBW or dBm
Minimum output power	Minimum system output power in W, dBW or dBm
Maximum output power	Maximum system output power in W, dBW or dBm
Frequency	Centre or reference frequency in MHz or GHz
Minimum frequency	Minimum system frequency in MHz or GHz
Maximum frequency	Maximum system frequency in MHz or GHz
Duplex spacing	Duplex spacing between duplex links in MHz
Tunability	Ability of system to tune between different frequencies
Tuning step	Tuning raster step in kHz or MHz
Frequency Tolerance	Variation around tuned frequency in Hz
Frequency hopset	Hopset for frequency hopping systems
Hop rate	Number of hops per second
Dwell rate	Duration of dwell in milliseconds
Occupied bandwidth	The bandwidth can be expressed as a single figure or it can be expressed as part of the power spectral density
Antenna gain	Gain of main beam of antenna in dBi or dBd
Minimum antenna gain	Minimum system antenna gain in dBi or dBd
Maximum antenna gain	Maximum system antenna gain in dBi or dBd
Antenna type	Category of antenna
Horizontal aperture	Horizontal aperture in m
Vertical aperture	Vertical aperture in m
Diameter	Diameter of a dish in m
Beam type	Type of beam emitted
Antenna polar pattern	Directional performance of antenna
Antenna gain response	Frequency gain response of the antenna. Normally expressed in dB down on the highest gain response
Antenna height/altitude	Height of the centre of the radiation pattern above local ground or altitude above sea level. Usually expressed in metres or feet
Antenna azimuth	For directional antennas, the direction of the main beam in degrees or mills reference grid or magnetic north
Antenna tilt	The electrical or mechanical vertical tilt of the main beam of the antenna. Usually expressed in degrees or mills
Power spectral density	The spectral shape of the transmitted energy. Can be used for interference analysis between dissimilar systems. Normally expressed as dB down from the main power
Antenna Polarisation	The polarity of the transmitted signal. Has an effect on propagation and also on interference
Time slots	For time based systems, the slot structure of the transmitted signal. Can be used for interference analysis. Normally expressed in microseconds, milliseconds or seconds

Table 18.2 (*Continued*)

Characteristic	Comments
Activity ratio	The ratio of activity when the system is transmitting. Again, used for interference analysis. Expressed as a percentage or in a range 0 – 1
Modulation scheme	The encoding scheme used to modulate the baseband signal into the carrier. Has an effect on receiver sensitivity
Modulation characteristics	Modulation parameters
Time slots	For time based systems, the slot structure of the received signal. Can be used for interference analysis. Expressed in microseconds, milliseconds or seconds
Activity ratio	The ratio of activity when the system is receiving. Again, used for interference analysis. Expressed as a percentage or in a range 0 – 1
Processing gain	Gain due to the use of a spreading code in a CDMA system. Expressed in dB
Spreading code rate	Rate of spreading code in Mb/s

will identify where the radars are, where the command and control points are and vulnerabilities that can be exploited.

The parameters that can be collected for communications systems are, as for the case of radars, those emitted by the transmitter. Many of these are shown in Table 18.2.

These parameters can be collected by direct measurement, but they can also be determined via a range of other methods, including analysing captured equipment, assessing likely parameters via simulation and from open source material.

18.4.5 Collection Methods

Collection of parametric information is normally carried out by observing enemy transmissions and recording the parameters using dedicated receivers. If these receivers are calibrated, transmission loss can be estimated in order to determine system output power. Other characteristics can be measured directly. Collection systems can be mounted on land, air or naval assets including submarines. Because the aim is to collect transmission parameters without decoding the content, such systems can be more sensitive than enemy system receivers. This allows data to be collected from a greater range than the detection or link ranges. Aircraft in particular can stand off while collecting data, making them less vulnerable to enemy counteraction or even detection. If the data is to be recorded during flight for later analysis, the collecting platform can be covert without making any transmissions during the task.

Aircraft can mount individual tasking missions over relatively short durations. Longer surveillance can be maintained by assigning several aircraft to the task, at high cost and with increasing probability of counter-detection as the activity goes on.

Land systems can be covert, but have lower capability to detect transmissions at long range. Special Forces can perform closer surveillance, but again with the risk of detection in enemy territory.

Naval surface platforms provide long-endurance collection tasks, but they are vulnerable to detection. Submarines can operate in enemy territory for long periods, but run a greater risk of detection by having an ES mast raised above the surface.

18.4.6 Collection versus Detection

An important consideration when conducting collection activities against enemy air defence or surveillance radars is the risk of counter detection or even engagement by enemy systems. However, if the radar parametric detection system is sensitive enough, it should be possible for the measurements to be taken without this risk. This is illustrated in Figure 18.13.

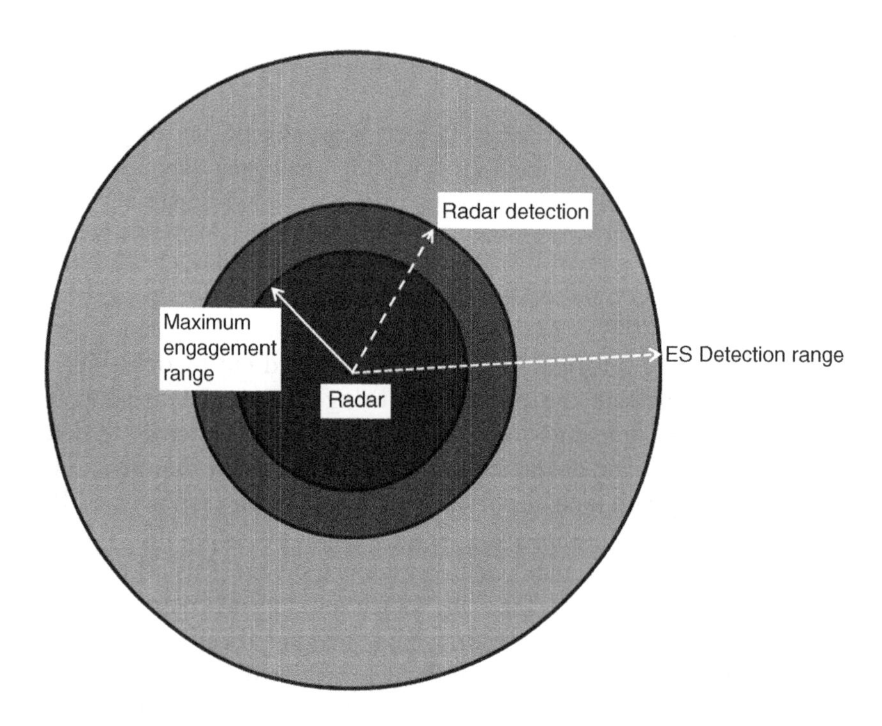

Figure 18.13 Range comparison between engagement, detection and ES detection.

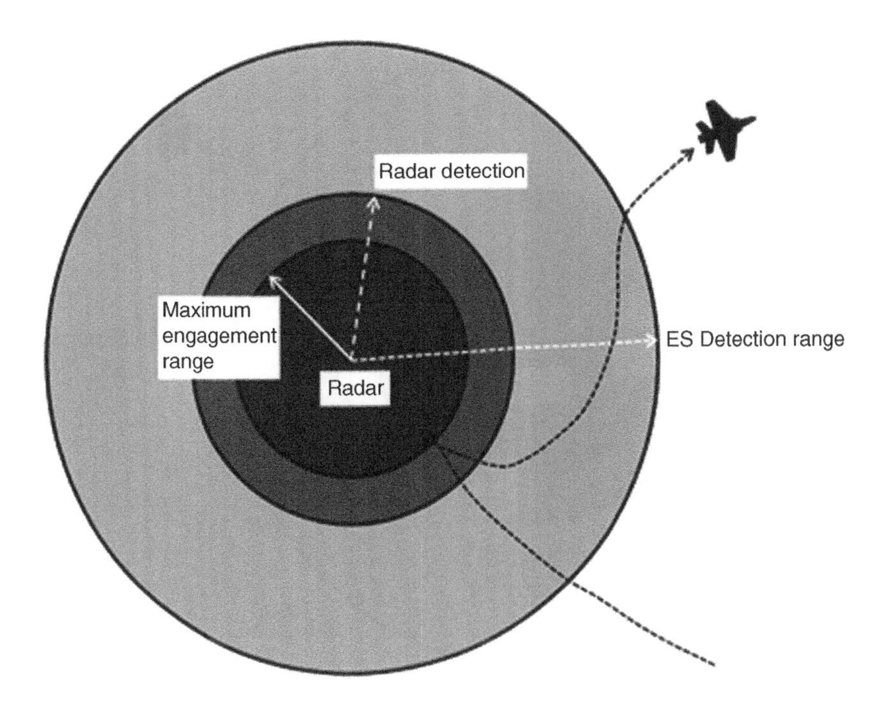

Figure 18.14 An aircraft used to trigger an enemy response and escape before it can be engaged.

18.4.7 Example Collection Activity

Data can only be detected off air if the enemy transmitter is activated. Many sensors are only activated occasionally, to ensure they work and for exercise purposes. Restricting their activity reduces the ability of the enemy to detect the vital parameters. This means that it may be necessary to pre-empt enemy activities to determine when these systems may become active. One example of this may be if the potential enemy declares an area of sea off-bounds for exercise purposes. This may indicate that missile firings are about to be carried out. Having collection assets in place to observe the exercise may reveal useful information.

An alternative in times of crisis is to trigger a response using aircraft, ships or decoys to penetrate enemy defences in order to prompt them to fire up their systems. Figure 18.14 shows an aircraft penetrating an enemy air defence system in order to trigger a response. Once the system has been triggered, the collecting aircraft have some time to collect the transmission characteristics.

Figure 18.15 shows the use of a decoy to achieve the same ends. In this case, the target aircraft can remain beyond engagement range and collect parametric information.

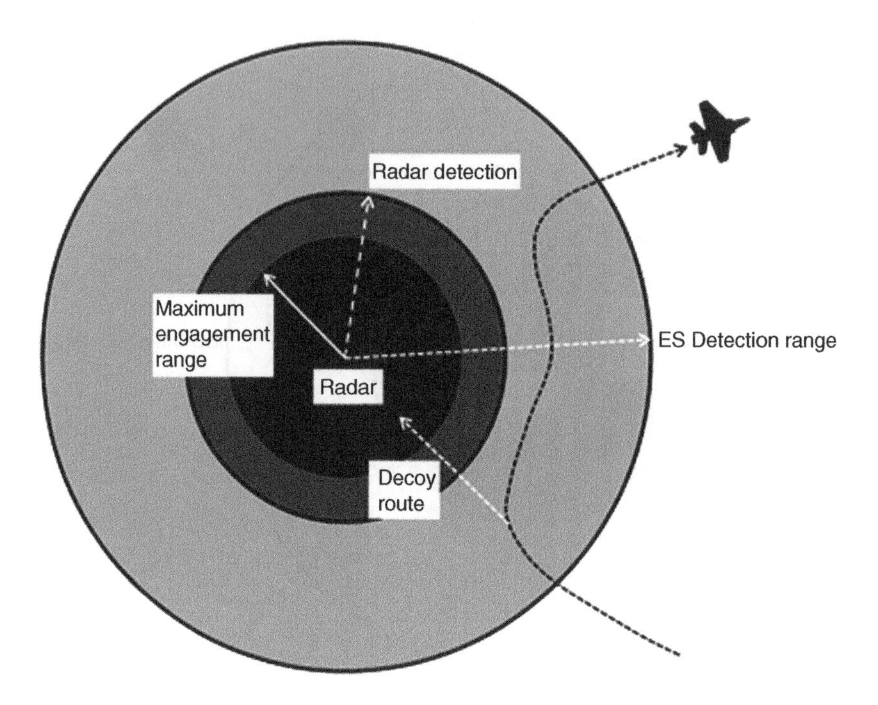

Figure 18.15 Use of a decoy to trigger enemy defences. In this case, the aircraft can stay beyond the engagement zone and still collect valuable data.

References and Further Reading

Peebles, P. Z. (1998), *Radar Principles*, Wiley-Blackwell, ISBN 978-0471252054.

Schleher, D. C. (1999), *Electronic Warfare in the Information Age*, Artech House, MA, USA, ISBN 0890065268.

Skolnik, M. (2008), *Radar Handbook*, McGraw-Hill, USA, ISBN 978-0-07-148547-0.

19

Countering Radio-Controlled IEDs

19.1 Introduction to IEDs

19.1.1 Asymmetric Warfare

Improvised Explosive Devices (IEDs) form one aspect of asymmetric warfare. Asymmetric warfare is where one side in a conflict resorts to different methods than the other. Typically, it is the weaker side that resorts to methods such as IEDs. IEDs have historically been used extensively in many conflicts, including Vietnam, Northern Ireland, the Middle East, Afghanistan and Iraq among many others. The type and sophistication of the IEDs used have varied according to the skills of the users and the context of the operational theatre. In many cases, unexploded munitions have been used as the source of the explosive material. In others, the explosives have been created from available materials such as those used in fertiliser-based systems.

The benefits of IEDs to the weaker side are that they can be used without exposing the operators to detection until too late, that they can pick and choose the time, place and target at will. If military targets are too well defended, then many other targets can be selected, including infrastructure such as particular buildings, bridges, roads and other targets important to the target organisation. They can also be aimed at specific people, transport systems or other targets the destruction of which is designed to reduce morale among the enemy and their people. Attacks can be aimed at slowing enemy forces down, tying up enemy personnel in defending locations and anti-IED activities, producing civilian unrest and causing disaffection to the enemy's population through generating rising and continuous casualties. In most recent conflicts, this has been the principal aim for those using IEDs; to reduce political will to fight, rather than defeating the enemy on the ground.

Communications, Radar and Electronic Warfare Adrian Graham

19.1.2 Types of IED

IEDs can be grouped via various categories, such as the delivery method, the type of device, the explosives used and by the initiation method. In general, there are three common types:

- command wire initiated;
- victim-operated;
- radio operated.

Command wire systems use a landline to the device trigger. This is linked to the commander, who will be present at some distance from the device, but usually within line of sight of it. When enemy forces approach, the commander can press a button or connect wires to a battery to set the device off. The advantages of the command wire type of IED are that there are no tell-tale emissions to alert the enemy. It is also a very simple system to set up. The disadvantages are that the command wire itself may be spotted. Also, there is a limit on the length of the command wire for it to be effective, meaning that the commander has to be fairly close to the device and is thus vulnerable to being detected immediately after the attack. After the device is activated, the command wire and potentially the triggering device will be left in position. This can be found by the enemy forces and used to collect evidence to determine who the culprit is.

Victim-operated devices are also known as booby traps. Once they have been set up, they can be left unattended, allowing the bombers to evacuate the area before the attack occurs. Any kind of trip wire, levers or other switches can be used. The device has to be triggered by the victim in order to detonate by an action such as picking up an object, opening a door, walking through a trip wire or some similar action. However, there is no guarantee that enemy forces will do this. The system is non-selective and can easily be triggered by civilians or even people from the bomber's own side. Once it is known or suspected that there are booby traps in an area, enemy forces can take far more care to avoid potential traps; for example by entering buildings via windows rather than doors.

Radio operated devices have become far more popular as suitable technology has become widely available. This replaces the command wire by a radio link. Devices can be triggered from a greater range and there is less remaining evidence after the attack. The trigger device can be reused as many times as required. The disadvantages are that the triggering radio signal is vulnerable to detection and jamming. Also, it may be possible for enemy forces to trigger the device early by transmitting likely triggering signals in advance of patrols or other potential targets.

19.1.3 Threat Mix

It is important to recognise that IEDs form one part of an organisation's arsenal or available tactics. They can be used alone, or they can be combined with ambushes to cause disruption to the enemy. For example, suspicious devices can be left on a road or patrol route. When these are investigated by the enemy forces, they may be exposed by, for example, leaving their vehicles or emerging from cover. At this stage, they are vulnerable to attack by an ambushing squad. Alternatively, an ambush can be prepared in an area where there is obvious local cover. The ambush can be carried out and if the enemy take cover in the place planned for by the bomber, then a pre-prepared device already hidden in that cover can be detonated. Other potential methods include having secondary devices planted ready for detonation after the primary device has been triggered. If the secondary devices are placed in an area, where it is expected that those under attack will re-group, then the effects of the initial attack can be magnified by the secondary devices.

Combined with guerrilla tactics such as hit-and-run attacks, ambushes and sniping, IEDs make life very difficult for enemy forces as can be clearly seen in Iraq and Afghanistan. The rise of the suicide bomber only serves to make the threat far more severe.

19.2 Radio Controlled IED

A wide variety of radio systems can or have been used to initiate RCIEDS. These include but are not limited to:

- radio control transmitters such as those used for model aircraft and boats;
- PMR radios in direct mode (often generically referred to as 'Icom' systems);
- pagers;
- GSM phones;
- custom built transmitters and receivers.

These systems cover a wide variety of frequencies from HF to UHF, posing a detection problem for forces attempting to counter them. They each have their advantages and disadvantages to the bomber.

Radio control transmitters are cheap and effective, and can have ranges up to several hundred metres. However, they are prone to noise and interference that may cause the device to be triggered in an uncontrolled manner. Their operational frequencies, typically crystal-controlled, are well known making transmissions relatively easy to detect. Typical PMR systems used operate in the VHF or UHF bands and there can be several hundred or more channels available. This makes a single transmission more difficult to detect, but since traffic is likely to be light, this is often not too much of a problem. Both pager and GSM systems offer the bomber a system in which triggering signals can be hidden within legitimate calls. This makes it difficult to detect suspicious traffic. However, of course, these systems can only work where there is existing coverage.

Radio transmitters and receivers can be readily made using standard components. Although this is more difficult than adapting existing equipment, it can provide greater flexibility in terms of the frequencies that can be used and the triggering signal used.

19.3 Basic IED Counter Methods

19.3.1 Detection of Enemy Activities

Countering IEDs can be achieved by a number of methods, all of which are designed to make it as difficult as possible for bombers to operate during all phases of their activities.

The first stage is gathering all information possible in order to produce detailed intelligence about enemy forces, structure, key personnel, recruiting methods, allies and supporters, operational methods and equipment used. This is a process that will continue to function and evolve throughout the operation.

Identification of the structure and key personnel allows intelligence agencies to focus assets on disrupting the structure and denying key personnel the ability to operate without taking extreme caution about what they do, where they travel and with whom they communicate. This makes logistics and organising operations more difficult. Identification of the recruiting methods and the type of people being recruited again allows intelligence agencies to optimise their surveillance and also gives them the potential to recruit agents or embed intelligence officers into the organisation. Determining supporters and allies is also important in a number of ways. Where practical support is being given by allies and supporters, it may be possible to disrupt these operations or to attempt to undermine the relationships, thereby denying the enemy these assets.

Identifying operational methods is of primary importance. The aim is to predict trends in order to predict potential targets and the methods that will be used. This means that likely targets can be protected and forces can be advised to keep alert for activities that may indicate that an attack is about to be made. There can be many different tell-tale activities, including prior surveillance of the target, changes to the way civilians are behaving, abnormal communications or movement of enemy forces.

The equipment used by the enemy is also a vital intelligence target. This can be identified by capturing enemy equipment or by using ES methods to identify transmissions. This again is a process that needs to be carried out throughout the operation, as enemy methods will evolve over time.

19.3.2 Non-EW Methods

Once enemy tactics have been identified, they can be countered by both technical and operational methods. Operational methods are aimed at preventing the enemy from functioning using their preferred methods. One of the primary methods is to prevent the enemy from operating in particular locations that are important to both sides. This is particularly true for the defence of fixed locations that may be regarded as targets. Static defences, proscribed zones and intrusion detection systems form the inner core. These can be protected by checkpoints and other methods designed to control access to these areas. Beyond this region, regular patrols can prevent the enemy from carrying out surveillance or setting up attacks.

Protecting moving assets such as convoys or patrols is considerably more difficult. The terrain and environment will change as these assets move through possibly unknown areas. This nullifies the advantage of being familiar with the location, as is the case with fixed defences. This problem can be reduced by using air reconnaissance prior to the assets arriving in areas; however, this may also alert enemy forces that a patrol or convoy is due. Another important aspect is to deny the enemy from being able to predict convoy or patrol routes by varying the time and routes taken.

Once a patrol or convoy has been mounted, it is also important to maintain alertness in order to identify visual or other cues that may suggest that an attack is about to be carried out. This includes looking out for signs of surveillance, changes to the behaviour of local people and of course, any unusual signs such as abandoned vehicles, bumps near the route or anything else out of the ordinary.

If there is a suggestion that there is a threat of attack, there must also be fallback options to go around threat areas or to react to an actual attack and its aftermath.

19.3.3 EW Methods

EW methods must be used as one part of the overall response to the threat from RCIEDs. While technical methods can be vital to countering such systems, without them being used within an overall tactical framework, they will be less effective in reducing casualties. This section examines the methods that can be used prior to and during an RCIED attack.

In most cases, RCIED jamming need only be effective over a fairly short range, typically within 500 metres or so. This should be remembered for the rest of the discussion.

Defensive positions can be protected by use of radio jammers used to prevent RCIED explosives from receiving their trigger signal. Barrage or responsive jamming can be used to achieve this. Barrage jamming will affect a force's own ability to communicate, however good the filters. Responsive jamming requires identifying the threat virtually instantaneously, which raises the possibility of failure at the critical moment. Therefore, both methods have their drawbacks. In both cases, however, their performance can be tested to ensure they provide adequate protection.

For the mobile case, vehicle mounted jammers can provide protection in the same way, and power can be supplied via the vehicle to the jammers. This can mean that enough power is available to successfully jam a wide range of potential

Table 19.1 Approximate parameters for a modern mobile jammer

Band	Power (W)	Potential target systems
20–28	25	Radio control systems; CB; analog phones and alarm systems
28–36	25	
36–43	25	
43–50	25	
66–88	75	PMR
135–145	50	VHF PMR
145–155	50	
155–165	50	
165–175	50	
300–312.5	50	UHF PMR
312.5–325	50	
325–337.5	50	
337.5–350	50	
350–362.5	50	
362.5–370	50	
370–387.5	50	
387.5–400	50	
400–415	50	NMT; car alarm systems; PMR
415–433	50	
433–450	50	
450–465	50	
465–475	50	CDMA450; PMR
475–490	50	
490–500	50	
500–530	50	
800–850	100	PMR
850–895	100	Cellular; PMR
925–960	100	Cellular GSM
1525–1670	100	GPS; satellite
1805–1880	200	Cellular DCS
1930–1990	150	Cellular GSM1900
2110–2170	150	Cellular UMTS, WCDMA
2400–2500	50	WLAN; Bluetooth; WiFi
Total	**2125 Watts**	

targets. Approximate parameters for a modern mobile jamming system are shown in Table 19.1.

If the available band power in the jamming system of Table 19.1 is spread over the available number of channels, we can work out the average power per band. Taking for example the 145–155 MHz band with potential channels spread at 25 kHz

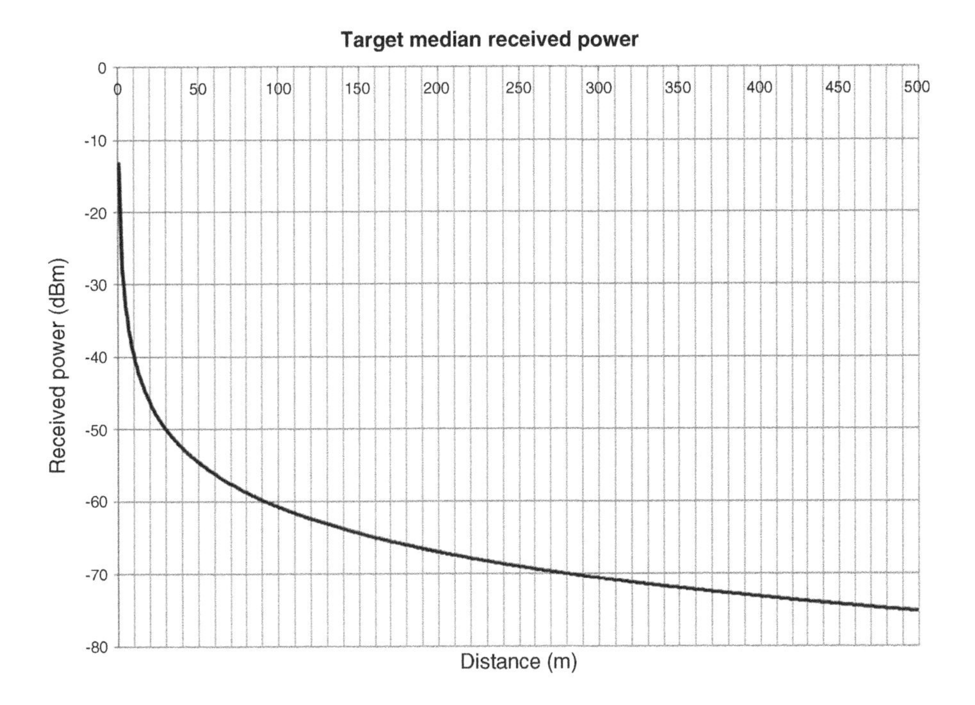

Figure 19.1 Median received power in the range 0–500 metres, the typical engagement range for RCIED scenarios.

raster spacing (although it may also be at 12.5 kHz) and assuming all of the jammer power is in the target bandwidth, we can calculate that there is 1W/channel jammer power available. If we assume this is targeted against a 1W mobile, adding 10 dB for fade margin and assuming a required J/S power of -10 dB, the two systems are equal in power.

A jammer of this power and with these parameters produces an equivalent power into an isotropic antenna as shown in Figure 19.1.

We can look at the jamming performance against targets at different ranges, based on the median received power as described above. This is illustrated in Figure 19.2.

It can be seen from Figure 19.2 that the jammer is nominally effective from half the range to the target. Given that they are of equal effective power under the conditions being considered, this is a not unexpected result.

It is, however, important to consider the effects of fading at these short ranges. Fading is not distance-dependent, so the fading effects are as pronounced at short ranges as they are at longer ranges. In this analysis, it is worth examining the short range scenario within which an explosive device might be effective. If we assume that for a small device this is approximately 50 metres when shrapnel effects are included, we can examine the effects of fading within this range.

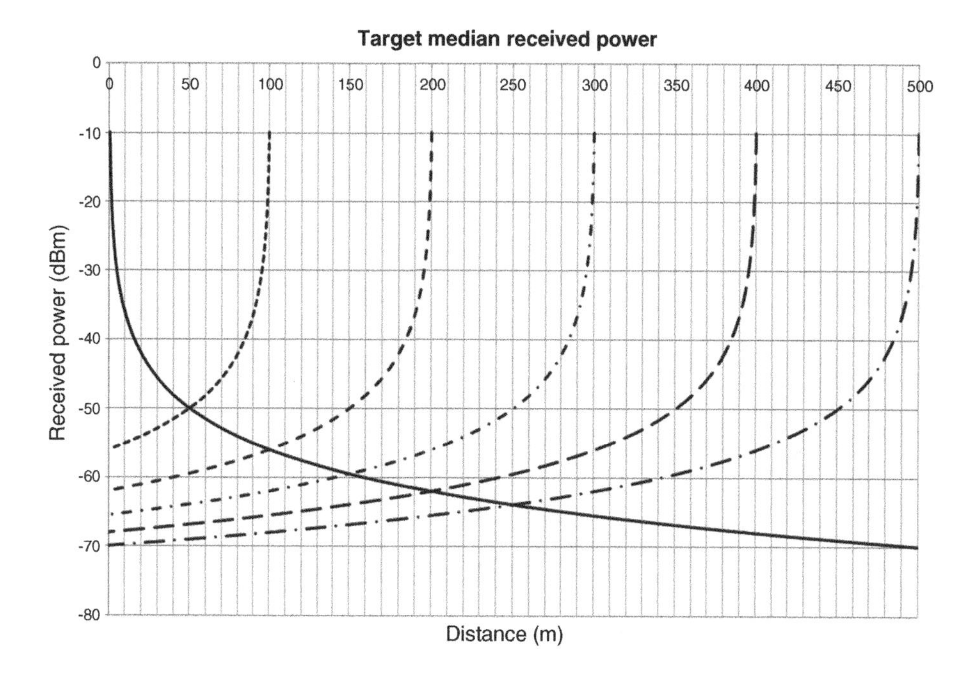

Figure 19.2 Performance against a target at ranges from 100 metres to 500 metres.

Figure 19.3 shows a predicted calculation over poor ground for a jammer of the same power as a trigger command on the same frequency, over the same ground, using a 10 dB margin. Although the median power of the jammer is higher than that of the trigger signal, fading causes the power battle to be lost at some ranges. This occurs even close to the jammer. This, as ever, is due to deep fades caused by ground reflections that affect the jammer at different points than the trigger signal. Please note that the distances will vary in particular situations; it is only the general form of the graph that is important. The grey areas show where the required 10 dB margin is not achieved.

The values of Figure 19.3 can be processed to show the percentage of the trigger signal jammed by range. This is shown in Figure 19.4.

Figure 19.4 shows something that commanders do not want to see, and engineers have difficulty in getting through to them; nothing in engineering is ever 100%. Even very close to the jammer, there is still some possibility that the trigger signal will win the power battle and trigger a device even despite the presence of a jammer.

This argument is based on a jammer and trigger of equal power, using omnidirectional or at least matching antennas on both sides. The percentage of jamming close to the jammer will be reduced if the jammer power is increased (substantially) or by using directional antennas for the jammer. There is however another way of reducing this vulnerability. This can be achieved for vehicle jammers but it is difficult to do for jammers intended to be used for dismounted patrols, where weight is important.

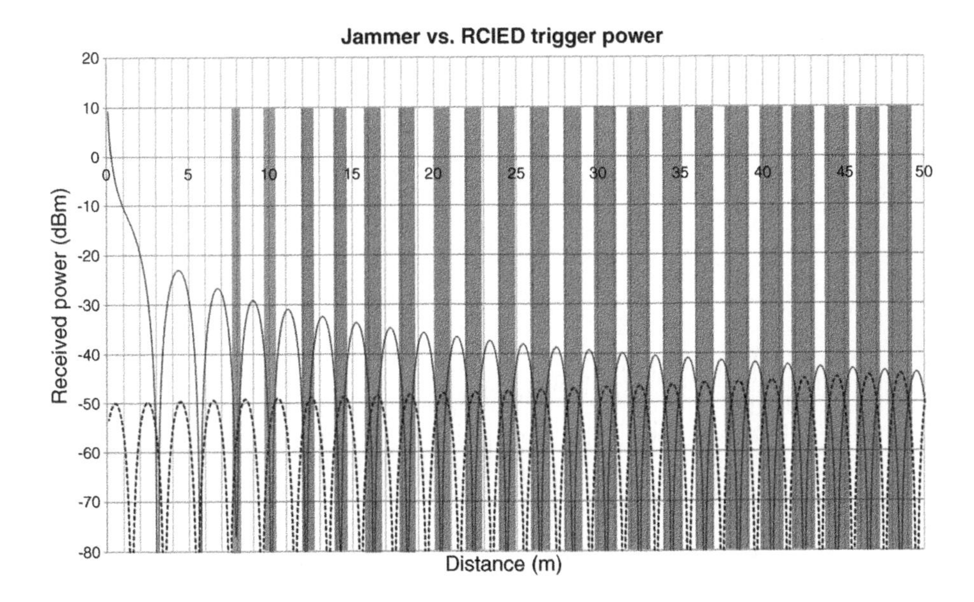

Figure 19.3 The effect of fading of both a jamming signal and a trigger signal in a different location such that the fades are uncorrelated. The grey areas are where jamming will not be effective.

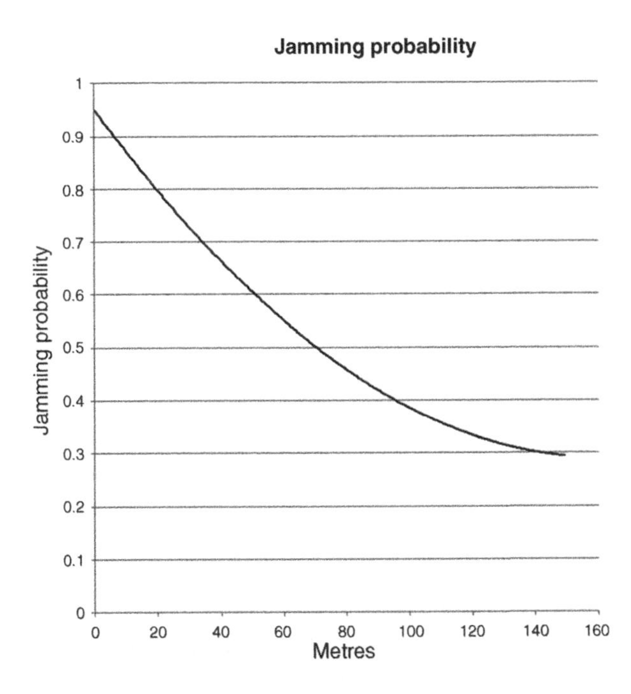

Figure 19.4 Probability of effective jamming by a jammer of equal power to the trigger signal when both signals are faded equally but in an uncorrelated fashion.

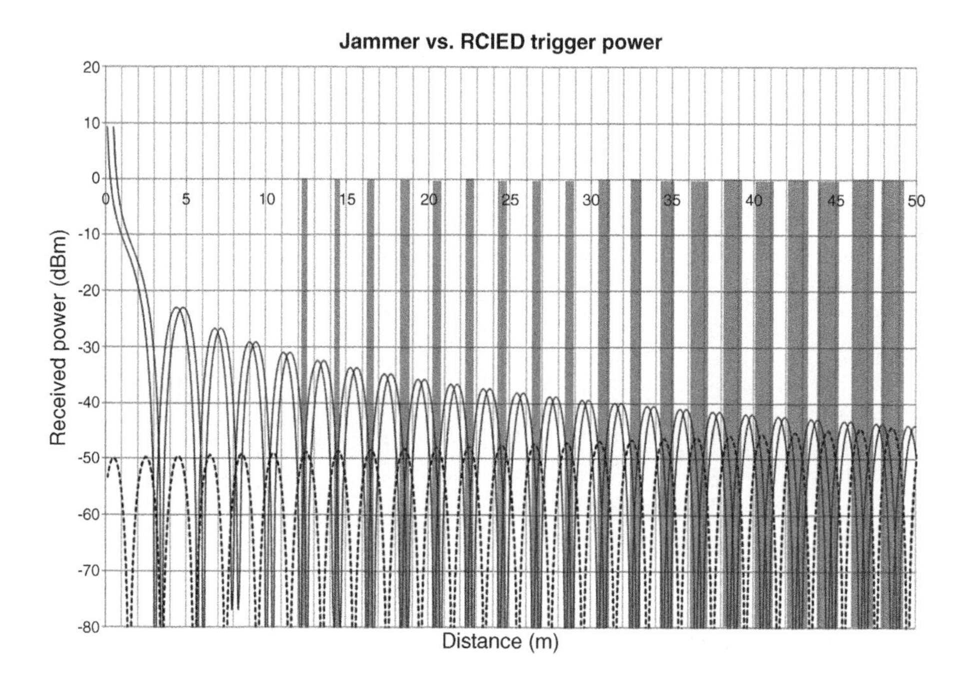

Figure 19.5 The use of two jamming signals to partially overcome fading effects.

The issue identified above is due to fading caused by ground reflections. This can be countered by using antennas with highly-directional vertical response – however, this runs the risk of missing signals outside of the narrow beam. A better way is to use diversity antennas for jamming. If they are spaced sufficiently far apart, the fading is uncorrelated and therefore statistically independent. There are two ways of achieving this; mounting horizontally diverse antennas on the jamming platform, or having multiple jammers transmitting continuously. Multiple transmissions may cause fading in the same manner as that due to ground reflections, but there are methods to overcome this, including slight frequency offsets.

Figure 19.5 shows the effects of two jamming antennas spaced a small distance apart. Again, the diagram is illustrative rather than giving an absolute representation of a particular scenario. With the two jamming signals, the percentage of jamming gaps is significantly reduced.

The improvement in jamming probability is shown in Figure 19.6. The improvement in this case is relatively small, however in many case it could be substantially better.

The fading problem can be addressed more effectively by having a number of jammers simultaneously operating, particularly for dismounted patrols. One option would be to replace the single patrol jammer with a smaller, lighter (and lower power) jammer for each member of the team. This would replace a 15–20 kg jammer of about 8W with eight 1W jammers. These jammers would be substantially lighter than the

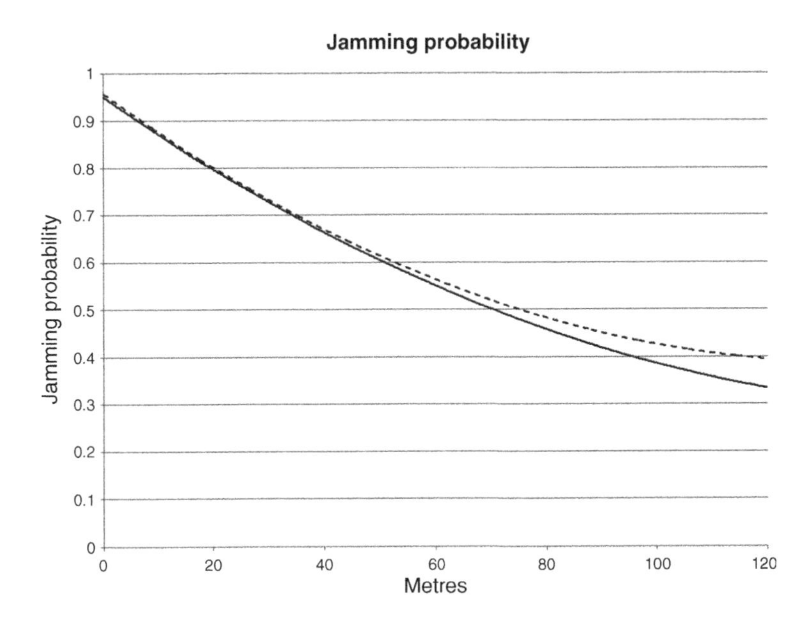

Figure 19.6 The improvement in jamming probability by having two simultaneous jamming signals. The more signals there are, the better the performance that can be achieved.

single patrol jammer, thereby sharing the overall weight as well as showing improved performance in overcoming fade problems.

Another method of protecting patrols is to send a high power jammer on an aircraft at low altitude over the patrol route immediately before it arrives. This can be used to prematurely detonate emplaced devices or, if the power emitted is sufficiently high, damage the device receivers by radio input overload.

20

Summary and Conclusions

I have tried in this book to balance the difficult task of providing a basic overview of these technologies while avoiding too much depth, which may discourage the reader. I have also tried to convey the view that radio is not a 'black art' and that most of the principles are fairly straightforward. The real complexity comes in realising systems and in positioning them effectively.

Readers who have a role in designing new systems should regard this book as a primer, and hopefully a useful repository of basic formulae and methods. Operators and managers will probably not need to delve too much deeper, except in the specific areas they are actively pursuing.

One other important factor I hope I have emphasised is that radio communications, radars and electronic warfare are all tools to be used in support of the operational objectives. Where possible, they should not be constraining factors preventing command from achieving their objectives. Additionally, they are a part of the overall challenge in modern conflict, and to be truly effective, operational techniques must be coordinated with the other activities being carried out.

I also hope that I have identified the practicalities of real systems. No system is perfect, no system is necessarily the best in all circumstances and, in engineering, no solution can ever be 100%. This is often ignored or rejected by non-technical people. This can be dangerous if those people are in charge. All that can be done is to attempt a continual process of education to ensure that as far as possible, pragmatism is adopted. When politicians and the public see casualties from RCIED systems, they often fail to appreciate that there are a variety of factors at play and that sometimes, systems will fail to perform at the crucial moment or in particular situations. Again, it should be emphasised that EW in particular is a continually evolving domain and that despite occasional failures, modern systems provide very good protection.

Finally, it should be recognised that professionals operating in the radio arena are dedicated in their work. The same is of course more than true for those members of the

Communications, Radar and Electronic Warfare Adrian Graham

armed forces and those supporting them in operational theatres. I hope that some of them will read this book and that it helps them to understand the concepts behind the systems that they use and that in doing so, they gain some comfort from the vast amount of effort being carried out by equipment suppliers, government procurement agencies and consultants to support them.

Appendix A: Working with Decibels

This section provides a brief refresher in working with decibels. Decibels are used in radio engineering to cater for the wide range of values involved and to make calculations much easier than the linear equivalent.

To convert from the linear form to the logarithmic, the equation is:

$$A(dB) = 10\log(A)$$

where *log* is the base 10 logarithm. It is vital to remember that this term is not the same as *ln*, which is the natural logarithm. To convert from the natural logarithm to base 10, the following conversion should be used:

$$A(log_{10}) = \frac{\ln(A)}{\ln(10)}$$

Ln (10) is approximately 2.303, so to convert to Log, divide the *ln* value by 2.303, and to convert from *log* to *ln*, multiply by 2.303.

To convert back to the linear form, use the following equation:

$$A = 10^{\left(\frac{A(dB)}{10}\right)}$$

Thus, by example:

$$30dB = 10\log(1000) \quad and \quad 1000 = 10^{\left(\frac{30}{20}\right)}$$

Converting formulae to their logarithmic form is also fairly straightforward. Consider the following linear equation:

$$x = \frac{A \cdot B \cdot C^2}{D^3}$$

In logarithmic terms, this becomes:

$$x(dB) = 10\log(A) + 10\log(B) + 20\log(C) - 30\log(D)$$

The squared and cubed terms are turned into multiplying coefficients when converting into the logarithmic form as

$$\log(A^x) = x \cdot 10\log(A)$$

We can work through the free space loss formula to consolidate these points. The linear form of the equation is:

$$L = \frac{(4\pi)^2 d^2 f^2}{c^2}$$

Converting to the logarithmic form, we get:

$$L = 10\log(4\pi^2) + 10\,Log(d^2) + 10\,log(f^2) - 10\,log(c^2)$$

$$L = 21.984 + 20\,log(d) + 20\,log(f) - 169.542$$

$$L = 20\,log(d) + 20\,log(f) - 147.558$$

However, this formula is in the form of metres for distance and hertz for frequency. We normally work in units of km and MHz. If we express the units in the preferred format, we can adjust the formula to a more useful form:

$$f(\text{MHz}) = f(Hz) \cdot 10^6$$

$$20\log[10^6(dB)] = 120\,log10 = 120$$

And similarly, converting between metres and kilometres, we get a value of 60 dB. Adding the 180 dB of the sum of both of these gives:

$$L = 20\,log(d) + 20\log(f) + 32.442$$

This is the free space loss normally used.

Values expressed in dB without units are ratios. Often, we will express dB values with reference to a particular set of units, such as dBm or dBW. Thus, we have a value of 100W, which is 20 dBW, or 50 dBm (dBm being 30 dB lower than dBW).

Appendix B: Common Conversion Formulae and Reference Tables

The following figures, tables and formulae have been collected in this appendix of ease of reference, so that the reader does not have to search through the text. See text for the variables if needed.

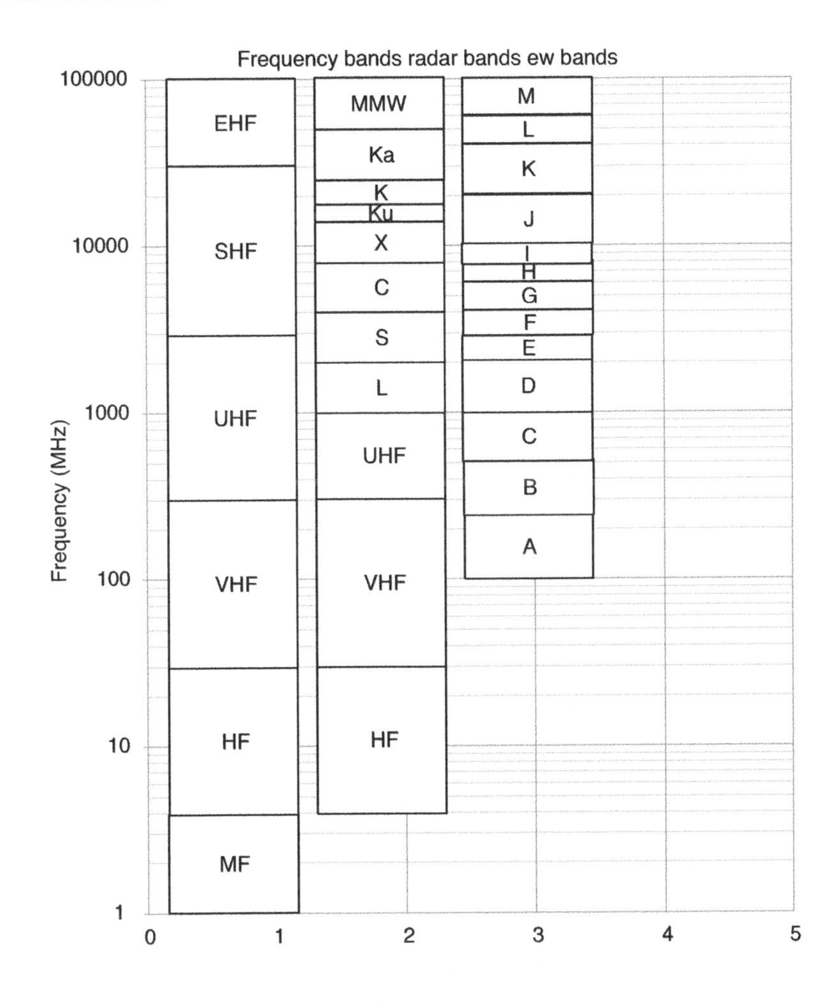

Communications, Radar and Electronic Warfare Adrian Graham

Power conversion formulae:

$$Power(dBW) = 10\log[Power(W)]$$

$$Power(W) = 10^{\left(\frac{Power(dBW)}{10}\right)}$$

$$Power(dBm) = Power(dBW) - 30$$

$$Power(W) = \frac{v^2}{R}$$

$$Power(dBW) = 10\,log\left(\frac{V^2}{R}\right)$$

Sensitivity conversions:

$$Sensitivity(dBm) = 10log\left(\frac{Sensitivity(\mu V)^2}{R}\right) + 30$$

Where R is impedance, usually 50, 75 or 300 Ω

$$Sensitivity(\mu V) = \sqrt{R \cdot 10^6 \cdot 10^{\left(\frac{Sensitivity(dBm)-30}{10}\right)}}$$

$$Sensitivity(dB\mu\mathrm{V}) = Sensitivity(dBm) - 107$$

$$Sensitivity(dBm) = Sensitivity(dB\mu\mathrm{V}) + 107$$

The complimentary error function:

$$Q(x) = \frac{1}{2} erfc\left(\frac{x}{\sqrt{2}}\right)$$

Power sum method of interference calculation:

$$P_i = 10\,log_{10}\left(\sum 10^{\left(\frac{I}{10}\right)}\right)$$

Radar range equation:

$$\frac{P_r}{P_t} = \frac{G_t G_r \sigma \lambda^2 F_t^2 F_r^2}{(4\pi)^3 R^4}$$

Simple propagation model formulae:

Free Space Loss

$$L = 32.44 + 20\log f + 20\,log\,d$$

Reflection Coefficient for a Two-Ray Model

$$\rho = \frac{(\varepsilon_r - jx)sin\psi - \sqrt{(\varepsilon_r - jx)\cos^2\psi}}{(\varepsilon_r - jx)sin\psi + (\varepsilon_r - jx)cos^2\psi}$$

Fresnel integral for an infinitely thin wedge:

$$J(v) = -20\,log\left(\frac{\sqrt{[1 - C(v) - S(v)]^2 + [C(v) - S(v)]^2}}{2}\right)$$

Basic form of an empirical model:

$$E_r = -\gamma \cdot \log(10) + K(P_{BS}, f, h_{BS}, h_{MS})$$

Okumura-Hata model:

$$L = 69.55 + 26.16\log_{10} f - 13.82\,log_e h_{BS} - a(h_{MS}) + (44.9 - 6.55\,log_e h_{BS})log_{10} d$$

$$Processing\ Gain(dB) = \frac{Spreading\ Bandwidth}{Signal\ Bandwidth}$$

Delta-h values for different types of terrain

Terrain Type	Typical Δh
Water or very smooth plains	0–5
Smooth plains	5–20
Slightly rolling plains	20–40
Rolling plains	40–80
Hills	80–150
Mountains	150–300
Rugged mountains	300–700
Very rugged mountains	>700

Conductivity and permittivity of common materials

Earth type	Conductivity (S/m)	Permittivity (ε_r)
Poor	0.001	4.0–5.0
Moderate	0.003	4.0
Average	0.005–0.01	10.0–15.0
Good	0.01–0.02	4.0–30.0
Dry, sandy, flat (typical of coastal land)	0.002	10.0
Pastoral Hills, rich soil	0.003–0.01	14.0–20.0
Pastoral medium hills and forestation	0.004–0.006	13.0
Fertile land	0.002	10.0
Rich agricultural land (low hills)	0.01	15.0
Rocky land, steep hills	0.002	10.0–15.0
Marshy land, densely wooded	0.0075	12.0
Marshy, forested, flat	0.008	12.0
Mountainous/hilly (to about 1000 m)	0.001	5.0
Highly moist ground	0.005–0.02	30.0
City Industrial area of average attenuation	0.001	5.0
City industrial area of maximal attenuation	0.0004	3.0
City industrial area	0.0001	3.0
Fresh water	0.002–0.01	80.0–81.0
Fresh water at 10.0 deg C (At 100 MHz)	0.001–0.01	84.0
Fresh water at 20.0 deg C (At 100 MHz)	0.001–0.01	80.0
Sea water	4.0–5.0	80.0–81.0
Sea water at 10.0 deg C (to 1.0 GHz)	4.0–5.0	80.0
Sea water at 20.0 deg C (to 1.0 GHz)	4.0–5.0	73.0
Sea ice	0.001	4.0
Polar ice	0.00025	3.0
Polar Ice Cap	0.0001	1.0
Arctic land	0.0005	3.0

Comparison of frequency and wavelength

Frequency band	Frequency range (MHz)	Wavelength (m)
HF	3–30	100–10
VHF	30–300	10–1
UHF	300–3000	1–0.1
SHF	3000–30 000	0.1–0.01
EHF	30000–300 000	0.01–0.001

Antenna separation formulae:

$$HI(dB) = 22 + 20\,log\left(\frac{x}{\lambda}\right)$$

$$VI(dB) = 28 + 40\,log\left(\frac{y}{\lambda}\right)$$

$$SI(dB) = (VI - HI) \cdot \frac{2\theta}{\pi} + HI$$

Index

Printed and bound by CPI Group (UK) Ltd, Croydon, CR0 4YY

16/04/2025

14658554-0006